D0007297

Managing Technology

McGraw-Hill Engineering and Technology Management Series
Michael K. Badawy, Ph.D., Editor in Chief

STEELE · *Managing Technology*

Managing Technology

The Strategic View

Lowell W. Steele

Consultant, Technology Planning and Management

McGraw-Hill Book Company

New York St. Louis San Francisco Auckland
Bogotá Hamburg London Madrid Mexico
Milan Montreal New Delhi Panama
Paris São Paulo Singapore
Sydney Tokyo Toronto

Library of Congress Cataloging-in-Publication Data

Steele, Lowell W.
Managing technology.

1. Technological innovation—Management.
2. Research, Industrial—Management. I. Title.
HD45.S7544 1988 338.'06 88-13471
ISBN 0-07-060936-5

Copyright © 1989 by McGraw-Hill, Inc. All rights reserved. Printed in the United States of America. Except as permitted under the United States Copyright Act of 1976, no part of this publication may be reproduced or distributed in any form or by any means, or stored in a data base or retrieval system, without the prior written permission of the publisher.

1234567890 DOC/DOC 89321098

ISBN 0-07-060936-5

The editors for this book were Betty Sun and Dennis Gleason, the designer was Naomi Auerbach, and the production supervisor was Dianne Walber. It was set in Century Schoolbook. It was composed by the McGraw-Hill Book Company Professional & Reference Division composition unit.

Printed and bound by R. R. Donnelley & Sons Company.

Information contained in this work has been obtained by McGraw-Hill, Inc., from sources believed to be reliable. However, neither McGraw-Hill nor its authors guarantees the accuracy or completeness of any information published herein and neither McGraw-Hill nor its authors shall be responsible for any errors, omissions, or damages arising out of use of this information. This work is published with the understanding that McGraw-Hill and its authors are supplying information but are not attempting to render engineering or other professional services. If such services are required, the assistance of an appropriate professional should be sought.

For more information about other McGraw-Hill materials, call 1-800-2-MCGRAW in the United States. In other countries, call your nearest McGraw-Hill office.

"What I have tried to tell the Chiefs of Staff is that their most important function is their corporate work . . . they are men of sufficient stature, training and intelligence to think of this balance—the balance between minimum requirements in the costly implements of war and the health of our economy."

DWIGHT D. EISENHOWER

Contents

Series Introduction

Technology is a key resource of profound importance for corporate profitability and growth. It also has enormous significance for the well-being of national economies as well as international competitiveness. Effective management of technology links engineering, science, and management disciplines to address the issues involved in the planning, development, and implementation of technological capabilities to shape and accomplish the strategic and operational objectives of an organization.

Management of technology involves the handling of technical activities in a broad spectrum of functional areas including basic research; applied research; development; design; construction, manufacturing, or operations; testing; maintenance; and technology transfer. In this sense, the concept of technology management is quite broad since it covers not only R&D but also managing product technology and process technology. Viewed from this perspective, management of technology is actually the practice of integrating technology strategy with business strategy in the company. This integration requires the deliberate coordination of the research, production, and service functions with the marketing, finance, and human resource functions of the firm.

This task calls for new managerial skills, techniques, styles, and ways of thinking. Providing executives, managers, and technical professionals with a systematic source of information to enable them to develop their knowledge and skills in managing technology is the challenge undertaken by this book series. This series will embody concise and practical treatments of specific topics within the broad area of engineering and technology management. The primary aim of the series is to provide a set of principles, concepts, tools, and techniques for those wishing to enhance their managerial skills and potential.

The series will provide readers with the information they must know and the skills they need to acquire in order to sharpen their managerial performance and advance their careers. Authors contributing to the series are carefully selected for their expertise and experience. While series books will vary in subject matter as well as approach, one major feature will be common to all volumes: a blend of practical applications and hands-on techniques supported by sound research and relevant theory.

The target audience for the series is quite broad. It includes engineers, scientists, and other technical professionals making the transition to management; entrepeneurs; technical managers and supervisors; upper-level executives; directors of engineering; R&D and other technology-related functions; corporate technical development managers and executives; continuing management education specialists; and students in engineering and technology management programs and related fields.

We hope that this series will become a primary source of information on the management of technology for practitioners, researchers, consultants, and students, and will help them to become better managers and to lead most rewarding professional careers.

MICHAEL K. BADAWY
Professor of Management of Technology
The R. B. Pamplin College of Business
Virginia Polytechnic Institute and State University
Falls Church, Virginia

Preface

This book places technology in the context of a total business. It attempts a balanced treatment of a subject that is usually presented from a narrower perspective, e.g., innovation, R&D, productivity, or new ventures. These more limited perspectives tend to oversimplify and distort the nature of the task of managing technology. The book is aimed at several audiences, even some not nominally associated with management at all. In addition to top management and senior technical managers, managers in other functions whose work is affected by the use or consequences of technology will also find the book beneficial—marketing, finance, strategic planning, business development, human resources, and production. Beyond managers, students of technology and of management more generally, as well as those concerned with science and technology policy and those concerned with economic vitality and international competitiveness, will, I believe, find the discussion helpful.

The book is deliberately structured so as to be useful to readers with these differing interests. It comprises four parts, each of which is preceded by an introduction that lays out the basic structure and themes of that part and summarizes the contents of each chapter.

Although the book has been carefully constructed to provide an integrated view of the full sweep of technology, most of the chapters have been written so that they also provide a free-standing exposition. The exception is the part on strategic management, in which the chapters are interrelated. A reader interested only in managing risk or evaluating a technical operation, for example, will not need to have read the rest of the book to find those chapters comprehensible.

The first part establishes the territory I will cover, laying out the basic concepts of technology and management as I will use them and introducing the major themes. The second part addresses some of the important aspects of operational management as they apply to technology. Part 3 provides a similar analysis for strategic management. Part

4 considers some crosscutting issues, such as achieving system integration and evaluating the effectiveness of a technical operation, which apply to both operational and strategic management of technology.

For a reader such as a CEO or general manager, the prologues, together with the introduction, provide a condensed overview. I believe the content of the first part, "Establishing the Territory," which elaborates more fully the basic themes and insights of the book, will also be of interest to senior managers in general. Chapters 13 and 14 on evaluating a technical operation and selling technology to top management focus directly on the interests of a CEO or general manager.

The book is intended to provide a comprehensive perspective. My view is corporate—the interplay between technology and a business, not the details of managing a technical operation. I focus on the interests of the general manager—are we working on the right things, and are we doing things right?—and the questions he asks himself endlessly: Is our competitive strategy sound? Are we capitalizing on opportunities in new technology and protecting against threats? Are we ensuring the continued profitability of our successful business and also providing opportunities for future growth? Perhaps most important, are we appraising risk realistically and doing sensible things to manage it? For the latter: Are we implementing our competitive strategy effectively? Are we meeting milestones and cost targets? Is our response to market signals timely? Do our development programs create products that provide value as judged by customers? Are we integrating technology with other elements of my business? Although I emphasize insight and understanding, the book contains much practical material that can be immediately applicable.

I owe readers an explication of my point of view. I believe that the view from the trenches, where decisions must be made and one must live with the consequences, is useful in understanding the practice of management. I also believe wisdom and insight can come from reflection on one's own experience. The principles so induced may not be scientifically validated, but the bedrock of every effective manager I have known is his own experience. The broader and more extensive the experience, the greater the opportunity to distill wisdom. I have had the fortunate opportunity of working for a very diverse and well-managed company, GE, supplemented by hundreds of interchanges with other executives from Europe, Japan, and the United States. Talking shop gave me the opportunity to compare practices, evaluate various approaches and techniques, and test the generality of my own insights. Consulting with companies in different industries and facing a diversity of challenges has extended and enriched that experience.

Involvement over an extended period of time provides a sense of perspective and of the dynamics of technology development which cannot be obtained any other way. I can look back on a mosaic of gratifying successes, pleasant surprises, disappointing failures, naive projections, mismanaged programs, and superbly well-managed ones. I believe my experience constitutes a data base from which to draw some limited but useful generalizations. Since I spent a substantial portion of my career with GE, readers should not be surprised to see that many of my examples draw on that experience. They are not used to suggest that the GE approach is a model for others, but rather to provide real-life examples of a point I am trying to make. Sometimes the same situation is referred to in a subsequent discussion to illustrate a separate point. The examples are not intended as even brief case histories. Nevertheless, I have taken care to ensure that the aspect I direct attention to is consistent with the entire incident as it unfolded. I have, of course, used many examples from other companies as well, some from my own experience and some from other sources, both public and private.

One additional observation: Every writer faces the problem of choosing language that is neutral with respect to gender without being excessively cumbersome. I have not succeeded in every case. I hope women who read the book will accept that my choice of words in no case reflects insensitivity to their nuances, but rather a desire to maintain reasonable felicity in expression.

Lowell W. Steele

Acknowledgments

How do I identify and thank the hundreds of people who have helped educate me? A lifetime of interacting with intelligent people, who recognize the value of new insights and new discoveries but are skeptical of validity, who insist on establishing the limits of generalizations but recognize their value, can only leave one grateful and humble to have had the privilege of knowing and working with first-rate scientists, engineers, and managers all his life.

Some I must recognize individually: Stan Neal for wisdom and perspective based on a career of effective management and a systems view of technology; Don Collier for valuable insight into innovation, a corporate perspective on technology, and constructive criticism, especially on strategic management; Bob Breen, Al Taylor, and Bill Gutzwiller, who helped me understand the interplay between product engineering, manufacturing processes, and information, and Bob Breen for constructive criticism; Bill Ehner, Dick Kashnow, and Ralph Hagan for helpful review and suggestions for additional examples; Lyle Ochs for thoughtful remarks and additional examples from Tektronix; Nico Hazewindus, Mikoto Kikuchi, and Eduard Pannenborg for review and valuable comments from a European and Japanese vantage point; Sam Tinsley for a perspective from process industry; Lloyd Harriott, Gunnar Thornton, and Don Marquis for providing generous access to a century of experience in product engineering and information technology; Bill Sommers and Hugh Miller for a broad perspective from management consulting and the National Academy of Engineering; and Alan Kantrow for reassuring support and guidance from his background as a professional editor and writer on technology and management.

Most important, of course, is my wife Jean, who not only encouraged me unfailingly to persevere and accepted the ensuing disruption to our household, but by her own devotion to endless typing and retyping, provided continuing incentive to keep at it. A change in hats turned her into a skillful editor, who contributed mightily to flow, structure, and whatever felicity of expression the book displays. 'Tain't easy wielding a blue pencil on your husband, but she did it superbly and gently. Thank you.

Introduction

Technology is a study in contradictions. Its importance and pervasiveness are universally accepted features of our world. At the national level, technology is the vital underpinning of national security and economic vitality. It is an indispensable ingredient of individual well-being and exerts a powerful influence on our standard of living and quality of life.

There is widespread agreement that the United States has been the dominant global player in technology since World War II. Its record of developing and using technology has been the benchmark against which others compare themselves. Our success in technological endeavors is a, perhaps *the*, key to our economic power and competitive position.

And yet . . .

Our effectiveness in generating and exploiting technology is cause for continuing concern. We fear we are not as effective as we were. Our relative technological and economic position has slipped, and competition is more effective. There is also virtually universal belief that we must do better; our future depends on it. Furthermore, there are countercurrents of concern that technology is not an unmixed blessing, that it must be guided and constrained lest its capacity for harm get out of hand.

At the level of the individual enterprise, there is also recognition that technology is the foundation of most businesses—it is hard to imagine one entirely devoid of technology. It is the principal source of innovation and the key to success and survival.

And yet . . .

We question management's commitment to adequate support. We fear too much emphasis on the short term. We worry about management's competence in managing technology. Why this ambivalence and anxiety? First, there is, I believe, general agreement that despite the dramatic advances since World War II, advances that have brought the importance and pervasiveness of technology to center stage, its role in business and international competitiveness will continue to grow. Electronics and information, which have been the two most visible symbols of technological dynamism, are now being joined by biotechnology. Still, metaphorically speaking, electronics is probably in young adulthood, information in adolescence, and biotechnology in infancy. Every function in a business, including finance and human resources, has an influence on, and is impacted by, technology.

The importance of managing technology effectively is not limited to its internal dynamics. The nature of competition has changed markedly, and the competitive significance of technology has grown apace. The concept of a domestic market or domestic competitors is becoming meaningless. Cost leadership and market opportunities must be appraised in global terms. The list of countries that aspire to participation in international trade—especially the U.S. market—and that have, or are acquiring, the capability to do so, now comprises a significant fraction of all the nations on earth. The idea of hiding behind so-called national advantages, such as land, location, a skilled and energetic work force, or natural resources, has become untenable. They are too widely available. And anyway, their competitive value is shrinking. The United States has no monopoly on the spirit of enterprise, willingness to work, level of education, and standards of skill that have played an important role in its past success. Some would assert it is now sometimes at a disadvantage in these features.

Perhaps most important, we have learned that we dare not take the survival of a nominally well-run business for granted. The seventieth anniversary issue of *Forbes* examined the present status of the 100 largest U.S. companies in 1917. Its headline read, "Dropout Rate for the 100 Behemoths of 1917: 78%. So Much for the Theory That Big Companies Are Invulnerable."* Failure to appreciate the dynamic nature of technological change and to recognize the fragile and transitory nature of technological advantage is now accepted as possibly the most important source of competitive decline. Self-renewal must become an explicit and measured aspect of managerial responsibility.

Two other attributes of technology contribute to our anxiety and fixation on its importance. First, it is rigorous and esoteric. Managers without technical training are distinctly uneasy with technology. The common wisdom is that laypersons cannot understand technology, that it is best left to the experts—and any technologist is by definition an

---

*July 13, 1987, p. 145.

expert compared with a layperson. The technical community has done little to discourage this perception, and the cost has been heavy. Second, technology is usually associated primarily with innovation, which is a complex, poorly understood process that is fraught with risk and uncertainty. A feeling that one must engage in such an activity in order to maintain competitive viability is calculated to generate anxiety.

It is ironic and somewhat puzzling that in the face of this widespread concern, the subject of technology has until recently been almost invisible in business school curricula. It is equally invisible in management development programs. I used to assert, without contradiction, that if one limited his or her knowledge of General Electric to the material covered at Crotonville (GE's management education facility), one would hardly know it was a technical company. Technical schools believe they already lack sufficient time to give engineers or scientists the basic substantive knowledge they need without diverting time to management considerations.

Technology also receives limited attention in the management literature, and even then, it is presented with an incomplete and distorted view. First of all, the term is rarely defined. Primary emphasis is devoted to research and development (R&D)—admittedly a critical component—and to technological innovation, which is closely associated with R&D. Although I do not deny the importance of these activities, they are by no means the only aspects of technology and of managing technology that are crucial to the success and survival of a business in a global economy. To make matters worse, the label of technology is sometimes used in discussions that in fact focus on R&D, and treatises ostensibly on R&D sometimes spill over into other aspects of technology.

These are the challenges to technology management. So far, we have had three responses. We have engaged in much introspection, asking, what's wrong with us? The questions have been most visible at the national level and have focused on questions of public policy and the national psyche: willingness to take risks, entrepreneurship, education, regulation, and antitrust, for instance. But managers of hundreds of companies have also been asking, what's wrong with us? Second, we have engaged in management bashing. Managers have become too short-sighted, they lack vision, they are fat cats, they are venal. . . . No doubt any of those apply to some managers, but they certainly do not apply to hundreds I have known in many different companies. For every manager who deserves opprobrium, there are thousands that are working very hard, experiencing heavy stress, and doing the best they know how! Still, there is widespread sentiment that U.S. managers must pull up their socks and do a more effective job. Finally, our third response has been to bring in new managers to wield a machete, to hack away at unproductive assets and people the business can no longer afford. *In extremis* such draconian measures may be the only recourse, but surely no one would argue they are the preferred solution.

The principal incentive behind this book is to show that we need a more comprehensive, more systems-oriented view of technology—what it is, how it fits in with the other activities in a business, what role it should play in a successful enterprise. Technology, as I believe the term should be used, requires a multidimensional perspective. In one dimension it comprises a broad sweep of activity all the way from the search for new knowledge to servicing a customer. I know many will assert that this view is much too all-inclusive, that it brings under one broad label a variety of activities—science, engineering, basic research, manufacturing processes, and so on—which should not be covered under one umbrella. But we need to change the way we think about technology, to view it more holistically over an extended continuum. I believe strongly that we need to pay more attention to the linkages among the various elements of technical work and to the effective functioning of the entire spectrum and less to the distinctive features of particular segments. The fragmentation that currently exists, both in the conduct of work and in our thinking about it, is a serious impediment to effectiveness.

Along another dimension, we need to broaden the understanding of the substance of technical work, i.e., to specify the technical disciplines which are included. We need to define technology as encompassing three fields of activity: product, process, and information technologies. These three different substantive fields of work have different traditions and different educational foundations. They are each recognized as fields of specialization, but are rarely thought of as an integrated entity that constitutes technology.

The penalty for excessive fragmentation is heavy. Product designs do not meet customer needs. Products have needless manufacturing or quality problems. Development cycles cannot respond to market dynamics. Engineering and manufacturing facilities are too inflexible in adjusting to external events in technology, markets, or competitive behavior.

In addition to the theme of the penalty for fragmentation and the need to view technology more broadly and more holistically, the book is based on other themes. All of management, but especially the management of technology, involves balancing the tensions between the rigor, discipline, and passion for stability necessary for effective operational management, on the one hand, and the change, risk taking, and innovation necessary for sustained viability and survival, on the other. Technology has two contradictory faces, and both must be accepted as essential for effective performance. Managing technology involves managing paradox, nay, means ensuring the paradox exists to be managed. To recognize only one face of technology—no matter which you choose—is to court disaster. The only questions are the timing and the speed with which trouble will strike.

My third theme is the influential role of local conventions in guiding and constraining managerial action and decisions *and* the power that technology offers to change those conventions. Much of the discussion of management focuses on the work of an individual manager or on the work of management as an abstract activity. Most of the real work of management takes place between those two extremes, i.e., it occurs in a particular context. A group of people must somehow constitute a team which can interact effectively and coherently in managing a particular enterprise. The character of that enterprise, together with a common set of norms of behavior and rules for action, largely determines how its management team will operate.

These common norms, assumptions, and rules of an enterprise cover what business it is in, how it competes, how it measures success, how it operates its plants or offices, how it makes decisions, and so on. Because many of these conventions are related to technology—for example, the cost and time required for particular operations, the amount of information that can be obtained and processed, or the time lag between an event and knowledge of it—technology can be a powerful lever for change. To exploit that potential, managers of technology must broaden their focus from particular products and processes to the entire business. Technological decisions that are thought to be narrowly focused on a product's performance or cost have, when aggregated over many products or processes, a profound effect on the entire business system. Effective management of technology means addressing these system effects, which are so deeply buried in assumptions that they are invisible.

Two somewhat more subtle purposes of this book also should be mentioned. The first is to demonstrate that addressing many of the critical issues and judgments about the management of technology does not necessarily require a technical background. I argue that nontechnical managers dare not leave technology to the specialists. Those without a technical background can and *must* participate in the decisions about technology as it relates to the rest of a business and to competitive advantage, and it is possible to do so effectively if they devote attention to learning how. I do not propose that economists or lawyers could be successful managers of engineering (although some probably could), but rather that in a corporate context, the issues which should be raised and the judgments which must be made can be dealt with effectively by a nontechnical general manager, with the help of his or her technical managers.

Competent technical people can be remarkably good expositors. Furthermore, they want people to understand. I long ago concluded that if people were having difficulty understanding a technical presentation and questions were not leading to enlightenment, probably the speaker was at fault—either the speaker was not very competent or did not want his listeners to understand. In fact, lack of a technical background can sometimes be an advantage. Technically trained people may feel they should understand something and be reluctant to expose their ignorance by asking questions. A person not technically trained need have no such qualms.

The second, less obvious purpose is to suggest that technology is a kind of leading indicator in the field of management. Managers of technology were pioneers in developing techniques for selecting, guiding, developing, and motivating highly creative professional people. As the use of professionals has grown and proliferated in other functions, those techniques have become more broadly applicable. The themes of the need to balance tension and the facilitating, yet constraining, role of conventions are also becoming more generally applicable, especially as the growing power of information technology is forcing reexamination of the foundations of managing a business. As we acquire capability to have more extensive and detailed knowledge of the real-time state of a business, we will need new conventions to guide behavior and decision making and the creative vision to recognize new business opportunities.

I believe we must change if we are to maintain global competitiveness. Managers do not deliberately opt for short-sightedness. They prefer an attractive vision if they can create one they can afford. They are products of their education, their experience, and their environment, just like people in other professions. They respond to the priorities and pressures they perceive with the tools they have at hand or can readily devise. Fortunately, there are productive avenues for change and guidelines for going about it.

I also believe the opportunities for improving our performance are great. We are probably underestimating the improvements that are occurring. However, as a manager and long-time participant in industrial technology, I was always puzzled and distressed by the difficulty of propagating the best management practices to other parts of an enterprise. Somehow the well-run businesses seemed to persist over successive generations of management, while the ineffective ones proved to be dismayingly resistant to improvement. Often they were not even aware of better practices in use elsewhere. This book attempts to provide a way of thinking about the role of technology in a business and to provide a frame of reference for competent general managers and their technical functions to work together more effectively. Much of what I present is already being practiced intermittently in bits and pieces by various businesses. I know of none that has succeeded in putting it all together. When one does, watch out!

Part

1

Establishing the Territory

This first part traverses the territory of the management of technology as I intend to address it. It sets forth the three themes composing the foundation of the book: (1) excessive fragmentation is seriously inhibiting effective management of technology; (2) nurturing and control of contradictory requirements (managing paradox) is central to the work of management; and (3) conventions are indispensable in enabling a management team to function, but they can become dysfunctional and technology should be a powerful force for change. This part argues that technology needs to be conceived in a more comprehensive and integrated framework: that excessive focus on territory and special status has led to so much fragmentation that the articulation of the entire system is suffering. One of the questions I am most frequently asked as a consultant is, how can the Japanese move so fast in introducing new products and responding to market dynamics? The answer clearly does not lie in access to or use of more advanced technology, because the United States still equals or leads the world in almost every important field. The answer lies in management.

Effective use of technology requires an adroit passing of the baton from one stage to another. Careful development of competence in individual stages will not lead to outstanding results unless the linkages are well understood and well managed. When these linkages are not well managed, a company is performing suboptimally, but even worse, the chief executive officer (CEO) frequently is not well equipped to lead the way to improvement. That leadership should come from the technical community. At present most CEOs cannot even get credible answers to the question, how are we doing in technology?

This part also argues that the dimensions of technology need to be expanded to include an integrated approach to product technology, process technology, and information processing and communications. These three dimensions have disparate roots. They have too often functioned as independent fiefdoms, and the price exacted is not only in product performance and cost but also in business performance—slow response, inflexibility, and ineffective use of capital and human resources. The opportunity to achieve better integration of these three dimensions has enormous potential for strengthening competitive advantage. As a subset of this discussion, I assert that the growing capability of information processing and communications is "rewriting the book" on all technology, and that whereas product technology has traditionally played the dominant role in most industries, information technology may soon become dominant in many cases.

A major theme of the entire book is the need to foster a dynamic tension between reliance on conventional technology and development of new technology to supplant it. This difficult task is the greatest challenge facing both managers of technology and general managers, who must incorporate technology strategy in their choice of directions and priorities. At a higher level, an analogous tension must exist between operating management, which strives for certainty, efficiency, discipline, and continuity, and strategic management, which addresses the continuing survival of the enterprise and thus must probe the need for change, the limits of successful change, and the implementation of plans for change. Thus the highest-order challenge for management is to achieve a productive, continuing tension between these two—to recognize that success and survival require wearing the perpetual hair shirt of wariness and concern over the balance between effective operations and strategic change.

Chapter 2 examines some specific aspects of technology that need to be understood by managers of technology and general managers as well. I take up the dynamics of technology life cycles, the changing nature of technology as it advances, and the requirement to modify management as a technology matures. A major theme that will recur repeatedly is the basic paradox in technical work: that saying no to discontinuous innovation would prove to be correct in a very large percentage of cases, but that the resulting failure to incorporate successful innovations could eventually wreck a business. I also argue that innovation is an elusive and often hostile objective, that it is important to heed the forces that determine its success or failure. Management cannot afford to chase every will-o'-the-wisp that appears, yet it must both seek to innovate internally and to respond aggressively to impending breakthroughs when they appear externally.

Chapter 3 examines the importance of management conventions in guiding behavior and decision making and in determining success or failure. It points out that every business is based on a few fundamental ideas or precepts. These precepts determine the role that technology will play in a business and the importance of different technologies to the business. It examines several categories of conventions and their relationship to technology. Many of them have historical roots in more primitive technology, but they are so internalized that their inadequacies are not seen. In particular, the chapter explores the role of conventions in operating management and the powerful impact that information processing could have on changing those conventions and thus on elevating operating effectiveness to a new plane. The costs of not doing so are seen in excessively long time cycles for product development, and in inefficient use of capital and especially human resources. Inevitably, the result is a response to market dynamics that lacks the timeliness and impact which could be achieved. Effective technology management begins with understanding these local conventions. Increased attention to focusing the emerging power of information technology on improving and tightening these conventions is a major opportunity that has yet to be fully grasped.

Chapter

1

What Is Technology? What Is Management?

WHAT IS TECHNOLOGY?

Technology. The term has become a global abracadabra. It's like beauty. Those who have it worry about keeping it. Those who don't feel deprived and strive to get it. In more prosaic terms, technology is universally accepted as central to economic growth and the creation of jobs; it is the foundation for improved health, longevity, and quality of life.

Viewed from another perspective, this wellspring of good has a dark mirror image. It threatens the loss of jobs and careers, degradation of the environment, frightening violation of life itself, and even destruction of civilization.

Irrespective of the point of view, technology is clearly one of the dominant features of the modern world. Whether one supports or condemns, there is universal agreement that managing technology effectively is critically important to the success and survival of individual companies and to economic well-being and growth. Somewhat surprisingly, despite the great advances that have been made and the dramatic contributions that technology has already provided, there is also virtually universal agreement that we must improve—that management of technology falls well short of what we must achieve. The "we" is universal; it is not limited to the United States, although that is the perspective most evident in this book.

Clearly, technology is too important to be left to the technologists—whoever they are. Politicians, diplomats, social scientists, ethicists—all see a need to help guide, constrain, support, or condemn technology. In part, this is because the impact of technology has become so all-encompassing. It is not surprising that those affected feel they have a right to exert influence. It is also in part because technology has always had social, economic, and political dimensions. Thus the directions and priorities of technology have always reflected perceptions of what is needed or desired for society. That being the case, it is desirable for these inputs to be from sources as diverse, informed, and responsible as possible.

Astonishingly, when collecting that input, when listening to the rhetoric, the exposition, the babble of kibitzing, and even the talk of those participating in technology itself, one is struck by the vagueness and diversity of the meaning ascribed to the term. The likelihood of mistakes, misunderstandings, confusion, equivocal support, and needless controversy is demonstrably greater under these circumstances. I believe it has become important to look for a better perspective, to be more systematic in defining what technology is, and to understand more clearly what its role needs to be.

A major theme of this book is the need for an integrated view of technology, which treats it as a closely linked system. This system spans the spectrum from creating new knowledge to servicing a product after it is sold. It includes the work to invent and develop products, the processes needed for their manufacture or delivery to customers, and the information processing inherent not only in all of these activities but also in the functioning of an entire business. Technology pervades all aspects of an enterprise, and effective management must recognize its pervasiveness and its crucial role in establishing competitive advantage and even in survival.

Defining Technology

Science, engineering, research and development, basic research, applied research, development, and *technology* are all terms used to cover some segments of a broad universe of technical activity for which there is no generally accepted single label. I have no desire to become mired in the task of distinguishing among these frequently overlapping and murky labels, even though clarity in defining this fine structure may be important to specific practitioners. It is worth noting, however, that most attempts to differentiate the various segments of this collection of activities are made by protagonists for a particular point of view, usually intent on establishing territory and preserving status relationships. Their intent is to make certain that the rest of us, whether we be managers or politicians paying for the work or just members of society, understand how important and unique *their* role is and to induce us to grant them preferred status compared with other similar-seeming activities.

So much time and effort have been devoted to identifying and gaining proper respect for the various pieces of the corpus of technical activity that we have lost sight of their relatedness. They are parts of a continuum. We must begin focusing on the whole instead of the parts, because they are in fact parts of a technical system with coupling requirements and feedbacks among the various elements. Failure to understand the linkages results in system performance that is far below optimum. Even brilliant performance in one portion of the spectrum, e.g., in science or invention, is of no avail if the total system is not effectively articulated. Many of the problems U.S. industry is facing in achieving international competitiveness in costs, quality, response time, and flexibility have their roots in this focus on the pieces instead of the system. I do not quarrel with the importance of being able to manage individual aspects of technical work effectively. And I would agree that there is a need to understand each of the pieces of the system better in order to manage it more effectively. Nevertheless, I believe we have focused attention too exclusively on this aspect of managing technology.

The perspective taken in this book is a holistic view. It is the view of the corporate funder or client who is not interested in the fine distinctions that may legitimately be sought by practitioners in various parts of the spectrum. It is the "corporate view" that President Eisenhower wished for so impatiently from his joint chiefs of staff with respect to national defense.[1] What he got was a parochial view from each service. All too often what a chief executive gets is a parochial view from his vice president of R&D or chief engineer or director of corporate information systems. Effective use of technology, however, requires the participation of every other aspect of a business. The connections with manufacturing and marketing are obvious, but technology affects people and pay rates and skills and relationships—so employee relations is involved. It also requires funds for investment and complex calculations of risks and returns—so finance is involved. Furthermore, the requirements of these other business functions are constraints on technology. An effective manager of technology must adopt a generalist's whole-business perspective. Furthermore, general managers or chief executives must be alert to the implications of technology for all aspects of their business. If the manager of technology insists on being a "pure technologist," the CEO cannot be blamed for feeling ill-served.

In adopting this perspective, I am not denying that there are important differences in the values and reward systems that inhere in different parts of the spectrum. Scientists place greater weight on gaining insight and understanding. Their culture emphasizes publication and sharing of information. Engineers are devoted to problem solving, even if why a solution works isn't fully understood. They accept greater restrictions on sharing information. But the good ones in both groups respect one another and value the role that each plays. Tony Nerad, one of the best engineers I ever knew, said Irving Langmuir, a Nobel Prize-winning physical chemist, was the best engineer *he* ever knew.

If one observes the way the term *technology* is used, two common themes emerge. First, it tends to be a crosscutting term, a term that is less easily pigeon-holed than *science* or *engineering* or *R&D*. As a corollary, it tends to be less polarizing. Although many scientists would probably not be happy to be called "technologists," the term probably evokes less unattractive status distinctions in their minds than others, such as "engineers." Second, the term is generally associated with capability—the ability to do things. In fact, one short and arguably valid definition of technology is "knowledge of how to do things." A somewhat more extended definition, and the one that will underlie all of this discussion, is that *technology is "the system by which a society satisfies its needs and desires."*

That definition is pretty sweeping. However, in trying to ensure that technology is viewed in a sufficiently broad light, I must be careful not to make it too all-encompassing. Some would assert that management is a technology. I am sympathetic, but stop short of that. The universe I focus on is the physical world and the realm of information processing. Thus the rules and principles of accounting are excluded, but the ability to store, retrieve, manipulate, and transmit data is included. When applied to an individual enterprise, it means the capability that enterprise needs in order to provide its customers with the goods and services it proposes to offer, both now and in the future. It also follows that technology has always been as indispensable to the functioning of an enterprise as the classical economic factors of production: land, labor, and capital. The requirement was so universal that classical economists simply did not recognize it. When GE was developing its system for strategic planning, the senior vice president in charge said, "We need a memo or position paper laying out the role of technology in strategic planning." After I prepared the paper, he circulated it to all corporate staff functions that were trying to help develop the new system. The finance people were astonished. They said, "Why, Steele says there is technology in finance!" To which the senior vice president replied, "Of course, there is. That's why I sent you the memo!"

Since "knowledge of how to do things" is the foundation from which a business satisfies the needs of customers, the choice of technology strongly influences the basic structure of the business. It is so intimately entwined with the very concept of an enterprise that one must consider them together. An enterprise with no technology is a virtually meaningless concept.

The Creation and Application Dimension

A company is not interested in capability in the abstract, in a generalized societal capability to know or to do. An enterprise is not necessarily even interested in creating capability; it *is* interested in having it and knowing how to use it for competitive advantage. But because the needs and desires of society are generally taken to be limitless, the creation of *new* capability has to be regarded as one aspect of technology. Thus one dimension of technology for any company is a creation-application spectrum. Science, or basic and applied research, is conventionally at the creation end of the spectrum. Research generates the new information that makes new capability possible. I argue that it is important to include this work as part of the spectrum—to emphasize the linkage, not the differentness. The extent to which a company should be active at the creation end of the spectrum can vary widely, but it dare not ignore the possibility of new technology, because that could threaten its survival. Efforts at creating new capability may be focused on better satisfying the needs already being addressed or on responding to new needs. In the latter case, an enterprise is seeking to create a new business—a different kind of management challenge. The other end of the spectrum is the applying or doing part. A portion of the effort is in applying a *new* capability, but a very much larger portion is in creative application of already *available* capability. Failure to do this job well can be just as disastrous as failure to develop or apply new technology. Unfortunately, it has received much less attention, and I am trying to redress the balance between the two.

I know some of my scientist friends may be dismayed at combining creation of capability (they would no doubt term it *research*) together with application into a single category of technology. I readily admit there are circumstances where it is desirable to emphasize the differences between the two. I simply assert that continued insistence on looking at individual pieces does not provide the integrated corporate view so badly needed. A business can no longer afford the parochial view of the advocate for new technology.

The spectrum of work I propose to include under the rubric of technology goes all the way from basic research to product service (Fig. 1.1). For any specific company this spectrum would be framed in terms

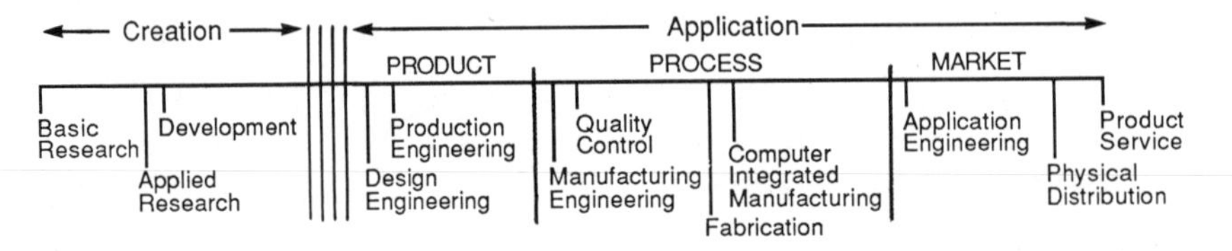

Figure 1.1 The creation-application spectrum.

of the technology that underlies, and in fact constitutes the basis for, the business.

Fragmentation

The interface between creation and application is shown as a band to suggest that the distinction cannot be sharply drawn. The management of the portion of technology beyond development exists largely in a shadow. But just look at the number of segments into which the spectrum is cut. Each represents a field of specialized activity. Each line represents an interface that must be crossed. Many of these activities have specialized languages and skills not easily understood by others. They operate in different time frames. No wonder we have a problem of fragmentation. If effective performance across the entire system did not require strong linkages, i.e., if the system were only loosely interrelated, fragmentation would be of little moment. Unfortunately, that is not true. Decisions in almost every portion of the spectrum affect the others. A considerable lapse of time may occur, but what is done or not done in creation affects all the subsequent activities. A breakdown in feedback from product service affects design. Design affects manufacture and vice versa. The linkage is crucial. And yet we've been concentrating on the pieces.

What we see is a badly fragmented body of activity in which even the participants in the various segments are frequently unaware of their impact on the other segments. It is not surprising that the general manager often fails to take into account the coupling between the various classes of technical activity when making decisions. The system does not make evident the cost or risk that may be incurred if the downstream and feedback effects are not taken into account. It is no wonder that perceptive CEOs are uncertain whether the technological needs of their business are even discerned, much less addressed. The penalty to an enterprise of this fragmentation is major: extended product development cycles, problems in manufacturability and quality, program recycling to "get it right," uncontrolled (even unseen) costs, and businesses in severe difficulties—even businesses going bankrupt.

Data on the magnitude of the total technological activity rarely exists at the company or even business unit level and never at the national level. The body of work I label technology is, in the aggregate, substantially ignored in the management literature and in engineering education, which concentrates on the creation portion of the spectrum and on engineering design; the other aspects of technology are learned on the job—if at all. The management techniques associated with technology are learned almost exclusively on the job.

Response to fragmentation

The position of vice president of technology or chief technical officer has come on the scene as an apparent successor to, or level above, the vice president of research or of research and development, vice president of engineering, and vice president of manufacturing. The title has become more conspicuous among the corporate representatives to the Industrial Research Institute. Corporate managers have come to believe that somehow the various dimensions of technology have not been receiving the appropriate emphasis, and even more, that the interfaces among the dimensions have not been integrated effectively. They are besieged with requests for more R&D to support product innovation while they are worrying more about costs. They see quality problems grow, with only finger pointing between engineering and manufacturing as a response. They see products poorly received by customers or product introduction dates missed with only a shoulder shrug of resignation that you must expect that when you attempt innovation. Corporate managers sense that something is not working right, but lack sufficient understanding to correct it. A typical response in such a situation is to create the new position of vice president of technology and ask the new incumbent to straighten out the mess.

In most cases the nature of the concern cannot even be articulated very clearly. The new vice president of technology expects that his purview transcends functional boundaries, but the nature of his position and his power vis-à-vis that of the traditional functions is rarely delineated carefully.

As a consequence of this uncertainty, the appearance of the new position vice president of technology has been more an indication of a problem than an agency for its resolution. Nevertheless, the action is a fumbling step in the right direction because it does put in place a person motivated to delineate the scope of technology and to attack the problems inhibiting its effective use.

Additional causes of fragmentation

The literature on technical management focuses heavily on the management of R&D and, more recently, on the management of technological innovation. In general, this literature is not misleading in the sense that it straightforwardly identifies its focus on R&D or technological innovation—especially big, dramatic innovation. The material often fails to make clear, however, that an additional universe of related work and managerial problems is excluded by the study being reported. Nor does it provide a hint of the downstream effects on the remainder of the application portion of the spectrum in Fig. 1.1. Thus one could easily be misled—however inadvertently—into thinking that the management of R&D is coextensive with the management of technology because the other elements are left virtually invisible. In fact, the rest of the corpus of technology is distinctly larger than R&D and arguably more important to the success and viability of an enterprise.

This emphasis on R&D tends to give a distorted picture of the total sweep of technology that an enterprise must employ in order to succeed. A focus on R&D, of course, stresses discovery or invention and conversion or extension to application. It also highlights the central role of technical risk in R&D. The long-accepted definitions established by the National Science Foundation (NSF), reinforced by adoption by international bodies and the Financial Accounting Standards Board (FASB), still are puzzling to people in industry because they make no reference to economic or market imponderables and cost constraints, which sometimes loom as huge barriers that technology must overcome. Programs facing these uncertainties are regarded as R&D by many (probably most) technical organizations, even when the technical risk is thought to be insignificant. Yet the key question addressed by the accepted definitions would appear to be, will nature permit it?

Uncertainty and risk are indeed present, but in general they are not literally of the "will nature permit it?" sort, but rather, "can I provide a configuration of product attributes at a cost that will attract customers?" This is not to say that *only* things which are already physically possible attract attention in industry and that industry devotes no effort to developing new capability, where questions of nature's cooperativeness are paramount. Rather, it is to emphasize that there is a huge body of effort that does not fit that rubric: a body of effort that if not done well can be disastrous for an enterprise. I suspect that more firms get into difficulty by neglecting to keep costs competitive, letting quality slip, and allowing product performance and features to get out of step with customer desires than by failing to create important innovations. The U.S. auto industry during the seventies is a well-known example.

The problems we face begin in technical education itself, which gives far too limited and fragmented a view of the nature of technical work in industry. Engineering schools feel that the requirements to produce a graduate with adequate skills in a single discipline are already being stretched. That pressure, plus departmental jealousies, results in graduates with an inadequate conception of the holistic, multidimensional character of technology. Engineering education (with the exception of chemical and metallurgical engineering) focuses on the skills that are most useful for work in research and development or product design. It provides little insight into problems of manufacturability, automated manufacture, sources of costs in engineering or elsewhere in a business, modification to meet customer needs, or design for easy field service.

The problems are exacerbated by technical career paths that emphasize functional specialization to the exclusion of breadth and by traditions and conventions that emphasize the special calling of technical work: complex mathematics, esoteric concepts, rigorous analysis, creativity, and an ability to live with high uncertainty. Unfortunately, these perceptions of the nature of technical work are easily translated to believing that one must be able to *do* science or engineering in order to understand them and their role in a business. I do not believe that. In fact, the fundamental premise underlying this book is that managing technology as an indispensable ingredient in the business equation is based on principles and insights that nontechnical managers can and must acquire and practice. Furthermore, a technical background and even experience in technical management do not necessarily equip one with the necessary understanding to manage technology from a corporate perspective. Managers, both technical and nontechnical, must work together for improvement without casting aspersions on the important contributions both must make.

An additional factor that contributes to the distorted and fragmented perception of technology is the relative isolation of the technical community—an isolation that is created by behavior, perceptions, and expectations of both technical and nontechnical people. It is common wisdom that ordinary mortals cannot understand technical matters and that one must trust the professionals. I have seen general managers' eyes glaze over when talk turned technical. I could almost see them thinking, "I will turn off until things get back to something I can understand." This attitude of the general manager is an abdication of responsibility for which he must be held accountable. Sadly, it is also unnecessary, as I hope to demonstrate. Not all general managers can, or even should, have a technical background—experience has amply demonstrated the fallacy of that position. But general managers, irrespective of their backgrounds, cannot call themselves general managers and evade responsibility for technology—and for acquiring the knowledge necessary to manage it effectively from a corporate perspective.

There is also an unfortunate corollary view: that technical people are not interested in or very skilled at business matters. Technologists cannot afford to limit their interests to just technology, and the rest of the management team dare not let them.

Naturally, these mutually reinforcing impressions lead to an environment in which technical managers are expected to make technical decisions—frequently with inadequate input from, and interaction with, other functions. And often they are only peripheral participants in other kinds of business activities. One member of a corporate executive office encapsulated this situation by referring to the vice president of technology as "the high priest of technology."

Most technical decisions cannot and should not be made in isolation. For the great majority of cases the key question is not, *can* we do this, will nature permit it? but rather, is it a good decision for our business, *should* we do it? With the limited funds and technical skills available, with the activities and plans under way elsewhere in the organization, with the dynamics of events in the marketplace, is this the best way to deploy our technical resources? Obviously, that set of questions cannot be answered exclusively by the technical community.

The Substantive Dimension

In addition to a dimension covering the spectrum from the creation of new knowledge to the application and blending of both new and conventional knowledge (Fig. 1.2), technology also involves a substantive dimension: what people actually work on. For what aspect of the business is the technical work being done? If people do not even share a common understanding of what constitutes the corpus of technical work—and careful listening to discussions of technology in different situations indicates that is indeed the case—the existence of an integrated system to perform technical work is very improbable. Here, we have a much more serious fragmentation problem than the one noted in connection with the spectrum from creation to postsale service. The

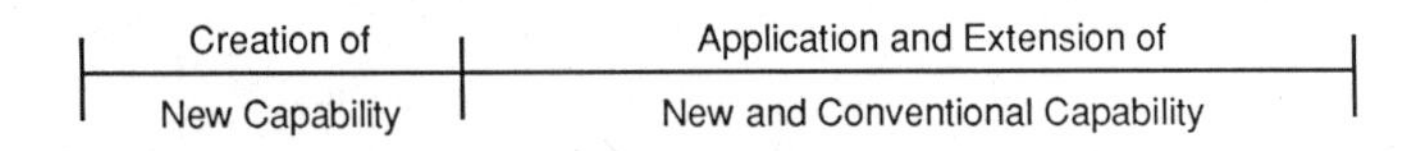

Figure 1.2 The creation-application dimension.

fragmentation covers both what is included and the way it is organized.

Product technology

The most widespread view tends to associate technology with products. Stated in organizational terms, it encompasses the body of work commonly assigned to the vice president of engineering, the chief engineer, or product engineering in an operating unit. If you ask most CEOs what they mean by technology, their first instinct will usually be to think of what their chief engineer does, i.e., the product dimension of technology.

Unfortunately, this limited concept of technology does not comprise all that should be included even in the domain of product technology. At least four subclassifications should be recognized (Fig. 1.3), two of them often associated with marketing rather than engineering:

1. *Product planning.* This is the work required to identify customer needs and to specify features and performance that will provide value at a price that generates the volume required to earn a profit. Often this work is located organizationally in marketing, but is done jointly with product engineering.
2. *Product engineering.* This is the work covering the entire spectrum from the development of new knowledge and new or modified products to the design for production of products that must meet performance, quality, and cost targets and that are producible, i.e., both design and production engineering. This includes activities to improve the process of design itself: more powerful analytical procedures, simulation, predicting performance and life, computer-aided design, and so on.
3. *Application engineering.* This is the work devoted to promoting sales or meeting customers' individual requirements by tailoring designs to their needs or by demonstrating the utility of a product

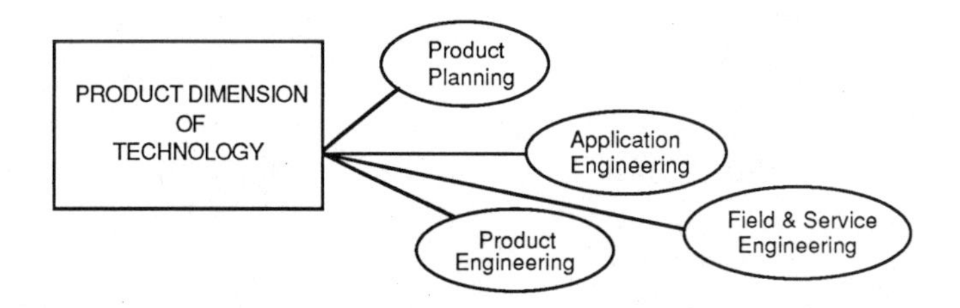

Figure 1.3 Components of product technology.

for their application. This, too, is usually found in marketing, but its intimate connection with product engineering is apparent.

4. *Service engineering.* This is the work developing systems and procedures to support field installation, maintenance, and repair. This activity includes installing and debugging new products, training operators, preparing technical publications and manuals, communicating with customers on the operation, storage, and treatment of products, analyzing complaints, and diagnosing field difficulties or failures. Again, this is often a marketing function and is all too frequently not identified as an indispensable element in the broad domain of product engineering.

The full extent of the domain of product technology as laid out above is rarely recognized as a single continuum. It is almost never placed in a single organization or aggregated as a single category of expenditure. Even worse, the communication among these four components is often inadequate, and the effects of activity in one area on the others are overlooked. Thus decisions made in product engineering to reduce costs may in fact increase the costs of maintenance or failure in the field and not even be noticed. Moreover, the fragmentation can lead to a kind of "product myopia": a focus on the product line itself and not on the fundamental sources of change or of competitive success, i.e., the underlying technology, evolving markets and customer needs, and changing competition. Nevertheless, recognition of the necessity for some kind of product engineering activity is well-nigh universal. A function exists to carry out the work, and in many cases the status of the function in the managerial hierarchy is quite high.

Manufacturing technology

A somewhat broader perspective of technology, and one which is finally beginning to receive the recognition it deserves, adds a second domain—manufacturing technology—the major components of which are shown in Fig. 1.4. This domain can be categorized into at least six subclassifications that are not necessarily separate subfunctions, but rather a set of disciplines:

1. *Materials: selection, vendor evaluation, processing yield, and fabrication.* This is the work to select materials with suitable properties, adequate availability, and favorable cost projections; to evaluate vendor capabilities in processing and quality control; and to select or develop processes for transforming, treating, or shaping materials into components needed in the final product.
2. *Equipment and tooling.* This is the work to develop or select

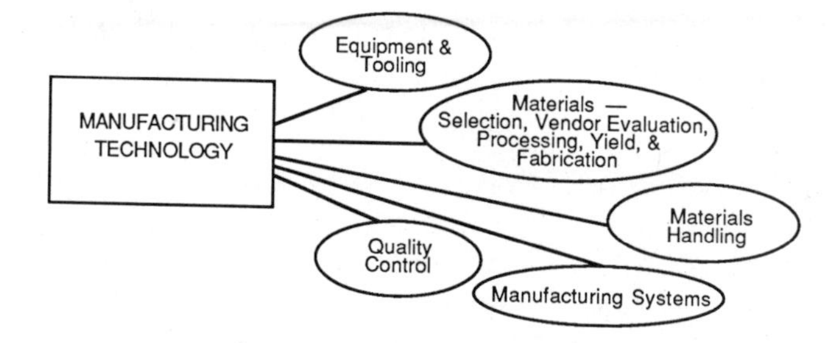

Figure 1.4 Components of manufacturing technology.

equipment and tooling needed to manufacture products and to provide optimum flexibility in manufacturing capability.

3. *Materials handling.* This is the work required to store, locate, identify, move, and orient materials and components. It operates at the macro level—in relations with suppliers, within the factory, or with a customer, and at the micro level—in assembling components into subsystems or finished products. Thus though the scale and precision differ greatly, the assembly of products and the movement of materials around on the factory floors are based on the same activity structure. At both macro and micro levels, the work includes automated materials handling and assembly.
4. *Manufacturing systems.* This is the work to develop the integrated information systems needed for parts explosion, procurement, scheduling, and production control.
5. *Quality control.* This is the work to provide systems and technologies for ensuring that materials, components, and subsystems have the dimensions and properties to meet final product requirements.
6. *Maintenance.* This is the work to understand the sources of equipment deterioration and failure and to devise preventive maintenance systems and techniques that provide optimum availability.

The work needed to develop and apply embedded control systems is included in all of these subclassifications. Typically, this means working with the hardware and software that are an integral part of the equipment (such as sensors, local memory, information processing, and signal processing) and are necessary for data acquisition and communication with actuators. Even though this kind of activity might ordinarily be thought of as information processing, it is, when embedded in a particular piece of equipment, usually developed or specified as part of the design process for the equipment. An artificial distinction is created by separating it from the machine in order to associate it with the development of the overall business or factory information system.

Information technology

Information systems are the third, and increasingly important, domain of technology. Since most of the work of management consists of the acquisition, processing, and communication of information, it is not difficult to see how central this domain is to management. It is just as important to include it as an integral domain of technology.

This area is so important and subject to so much confusion that it is necessary to pause to clarify the way in which I propose to use the term. Information processing and communications (the two initially separate fields are rapidly merging into one) is a clearly identified field of technical expertise, but the term is used by different speakers to refer to different things. One should start by realizing that information processing is not a new function per se. The medieval craftsman had to process information and communicate all for himself. The challenge now is to accomplish a level of system integration equivalent to his, but using tools of incredibly greater power in more complex environments. The various fields included under this rubric can be thought of as a pyramid resting on a foundation of basic computing power and communication capability, as shown in Fig. 1.5 and as described below.

Information hardware and software cover the physical principles and intellectual constructs that underlie the advances in memory, processing speed, and power in information processing systems and that lead to more powerful computing "engines" and communication networks.

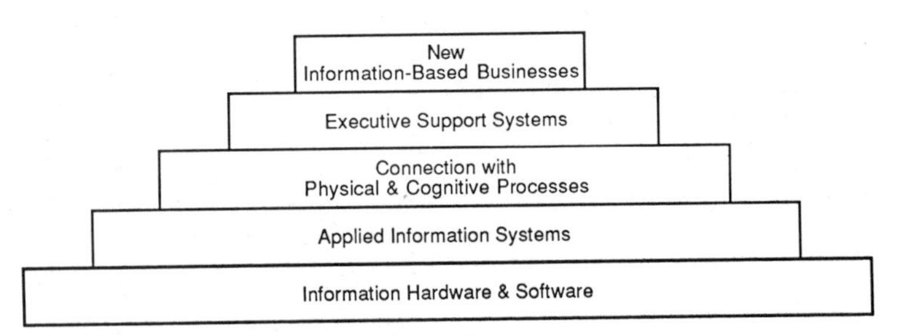

Figure 1.5 Fields included in information systems.

Applied information systems offer the system configurations, hardware, operating systems, application software, and communications protocols that are designed and operated for particular applications.

Connection with physical and cognitive processes includes the hardware, software, and physical insight that permit coupling of physical events and transactions with information systems (i.e., people, machines, and material) and work with artificial intelligence, expert systems, or other simulation of cognitive processes, no matter what the label.

Executive support systems can convert this information processing power into strategic competitive advantage.

New information-based businesses seek to develop new goods or services that may be possible through creative application of new information processing power.

Information hardware and software. Improvements in the power of processing capability are primarily in the hands of academics and computer and communications companies, such as IBM, DEC, Cray, or AT&T. Although the linkage between this level of activity and the application of information and communications systems is obviously important, I will exclude it from consideration in this discussion.

Applied information systems. The second level of activity, applied information systems, is what most people think of when the term *information systems management* is used. It is concerned with the configuration of the system and the hardware and software that are chosen and installed, the application programs that are written, and the data that are recorded and manipulated in actual systems. In effect, specialists in this field say to the managers of a business: We cannot determine what kind of information you need, who should get it, and how it should be used, but tell us what inputs you want, how you want the data manipulated and provided as an output, who you want it made available to and with what time lag, and we will design, install, and operate a system that will do it most cost effectively. We will also tell you what new capability is coming along and what its implications are for your system in terms of capacity, processing power, cost, and time response. In a sense, the information systems expert develops the brain of an enterprise, but not its mind.[2] The responsibility for deciding what the system should do, who should have access to it, and how the information should be used rests clearly with the management team.

Connection with physical and cognitive processes. The third field of activity, which is the principal area of interaction with scientists and engineers, concerns joining physical reality or cognitive processes with information processing capability. This need to connect physical items, events, and processes with data management cuts across many endeavors, including product engineering, manufacture, and product service.

Engineering. In product engineering, information technology began with using the computational power of computers to perform analytic manipulations for design, namely, programs to calculate heat transfer, perform stress analysis, examine fluid flow, calculate magnetic field configurations, and so on. Extensions of graphics capability led to computer-aided design, which first permitted graphic representation and then began to integrate it with the analytical capability for evaluating the integrity and effectiveness of a proposed design. Subsequent developments are leading to expert systems and artificial intelligence, which bring to the task at hand the cumulative knowledge and wisdom of the best authorities in the field. Engineers have shown foresight and aggressiveness in exploiting this aspect of information technology.

You will note, however, that this sequence of advances, spread over thirty years, is focused on work within what might loosely be termed the product engineering community. Transactions with other functions, especially manufacturing, have until recently received much less attention. A vision now exists, and is slowly coming into reality, in which traditional drawings disappear and engineering designs and specifications automatically extend into production instructions for guiding machines and even mold manufacturers. This vision of "art to part" requires an unprecedented melding of engineering and manufacturing lore with engineering principles and techniques.

The impact on time, cost, and performance can be immense. The Boeing 767 was designed at seven different locations in four countries, and yet the overall development program came in below budget, mostly because conventional allowances for mistakes and rework turned out to be much too large. Computer-aided design eliminated many of the errors and omissions that formerly were regarded as inherent in such a massive task. The dispersed development, however, was possible only because a massive information processing and communication system had been developed and installed. The system included common protocols for notation and communication, which permitted different specialized groups to work in their own language and yet interchange enormous amounts of data in usable form and with minimal errors. The design, installation, and operation of such a system is a daunting management challenge. A fundamental tension exists between systems visionaries who want to establish the truly optimum system and the "operators" who must be induced to learn the skills needed to use the system.

To the system purist, the information system used for the Boeing 767 might seem like a kludge—a structure that was permitted to grow unrestrained, with far too much variety in kinds of computing equipment and application software being used. Boeing understood that success depended on establishing a sense of "ownership" of the system among the thousands of engineers who had to be induced to use it. Consequently, it chose an approach of maximum autonomy for individual technical groups. It said, in effect, you select the computing equipment and the application software you believe best suits your needs. The only thing we require is that we all use a single system of notation and measurement and we all use the same communication protocols. Beyond that, we at the system level will figure out how to integrate everything into a workable system.

The principle of system design that emerged is that individual users may be free to choose the "computing machine" best suiting their needs, while the system provides an environment permitting the necessary communication and interaction among different users.

Manufacturing. In the meantime, an analogous activity, focused on a different set of needs, has been under way in manufacturing. It is useful to begin by noting the few simple to state, but profoundly difficult to answer, questions that manufacturing must address. They can be classified under planning and operations. Under planning:

1. What instructions have we received from engineering and marketing regarding designs and specifications, order volume, and delivery dates?
2. What do those instructions require in terms of materials, people, and plant and equipment?

3. Are those resources available, and if not, what is required to make them available?
4. What sequence of events and transactions is optimal for carrying out our instructions in terms of cost and time?
5. Given all the above, can we in fact carry out our instructions?

Under operations—for every part and component of the product or intermediate in a process:

1. What is it?
2. Where is it in the sequence?
3. What is being done to it?
4. Is this operation within limits?
5. Where is it to go next?

The first set of questions involves converting information into greater and greater detail until it gets down to individual units of people and machines and raw material or components. The requirements for information handling are immense, and most information processing work to date has been devoted to developing systems that can cope with this incredible amount of detailed information. The level of progress is illustrated by the fact that an automobile manufacturer can now know within thirty minutes when the new car you have ordered will be assembled.

The second set of questions is more difficult because it involves closer coupling, in real time, of physical reality with information processing capability. First, it is important to note the dual requirements of manufacturing—physical processing and information processing (Fig. 1.6).

The conventional approach has been to develop an information system for an existing factory—usually the only available step. This means, however, that the degree of coupling that can be achieved is limited by what is possible in the factory as it was planned and built. Configuring the factory and the information system at the same time

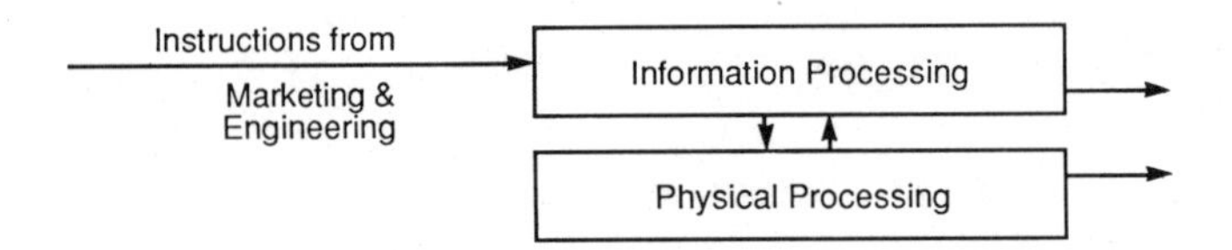

Figure 1.6 Dual requirements of manufacturing.

is more effective, but often is not practicable. Even when a "green field" installation is contemplated, the kind of system that people envision as ideal is not yet possible. If engineering instructions cannot be communicated to a machine, a vital link in improved efficiency is lost. An even bigger barrier pertains to the materials-handling questions that drive operations. Dealing with millions of parts, distributed over a large area, makes questions of *what is it?* and *where is it?* profoundly difficult to answer precisely and in anything approaching real time. These questions, which are central to information processing in factory operations, are only slowly leading to productive answers, but they basically cannot be addressed by the information systems expert. Once physical reality is converted to an information input, the information systems expert can design and operate a system to process the information effectively. The coupling between the two, however, lies in the domain of the product and process technologists. Often what is needed is a kind of information transducer that senses some aspect of physical reality and converts it into a signal that can be manipulated by the information system. The widely publicized flexible automation system of Allen Bradley[3] accomplishes the automatic manufacture, as though on a mass-production line, of one-of-a-kind products. It rests on two simple but critical advances: development of a coding system that provides unique identification of every material, part, or component, and a sensing system, analogous to vision, for recognizing parts.

Despite the progress that has been made, this melding of the physical and informational worlds remains as one of the great unexploited opportunities in technology. Present limitations continue to require huge safety factors in all planning and factory operations to provide insurance against lack of knowledge about the state of the manufacturing system. I can recall old-timers in the Schenectady plant of GE, a large-apparatus factory complex, who laughed about spending the first half day on a new shop order trying to find the part to be worked on out in the storage yard somewhere. We have come a long way since then, but we still have far to go.

The final stage of information processing in manufacturing is quality control, which ascertains whether an output conforms with configurational and dimensional requirements and whether specified product attributes are indeed achieved. This field has been treated in a fashion similar to that in product engineering and in the other phases of manufacture. In other words, inspection, testing, nondestructive evaluation, application of experimental design, and statistical quality control have been pursued more or less in isolation from the other fields, and even though progress is being made, are still far from what would be optimal.

Distribution and service. In addition to product engineering and manufacturing, physical distribution and service represent a third sphere requiring the connecting of the physical and informational worlds. The circumstances here are similar to those in plant operation, even though the distances and quantities involved and the requirements for real-time control are very different. Again the crucial questions are

What is it?

How much is there of it?

Where is it? (In the case of service, is it performing satisfactorily?)

Where is it supposed to go next? (Or, what should be done to repair it?)

Just as in the case of plant operation, a lack of timely knowledge about the state of the logistical system, together with inadequacies in the field diagnosis of problems, forces the creation of allowances to cover contingencies and slows the response in serving customers. And once again, the present level of integration is primitive compared with what is possible. Opportunities for remote diagnosis of difficulties, for practically instantaneous determination of product availability, and for smaller finished goods inventory and improved spare parts logistics offer attractive avenues to lower cost. Even more, they can be effective competitive weapons in giving greater value to a firm's customer.

Cognition. The level of information processing work I call cognition is difficult to incorporate in my schema. Work on expert systems, artificial intelligence, simulation, and so on, cuts across some of the other fields, but it is too important to bury in that fashion. The potential applications lie in almost every area from basic work on information science to troubleshooting equipment in the field and to executive support systems. At this early stage of its development, it is hard to do anything more than note its existence and its promise; and at some point, it may require designation as a separate field.

Executive support systems. The fourth field of activity in information processing is executive support systems. These systems comprise three major elements. First is the ability to reconfigure budgets and strategic plans under various assumptions about the level of business activity and the economic environment—massive "what if?" capability that in effect permits a response in hours instead of weeks. With this capability, top management can examine more alternatives and issue guidelines to operations more quickly. To those who sometimes did not know what their exact budget was until the end of the first quarter because it took that long to work out the details, this can seem like nirvana. The capability for providing this kind of support is slowly being installed, but learning how to use it is still in its infancy.

The second element is the development of business models for examining the effect on the total business of a proposed change, say, in product design, product mix, or operating procedure. Accounting and reporting procedures and systems have their roots in techniques designed to enable one to determine profitability and to control operations. They have limited utility when used as guides to decision making for product development, technology, investment, and strategic planning, for instance. They assume that the configuration of a system is unchanged by a proposed course of action. For example, they cannot tell you the effect of the introduction of a new product on the whole system, e.g., on inventory requirements, worker training costs, material utilization, or documentation complexity. Consequently, projections on costs, investment requirements, and return on investment are suspect and sometimes badly misleading. At present, the surface has barely been scratched in developing and applying this kind of capability.

The third element of executive support systems concerns access to, and interaction with, external and internal data bases in considering the strategic consequences of various alternatives. This area has received considerable attention, but usage is still limited. Many general managers have only a fragmentary perception of the role information processing can play in attaining strategic dominance. For example, the complex global relationships that are becoming important are possible only because of the remarkable information processing power now available.

New information-based businesses. The final level of activity in information processing is the bailiwick of entrepreneurs: people who have the vision and insight to see in this emerging information processing power the basis for a potent new capability, offering new products and services to the customer. Completely new businesses, such as overnight delivery, multidimensional financial services, and hospital supply, have been created. Powerful new strategic advantages, such as sophisticated airline reservation systems, have been established. These systems not only improve customer service but also help provide product differentiation, build customer loyalty, and constitute a valuable tool for market research and promotion. And we have just begun. No comparable entrepreneurial impact has yet been felt in manufacturing.

The challenge that management will face in coming years is to exploit the burgeoning power of information technology. The truly mind-boggling question that should be addressed is not yet receiving attention. If the cost of information becomes so low that it is no longer a significant constraint in designing information systems, *what should one choose to know* and with what time lag? The easy answer of "everything, and in real time" is clearly not correct. There is good psychological evidence that too much information can overwhelm and substantially incapacitate a decision maker. But if a manager *could* know practically anything he wanted about the state of his business, the choices he might make—or fail to make—about *what* to know and *when* would have a profound effect on the competitive viability of his operation. Faster response, more reliable performance, more consistent quality, greater adaptability, and better market segmentation are the stuff of powerful competitive advantage. They lead to smaller investment, fewer people, higher returns on investment, and more stable operations—all the attributes so highly prized by managers and investment analysts.

Fragmentation of the substantive dimension

Each of these three domains—product technology, manufacturing technology, and information processing technology—developed independently. They have different academic and industrial traditions and are at different stages of maturity. Engineering (really product engineering in the United States) has an academic and professional tradition that requires college- or even graduate-level education, but industrial practice has also contributed greatly. For example, GE began a graduate-level internal engineering education program in 1924. Managers of engineering almost universally rise from the ranks of those practicing engineering.

In contrast, manufacturing's traditions are more closely associated with the shop floor. Manufacturing management in the United States in turn has its roots in craft skills and shop floor supervision. It was almost invisible in academia until the midseventies. Over the last decade a number of companies have actually encouraged colleges and universities to devote more attention to manufacturing. Within industry, organized efforts to encourage manufacturing management are now beginning to appear.

The background of information processing professionals is more diverse. It includes a mixture of electrical engineering, communications engineering, mathematics, operations research, and a smattering of other disciplines. People with a financial background are strongly represented in applications and business systems. Perhaps its most important feature is the large number of people who converted themselves to this specialty after training in other disciplines. There is also an important cultural difference between engineers or scientists who become involved in information technology and business systems people. The former tend to be problem-centered, sort of bottom-up in viewpoint. They use information processing capability to solve their specific problems. The latter tend to be top-down people who think in terms of common data bases, control of access, and corporate directives. These differing viewpoints permit, almost encourage, people to pursue their individual interests with little interaction.

Are there signs that these three domains of technology function as an integrated whole? Unfortunately, work requirements have not provided sufficient impetus for bringing together these three domains with such disparate traditions. Consequently, they function largely as independent entities.

The degree of coupling differs among industries, and also among countries, as we shall see shortly, but the problem in the United States is particularly severe. The process industries—chemicals, petroleum, and pharmaceuticals—as might be expected, have achieved the greatest integration of product and process technology, and batch-output discrete parts manufacture, the least. In no industry has optimum integration among the three domains yet been achieved. In part, this results from localized deficiencies in various aspects of technology: there are many pieces missing. The dream of a fully automated flow of information from "art to part" is still in its infancy. Many analytical techniques (e.g., fluid dynamics) lack the precision and generality that are needed. But to a much greater extent, inadequate integration indicates a failure to perceive the enormous opportunity that is emerging and to focus effort and talent on it. Each function has pursued its parochial interests, and nobody has addressed the opportunities for the total business. It seems to me that the impetus has to originate in the technical community. Who else can do it? The general manager obviously must be on board, but the vision should come from the technologists.

In the midseventies I did an informal survey of managers of product engineering to ascertain the most important technical barriers to innovation in their businesses. I hoped to use the information to help guide program planning in corporate research and development. The overwhelming response was that the cost and time required to obtain tooling was the largest single barrier to innovation. Tooling usually required 18 to 30 months when the market window dictated 9 to 12 months; it cost $500,000 and provided 500,000 cycles, when often what was needed was $50,000 and 10,000 to 50,000 cycles.

I was totally ineffective in eliciting interest in this opportunity, in part because the problem was not regarded as technically "interesting." An effective attack on the problem required melding skills in materials, processing, and fabrication together with those in computerized analytical procedures to validate designs (e.g., in heat transfer and stress analysis) and in computer-aided design and automation. This required an integration of the three domains of technology, which was at the time beyond the ability of the laboratory, primarily because of internal territorial parochialism.

The degree of linkage that has been achieved in information processing between functions is particularly weak. Some of this can be attributed to the state of information technology, for despite the remarkable advances that have been achieved, deficiencies in both hardware and software continue to make integration between functions difficult. Even within a relatively advanced function such as engineering, human intervention is still too often required to enable different application programs to interact effectively.

To a major extent, however, the inadequate integration demonstrates a lack of attention on the part of managers of technology. Achieving greater integration is possible with present technology, and there is little effort to stimulate work on the missing elements. Even more, the view of these managers is still largely parochial—they do not see the total business leverage that can come from an integration of the three domains of technology. In fact, territorial warfare still goes on. On one consulting assignment for a major company, I soon became convinced that a strong motivation or hidden agenda in hiring me was engineering's concern over the growing power and influence of manufacturing. I was supposed to supply ammunition for its territorial battle.

In a world where global competitive pressures are creating inexorable demands to work assets harder, as well as to achieve faster response time and greater flexibility, the penalty for failure to integrate these three domains will be heavy indeed. Said positively, the companies and countries that manage to do so first will create a powerful competitive advantage.

Impediments to integration

The impediments to integration are multidimensional. The situation in which we now find ourselves can be represented graphically in Figs. 1.7 and 1.8. Most companies (especially manufacturing companies, but increasingly others as well) have an identified function called engineering. As we have seen, not all of the activities associated with engineering work on a product are in a single function, but

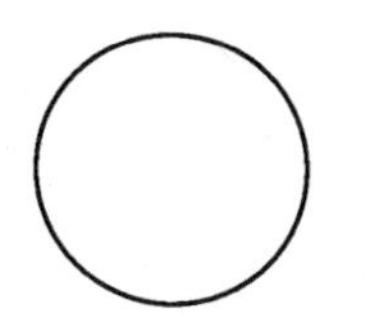

Figure 1.7 Conventional view of technology.

the function does exist and someone is held accountable for its effectiveness. In similar fashion, most industrial operations have a function called manufacturing for which someone is held accountable.

In this conventional view of technology, the two circles representing the two dimensions in Fig. 1.7 are separated—reflecting the integration gap—and manufacturing is somewhat smaller. In a great many (I am tempted to say most) U.S. companies, the manufacturing function has been placed in a junior position. It has been regarded as lower in status than engineering, its compensation levels have been lower, and its people have been deemed less technically competent. This difference has been demonstrated dramatically in the written reports of interviewers visiting college campuses for a major company. An occasional interviewer would note, "Not good enough for engineering, suggest manufacturing." The difference in status was exacerbated by attitudes on campuses. For essentially all of the post–World War II era until the late seventies, engineering schools almost completely ignored manufacturing. Chemical engineering was an important exception, providing thorough grounding in the basic unit operations and unit processes of chemical processing. A degree in manufacturing per se was virtually unobtainable. The fields that might have included some attention to manufacturing, especially mechanical engineering, carefully avoided it and instead pointed graduates toward careers in product engineering. The field with the obvious connection with manufacturing, industrial engineering, was clearly low on the engineering totem pole.

The situation in Japan and to some extent in Europe, especially West Germany, has been different. Many engineering graduates expect to go into manufacturing. The need for experience on the shop floor as a mandatory component of engineering training is widely recognized in Japan. For some years the senior technical officer of Siemens, the giant West German electrical equipment company, was a man whose background was in manufacturing and plant management. This situation was regarded as so normal as to occasion no comment. It would be viewed as bizarre in most American companies—at least in the era since World War II.

The history in information processing is somewhat different but leads to even more fragmented results. Work in information processing usually originated separately in engineering and in finance or accounting for use in payroll, billing, payables, and general ledger accounts. Prior to the advent of computers, financial people were using card sorters for classifying and aggregating inputs (in fact, punched cards continued to be used as the principal means of data entry for many years and are still being used for recording time worked). Consequently, for these people the introduction of computers was more an evolutionary step. Finance usually purchased the first computer and established territorial rights to its use. Batch processing was universal. Engineering employed early computers to develop numerical solutions for complex analytical problems. As engineering interest in computers grew, the typical response from finance was, use ours—at night. The machines were expensive and pressures to increase utilization were great. As capability for interactive use emerged and engineering needs became more specialized, engineers turned to dedicated minicomputers—sometimes using artful labels, such as programmable controller, to circumvent territorial feuds with finance. Kenneth Olsen[4] of Digital Equipment says they called their early computers "modules" to avoid the use of the term.

Other functions gradually developed their own specialized application programs and languages as their sophistication grew.

The result in our graphic representation (Fig. 1.8) is a tower of Babel. Work on the development of information processing is being done all over the place. Unfortunately, rarely is anybody charged with responsibility for the overall system needs of an entire business. In other words, nobody has the charter to develop and manage a businesswide system analogous to the circles for engineering and manufacturing.

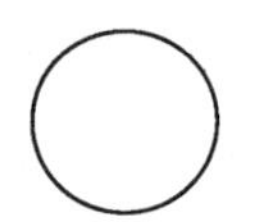

Figure 1.8 Current typical view of technology.

One of my responsibilities at GE was to review the technological dimension of strategic plans from operating components. In general the product engineering portion was easily identifiable and in acceptable shape; manufacturing was less satisfactory, but showing steady improvement; the information dimension was usually nonexistent in a strategic sense. That is not to suggest that little was being done in information systems. In fact, major, competent, and successful efforts were being made in information systems for manufacturing and distribution. The various functions would make reference to their own parochial needs, and finance would talk about system (but really financial system) plans. But nobody addressed the opportunity for the entire business. A moment's reflection indicated that no function had that responsibility.

Finance might argue that it has such a responsibility, but finance has no requirement for real-time interactive systems of the sort needed in engineering and manufacturing; nor does it have any feel for, or interest in, the coupling with physical reality that must be achieved. The pressures for achieving greater flexibility and increased automation call for information systems that finance is poorly equipped to lead in developing. The critical problems are in system architecture, in software, and in linking mechanisms for the physical and the managerial worlds. Perhaps most important of all, as said before, is the need to rethink how one might choose to manage the business in a world of virtually unlimited information.

The information systems specialist tends to be just that. He can configure systems and develop software to process information, but he has no competence in relating that capability to operational or strategic opportunity; nor is he equipped to develop the coupling with physical reality. Some companies are responding to this situation by installing a chief information officer to focus on the meaning and use of information as well as on its manipulation.[5] A major premise of this book, however, is that even that step is not enough. Effective exploitation of technology requires integrating its three domains. Only in this way can the full potential of information technology be achieved. A chief information officer would not even have a purview of the other two domains—product and manufacturing technology. Consequently, he would be unable to direct all of the work that is needed, even if he had the vision to recognize what was needed.

The loss in time and increase in error that result from the manual retranslation of engineering outputs into manufacturing drawings and instructions are obvious. Less obvious are the attitudes in both functions that lead to difficulties in manufacturing and quality assurance: ineffective scheduling, engineering changes that cost more than they are worth, and unnecessary expansion of in-process inventory. Finally, failure to achieve adequate integration means that the entire system both slows down and builds in safety margins in human resources and materials to protect against the lack of real-time information about the state of the system.

If one examines the problem among the three key operating components—engineering, manufacturing, and marketing—the cost of inadequate integration becomes more obvious. A partial list of the operational activities affected by inadequate integration would include orders processing, capacity, overtime costs, inventory, material availability, line rebalancing costs, substitutions, customer status, payables, and engineering workload.

The interdependence suggested graphically in Fig. 1.9 represents the state we must seek. In fact, the eventual equilibrium for many companies may look more like Fig. 1.10, even in manufacturing industries. I was intrigued at the recent comment of a vice president of research for a major consumer products company: "Information is now our most important technology."

Achieving the more comprehensive, integrated view I advocate requires the active involvement of technologists and those in other functions as well. Most of the important decisions about technology cannot be made in isolation by the technology managers. Conversely, decisions in other areas that ignore the interaction with technology which is implicit in those decisions are unlikely to be optimal. Since no company can run exclusively with people who are technically trained, those without a technical background can and *must* participate in such deliberations, and I believe strongly that this is possible if they but devote enough attention to learning how.

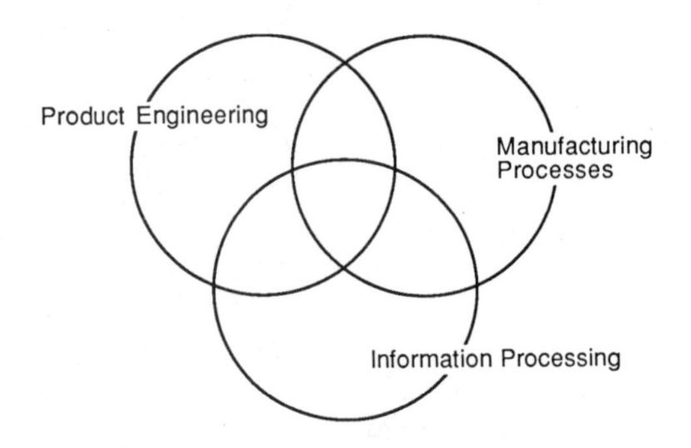

Figure 1.9 Integrated view of technology.

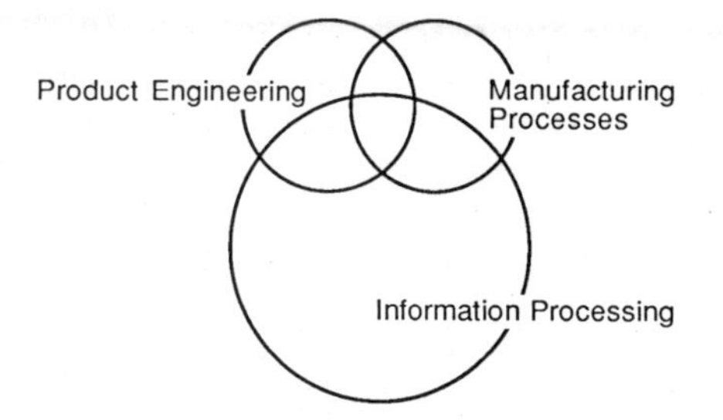

Figure 1.10 Will information dominate?

WHAT IS MANAGEMENT?

Balancing Tensions

Having made an extended examination of the boundaries and inner structure of technology, let us now turn our attention to management. Although there is greater consensus on the meaning of the term, it is important to clarify my perspective. Throughout this book, I will focus on the cognitive dimension, not on the equally vital dimension of leadership. As the *Business Week* follow-up[6] of the superior companies identified in *In Search of Excellence*[7] showed, sound business decisions are also indispensable ingredients of excellence in management.

The classical view of management emphasizes decision making and problem solving. Managerial work tends to be presented as though it were episodic. Over time, managers periodically make decisions that set directions and then solve problems that get in the way. Furthermore, decisions tend to be presented as binary choices—yes or no. If one observes the behavior of managers over a long period of time, it takes on a different coloration—especially if one examines the forces that drive them. Most of the work of management is orchestrated by a standing agenda that permits monitoring the state of the business. The skill and ingenuity with which a manager establishes the items on that standing agenda determine the sophistication and vigor with which he is able to control the operating performance of the business.

Most managers most of the time, and many managers all of the time, concentrate their attention on this kind of work. In practice, it can be thought of as a sophisticated way to monitor the state of the system and then focus attention on perturbations that threaten to take it outside prescribed limits—e.g., orders received, sales, deliveries, quality, cost, profitability, and cash flow. In other words, this is the classical management by exception. The perspective of this aspect of management, which focuses on operational performance, can be thought of as the creation of certainty—trying to preserve order in a world that persists in trying to come unglued. Managers are expected to establish challenging and financially attractive targets and then achieve them. Harold Geneen's treatise on management focuses almost exclusively on this aspect.[8] The investment community always maintains a sharp scrutiny of this facet of management. It values evidence that management is in control, that it achieves the performance it targets. Surprises, as much as failures, are regarded as evidence of poor management, a lack of adequate control.

Managing paradox

It would be convenient to say that management responsibilities stop with operational management, that is, with achieving effective or efficient performance. Unfortunately, managers cannot escape so easily. They are fated to have to nurture, and yet keep within constructive bounds, incompatible requirements. The continuing survival of an enterprise requires both operational management and strategic management—the former seeking near-term efficiency, the latter, the desirable change necessary for survival. I define survival as achieving a level of performance that continues to attract capital and that provides protection against raiders ferreting out underperforming assets to "restructure." This duality generates a paradox that is perhaps the greatest challenge in management. Thus my theme of the need for effective management to live with—nay, to nurture—paradoxical tension.

Managing for survival may seem melodramatic and overly stark, especially when applied to established businesses. The process may be described as revitalization or renewal, but the goal is clearly survival. Historical studies demonstrate that survival is, or should be, a matter of continual concern for every management team. And it cannot be taken for granted, even by firms as dramatically successful as Xerox.[9] The mortality among young enterprises is well known and accepted as part of the winnowing process for determining those that have the "legs" to continue to survive and prosper. Survival cannot be taken for granted by any enterprise. Table 1.1 indicates the fate of the 25 largest industrial corporations in the United States in 1900.[10]

Of the ten most profitable companies fifteen years ago, three no longer exist as independent companies, and four had returns on equity roughly equal to bank certificates. Of the forty-three companies cited for excellent management by Peters and Waterman in their book *In*

TABLE 1.1 Fate of 25 Largest U.S. Industrial Corporations in 1900

Surviving intact and still in top 25	1
Merged into a new top 25 entity	7
Bankrupt	2
Merged, but no longer in top 25	3
Still in business below top 25	12

Search of Excellence, fourteen had lost their luster within two years.[11] Obviously, it is not enough to take good care of the short run—that's one of those necessary but not sufficient conditions.

Strategic vs. operational management

Although strategic planning is usually initiated to improve performance or to circumvent a threatening vulnerability, it has provided a powerful impetus to managing for survival by encouraging managers to project the present trajectories of their businesses and ask themselves if performance is adequate to ensure continued viability. If strategic planning evolves into strategic management, as it must if it is to survive, it requires different skills and priorities from those for operational management. Strategic management becomes the countervailing force that creates the paradoxical tension requisite for survival.

Where operational management prizes stability, imposes structure and discipline, and resists change, strategic management asks: What about future markets? What are continuing prospects for growth? Will our current technologies be replaced? Will present modes of competition continue to be effective? Will our management style and corporate culture continue to be efficacious?

Thus strategic management addresses fundamental questions of strategic change, and change of this magnitude is antithetical to operations management. It is disruptive of careers and relationships and values. Change creates anxiety. One cannot be sure that the change itself or the path chosen to it will be productive or even necessary. Change requires both financial and human resources—resources that otherwise might be used to improve other operations. Sound operations management recognizes the need for incremental change, but it regards strategic change as something to avoid.

Questions of change and survival are not easily addressed. They are murky, unstructured, controversial, and threatening. Operations-focused management is unlikely to face up to them. Unless the issues of strategic management are assigned as an explicit responsibility to parts of an organization that will not be diverted by daily operating problems, they are unlikely to be confronted effectively. Fortunately, the human and financial resources required for these activities are tiny compared with those for operations. Nevertheless, the work is vital to the survival of an enterprise.

Invention, the starting point of innovation, is considered by many as the central responsibility of research and development, a major facet of technology. But I and others have long argued that innovation (the sequence required to go from invention to use) must be seen as a responsibility of an entire business—it cannot be accomplished by R&D alone. By that same token, the introduction of the changes needed to ensure the survival of an enterprise cannot be thought of as limited to technological innovation of new products or processes. Survival may demand significant changes in the business portfolio, modes of competition, or corporate culture—any or all of the above. Changes of this sort are not conventionally thought of as innovation, but the process of accomplishing them faces many of the same barriers. Thus a management cannot assert that it has met its responsibility for innovation (even just for *technological* innovation) by establishing an R&D component. Nor can it succeed by believing that only success in technological innovation is essential for survival.

It follows, quite obviously, that strategic management is often in conflict with operations management. It is in competition for resources, and its point of view is contradictory. There is an inevitable tension between the two. Successful management requires, first of all, recognition that there *must* be such tension, then assigning responsibility and allocating resources so as to have effective protagonists for each kind of activity. Continuing profitable survival depends on balancing the tension. Management is condemned to a perpetual hell of ambivalence. The most critical aspects of successful management do not admit ultimate solutions. To the contrary, successful management involves a continual balancing of contradictory alternatives, none of which can be dispensed with, and then a ceaseless monitoring of the balance to ascertain whether it is still optimal, given the changing external world and the maturation of the business. I liken the successful manager to a skilled piano tuner, who is continually checking to see whether all the tensions on the strings are still correct, i.e., whether the instrument (enterprise) can still create good music.

This aspect of management is not episodic. It does not solve problems, put them behind, and go on to other problems. Rather, this face of management recognizes that it is important to keep some issues alive and healthy and cause for relentless concern. The apparent disappearance or assumed resolution of an issue is, in fact, a danger signal. Companies dare not act like lions, lords of all they survey, too powerful and successful to be attacked. Instead, the price of long-term survival is a certain level of discomfort—a hair shirt that management must actively seek rather than deny.

Implications for Managing Technology

It is clear that this perpetual contesting of contradictory forces has important implications for managing technology. Our discussion of technology highlighted the importance of the creation-application spectrum of technology. The creation and introduction of new technology forces change. The requirements for the successful application of existing technology resist change. The tension between these two is inevitable.

Intolerance of error

What are the requirements for effective operational management of technology? The first is that it be intolerant of error. Technology must work as specified all the time. That ideal may not be achieved, but it remains the ideal. For technology to perform as society demands, the primary criterion in its choice must be conservatism—there must be substantiating data that the technology will perform as specified. Consequently, there is a strong preference for thoroughly proven technology with massive data to delineate the boundaries of its capability. I recall the wonderment of one of my associates who transferred to operations to introduce a new process. He said, "Son of a gun, when my boss says, 'I want it to work,' he means 24 hours a day, 7 days a week, 365 days a year!"—and he might well have added, "with an idiot running it."

The second criterion is that the mode of application be as constant as possible, that features and requirements be fixed and not casually changed. Standards and disciplined routines should exist for design and manufacture. These constraints reduce the likelihood of failure and they permit repetition, which is the foundation of the learning curve that reduces cost.

Cost

Cost introduces the second requirement for technology: it must be cost-effective and competitive as well as certain. Generally, more than one technological solution is available. Competent technology management dictates that the most cost-effective solution be chosen. This is true even though the parameters defining costs and benefits are uncertain and are strongly influenced by convention and tradition.

Certainty of performance and cost-effectiveness are the twin demons that haunt managers of technology. In our rush to proclaim the virtues of innovation, we tend to forget the fundamental mandates for technology. Successful managers must be hard-nosed and demanding of data to support choices; they must place a premium on certainty; they must be conservative in their selections; they must urge standardization and structure to guide and constrain decisions. But they must also demand cost-effective solutions and insist on minimum acceptable performance. Otherwise, a customer may decide that he is getting more capability than he is willing to pay for. "Good enough is best" and "the best is the enemy of the good" are appropriate aphorisms.

Unfortunately for the manager, the twin demons of certainty and cost do not push in parallel, leading to another source of perpetual tension between the two. Very often, the most certain technological solution is more expensive. Managing technology would be much less difficult if cost constraints could be ignored. Technology not only must be cost-effective in terms of perceived value to the end user, it also must be cost-effective compared with competing solutions.

Endless improvement

Technology has a third requirement: relentless improvement. The next generation of products and processes—and there must be a next generation—must be better or less expensive than this generation. Technology is a crucial weapon in competitive success, and it is a foolhardy manager who assumes he is safe with present levels of cost or performance.

Even the skilled operations manager is not permitted the luxury of the status quo. He, too, must strive for "more" in terms of revenues and profitability, but in the search for incremental improvements he will reduce certainty.

Readiness for revolutionary change

The final requirement for the management of technology is a readiness for revolutionary change—an ability to utilize a discontinuity in performance, as opposed to incremental progress, that makes conventional capability obsolete. In the vast majority of cases this new capability is created by people possessing quite different skills and even pursuing disparate objectives. Even endless effort in pursuit of perfection with conventional technology leaves one vulnerable to a big leap forward coming from elsewhere. The maser, electronic predecessor to the laser, provided the very low noise amplifier that has made space communication possible. One of my physicist friends pointed out that no amount of effort to reduce the noise of conventional amplifiers would have produced the same result. And yet the inventor of the maser was not seeking a low-noise amplifier—although he recognized it when he saw it!

Thus the effective manager of technology shares, in extreme, the inherent tension that is central to all of management: balancing, but also nurturing contradictory requirements. He must apply the proven state of the art to create value for the user. But he must also extend that conventional technology to achieve continually better performance or lower cost. The operative criterion in product design is "good enough," in technology development it is "more," and in strategic change it is "new" to supplant the old.

Even at this operational level, the effective manager of technology has one final contradictory force he must both foster and balance: he must decide whether to rely on internal resources to apply the state of the art or to use technology developed and demonstrated externally. In this capacity, which is the analog of the "make-buy" decision in manufacturing, the manager of technology faces a unique handicap. All his professional instincts lean toward "do it yourself." The value

All his professional instincts lean toward "do it yourself." The system of scientists and engineers is biased toward the application of individual creativity—not toward the active search for somebody else's solution. The "not invented here" syndrome is deeply rooted in the values and traditions of most technically trained people, and it is a bias against which the effective manager of technology must always be on guard.

The tension between conventional and completely new technology also includes special hazards. As we will see in more detail subsequently, most candidate new technologies prove inadequate. The manager of technology dares not chase every will-o'-the-wisp that comes along. Furthermore, with the very high probability that the new technology will require skills that are foreign to his in-house capability, he lacks the ability to pursue them effectively anyway.

What is he to do? If his organization is too small to permit this ultimate step in specialization in the responsibility for pursuit of conventional technology and creation of new technology, he has little choice except to stay nervous. He *can* remain very conscious of his vulnerability and maintain a ceaseless surveillance of external developments. As one of my perceptive associates used to say, "If you run with your head down, pretty soon you may find yourself out in the wilderness." If the operation is large enough, and profitable enough, to support the diversion of resources from operational work, then the effective manager of technology (and the perceptive general manager) will establish a separate operation charged with creating and introducing new technology.

Conclusion

This chapter has laid out two of the themes I believe are fundamental to effective management of technology. We must approach technology as a system and seek to understand and strengthen the linkages between its various elements and dimensions. We must recognize that effective management (all management, but especially management of technology) requires nurturing and balancing tension between contradictory requirements—particularly the requirement for rigor and continuity and the requirement for change.

Before turning to the final theme in chapter 3, the complex relationship between technology and management conventions, I will pause to explore the dynamics of technological change. We must have it, but it is a complex, anxiety-ridden process.

Notes and References

1. Dwight D. Eisenhower, in a 1956 White House letter reproduced in the *New York Times,* March 17, 1985, Section 4, p. 23.

2. Dr. Donald Collier suggested this analogy.
3. "A Breakthrough in Automating the Assembly Line," *Fortune,* May 26, 1986, pp. 64–66.
4. "America's Most Successful Entrepreneur," *Fortune,* October 27, 1986, pp. 24–32.
5. "Management's Newest Star," *Business Week,* October 13, 1986, pp. 160–172.
6. "Who's Excellent Now?" *Business Week,* November 5, 1984, pp. 76–88.
7. Thomas J. Peters and Robert H. Waterman, Jr., *In Search of Excellence: Lessons from America's Best-Run Companies,* Harper & Row, New York, 1982.
8. Harold Geneen, with Alvin Moscow, *Managing,* Doubleday, Garden City, N.Y., 1984.
9. Norman R. Augustine, *Augustine's Laws,* Penguin Books, New York, 1987, p. 5.
10. Ibid, p. 6.
11. "Who's Excellent Now?" *Business Week,* November 5, 1984, pp. 76–88.

Chapter 2

Technology Maturation and Technology Substitution

The first chapter examined at considerable length the comprehensive and holistic view of technology that must become a management imperative. It also argued that establishing, and modifying over time, the appropriate balance between the application and extension of conventional technology and the development and introduction of new technology was the most important and difficult task in managing technology. This chapter will explore additional characteristics of technology, its life cycle, and factors that influence its introduction and use, which we must understand and take into account in establishing this critical balance between conventional and new.

Most candidate new technologies will prove unworthy. Moreover, a manager of technology lacks the in-house skills to pursue every candidate that comes along. He must understand the dynamics of technology, how it advances, and what hurdles new technology must pass in order to be accepted. The endgame is application. I learned long ago that R&D managers do not and should not get Brownie points for elegant new technology that somehow does not make it. One CEO, in instructing the new vice president of R&D of a laboratory that had this problem, said, "Get that damned place hooked to the company!"

Technology Is Like a Jigsaw Puzzle

A particular technology does not function in isolation; it is part of a complex system of knowledge of how to do things. In considering whether to continue to rely on present technology or to opt to develop new technology, it is important to understand the process by which a new technology enters the system. I find that it helps to think of the edifice of technology as a complex jigsaw puzzle—a puzzle that has literally been created over centuries. The fit between the pieces, each representing a particular technology, is exceedingly intricate and precise. Occasionally, an invention, e.g., instant photography, antibiotics, lasers, or television, leads to a new capability which actually extends the puzzle, but this is a rare event that cannot form the basis for management principles and priorities. Even the continuing advances in microelectronics, information, and biotechnology, which in the aggregate will have enormous impact and extend the puzzle, will as single entities have a limited impact on the management of most businesses. The difference between a micro- and a macroperspective is important. An individual management team cannot act as though all possible dramatic advances will be important to it.

Even the rare discontinuities, however, cannot be brought into use independent of the existing body of technology. They will have to be made of materials that can be produced in adequate quantities and fabricated with tolerances that can be achieved. They must not introduce undue hazards or side effects and must be amenable to operation by ordinary humans.

A more typical situation, however, occurs when a potential new technology provides a capability that overlies that of one or more pieces in the puzzle. The present corpus of technology represents a capability of enormous power, which is by no means yet fully exploited. Therefore, the new piece (technology) must enter by demonstrating attributes that are in some way superior to those of the present pieces. Even such dramatic advances as the transistor and jet engine had to run the gauntlet of competitive evaluation.

The existing puzzle is both intricate and unforgiving. A new piece must fit the present configuration with great fidelity or else demonstrate enough promise to energize a reconfiguration of the puzzle—a requirement that few advances satisfy. Unfortunately, the fit cannot be determined by inspection, analysis, or simulation—it actually must be tried. This demand imposes so many constraints on a new technology that most prove deficient in some crucial respects. Even more unfortunately, the new technology may exhibit a potential fit with several different pieces of the puzzle, but only trial will demonstrate which, if any, can in fact be replaced.

The situation is even more intractable because several different new pieces may appear at roughly the same time—witness the recent competition in laser disks between electronic and mechanical scanning or the earlier competition between Kevlar, steel, and fiberglass in radial-tire belt cord. None may prove successful in entering the puzzle, but the first one to succeed will usually foreclose the acceptance of any other contestant. (The success of the VHS system of video recording over Betamax is an interesting example of an exception.)

Finally, the puzzle itself is not passive. It recognizes the potential threat of new technology and busily tries to improve its existing structure so as to resist replacement. The manufacturers of photo flashbulbs managed to delay the adoption of electronic flash for widespread amateur use for approximately twenty years by creative invention of lower-cost, more convenient flashbulb systems. The threat to their business proved to be a powerful stimulus to innovation that succeeded in holding off electronic flash despite continuing important advances in the latter. Vacuum tubes were thought to be doomed by the transistor over twenty years ago, but by virtue of emphasizing cost reduction and identifying niches where they had special advantages, they managed to remain profitable long after they had been written off as finished.

Thus new technology is chasing the constantly moving target of conventional technology, which is goaded to accelerated improvement by the threat. Frequently, the new technology never catches up. In fact, I would argue that one of the most important economic benefits of a high level of innovative activity is the stimulus it provides for improving conventional technology. The continuing advances in magnetic tape for audio and video recording prevented thermoplastic recording from ever becoming established, and analogous advances in magnetic disks represent a dynamic challenge to optical memories. In a similar fashion, advances in existing technologies for interrupting high-power circuits proved to be an insurmountable barrier to vacuum interrupters despite their technical elegance and many attractive features. Steel manufacturers, threatened by aluminum, found ways to resist corrosion and use thinner sheet and lighter shapes.

Videotape and videodisk are examples of competing new technologies. The advances of the tape cassettes in recording length, ease of use, and miniaturization, together with the advantage of home recording and reusability, proved a formidable—and moving—obstacle to the success of videodisk.

Fortunately, the circumstances that discourage the incorporation of a new "piece" also stimulate an active search for additional applications once a new piece is accepted. I have seen a number of situations in which the most enthusiastic supporters of a fledgling technology failed by an order of magnitude to foresee the success it would achieve over the next ten to fifteen years.[1] Many of the new applications were ones they never even imagined.

The pioneering user of a new technology runs two risks. The technology may not be successful, but that can happen with conventional technology too. With a new technology, however, the pioneer can look silly as well, and be criticized for having tried something unproven. Once the technology has been demonstrated to perform successfully, the second risk evaporates. After the credibility of a new technology has been established, one may have thousands of "helpers" seeking additional uses. Their aggregate impact is frequently grossly underestimated.

The Technology Paradox

Given the power and complexity of the existing edifice of technology, it is not surprising that most attempts to replace existing elements are doomed to fail. Evolutionary advances that require adjustments among present pieces are much more likely to succeed. Since the failures usually disappear quickly, while the successes receive wide attention, we often underestimate the failure rate. Starting with the beginning step—a bright new idea—those that survive all the way to successful application are probably less than 1 percent. The director of one lab in a company renowned for its success in innovation says you need 2000 tries to get a hit.[2] Even those that survive long enough to compete for entry probably fail 95 percent of the time.

Fortunately, human nature manifests the duality of seeking security and predictability, but at the same time, of being fascinated and excited by change and the uncertainty of the unknown. Managers face a perpetual trap in technology. If you say no, most of the time you will be right, but that other 5 percent can kill you. I repeat, the price of survival is to wear a hair shirt of wariness.

Thus a manager of technology is left with a threatening paradox. If you examine the history of technology, you are forced to conclude that all technologies are fated to be replaced—eventually; however, most attempts to replace them will fail. That "eventually" is like Russian roulette: when will it fire?

Achieving a sound balance between continued concentration on conventional technologies and diverting effort to bring on a new technology means understanding the process by which a new technology arrives and appreciating the barriers to introduction that an aspiring new technology must overcome. One must remember that a perception of the conditions which foster innovation comes from observing failures as well as successes and that failures are more common, though less amenable to study.

Technology Life Cycles

Interestingly enough, despite the intense interest in technological innovation, much insight has been gained by studying the process of technology maturation. If we understand how and when a technology becomes vulnerable as well as the issues involved in technological substitution, we are more likely to achieve a sound balance of effort.

Understanding the maturation of technology is important for three reasons. The kind of advance that is made in a technology tends to change as it matures, and work that continues to focus on objectives which were appropriate at an earlier stage is likely to become progressively less productive. Second, this change in the nature of technical progress is a signal that a technology is maturing and therefore may—I emphasize, *may*—be becoming vulnerable to attack by a new technology. Third, accompanying the maturation of the technology is a change in the nature of the management focus that is needed and in the business strategy that should be pursued.

Maturation of industries

Two lines of inquiry have given much better insight into the dynamics of maturation. One line of study has examined the history of particular industries from birth to maturity. The other has looked at individual technologies, their characteristics during the early years, how those characteristics shift over time, and the events that occur as they are replaced by new technologies. Although the specifics of the two types of studies differ, their findings are compatible.[3]

Study of industries indicates that at the beginning an industry is characterized by a great variety of product features and a large number of product offerings from many different suppliers. Gradually, a dominant configuration of product features and attributes emerges. Figure 2.1 shows this change graphically.

As this occurs, the driving forces that shape the industry begin to change. Initially the rewards go to those innovations that contribute to the further proliferation of products and features in the search for the dominant survivor; i.e., successful innovations tend to keep an industry in flux. Eventually, the rewards go to those innovations that stabilize the industry and contribute to the strength and penetration

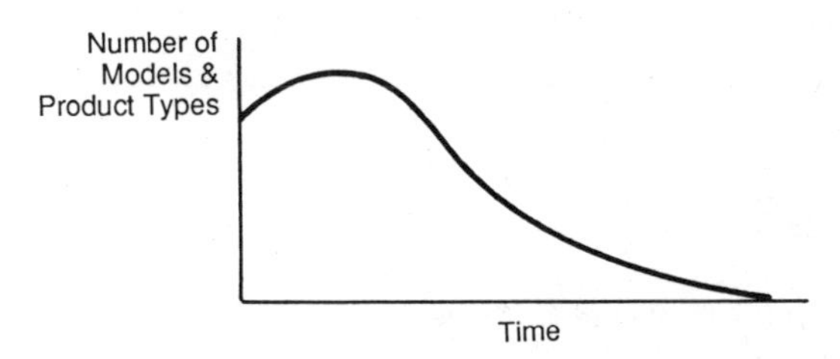

Figure 2.1 Industry maturation of product and model diversity.

of the dominant configuration. As the sequence progresses, extensions and refinements of the dominant attributes and configurations become more and more prominent. As the product stabilizes, process improvement and contributions to improved productivity become increasingly important, as shown in Fig. 2.2. As the industry moves toward maturity, product differentiation becomes more difficult to establish, and total business effectiveness largely determines competitive success—breadth of product line, strength of distribution, quality of service, and so forth.

Persistent efforts to introduce destabilizing innovations when a young industry, for example, personal computers, is moving toward a dominant configuration—as Apple attempted with its Apple II, Apple III, Lisa, and Macintosh, each of which largely ignored prior history—are unlikely to be effective. Conversely, when an industry is far advanced in maturity, its modes of operating and of competing become a kind of security blanket that is increasingly vulnerable to a destabilizing attack—remember the problems encountered by the U.S. auto industry and the tire industries' persistence in ignoring radial tires. This process is not immutable, however. Sometimes an industry headed into maturity can be revitalized by the introduction of new technology or new design concepts that greatly enhance its value. The introduction of the self-cleaning oven is an example of new technology that induced many owners of gas or electric ranges to buy a new product. In analogous fashion, in-the-hood microwave ovens markedly enlarged the market beyond that for countertop ovens.

Maturation of technologies

The life history of particular technologies reflects an analogous pattern. When a technology first emerges, there is a sense of wonder about the new capability, e.g., jet engines, semiconductors, or integrated circuits. At that point the technology is relatively primitive

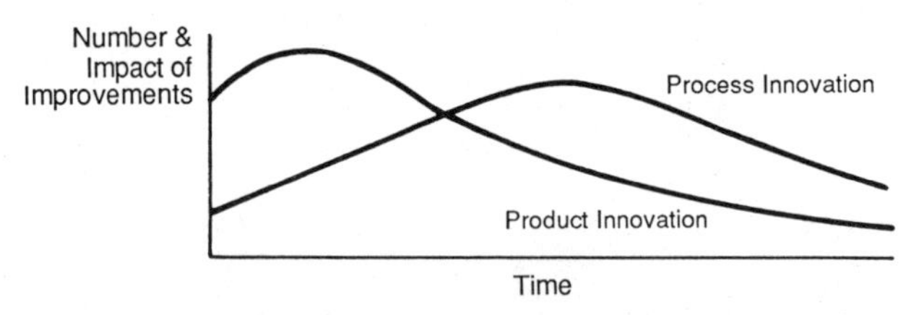

Figure 2.2 Industry maturation. (*Source: James M. Utterback, "Innovation and Industrial Evolution in Manufacturing Industries,"* Technology and Global Industry, *National Academy of Engineering, Washington, D.C., 1987.*)

and many of the paths for improvement are evident to those working in the field. Since no inventory of the product is in the field, there are almost no constraints on what can be tried. During the early years, progress is rapid and nearly all the effort is focused on improving the raw physical capability of the technology. Also, during those early years a business seeking to capitalize on the new technology is dominated by engineers. A sophisticated understanding of the potential of the new technology and an aggressive pursuit of key advances are critical to success. An engineer whose career coincided with the development of jet engines commented of the early days, "Hell, it was a miracle that the things flew at all. Nobody had any preconceived expectations, so you felt free to try anything."

As the technology is applied, constraints begin to emerge. Advances in technology must be applied in ways compatible with equipment already being used. Basic geometries and configurations become standardized. Preferred materials and components begin to dominate. Unrecovered investments in facilities encourage adoption of advances that make use of those same facilities. In many respects, the price of progress in a technology is to take the fun out of it. In part, this is because nature does impose limits. As an example, the melting point of tungsten has been a constraint on incandescent lamps for more than 75 years. As the gap between natural limits and reliable practice narrows, the effort to achieve additional advances becomes more difficult, and advances are likely to be smaller and less frequent. More and more attention shifts from improvements in capability to improvements in processes that lead to lower cost. Thus manufacturing effectiveness rises in importance and visibility for management as well.

This neat sequence is more complicated in cases where product and process technology must be developed together. This has been obvious in the so-called process industries, especially chemicals and petroleum, for many years. It is becoming more prevalent in other industries as well, e.g., instant photography, integrated circuits, liquid crystal displays, and compact disks. In each of these cases, there could have been no product without parallel sophisticated process development. Furthermore, the success of the new product was dependent as much on advances in the process technology to reduce the initially high cost as it was on advances in product features and performance.

As processes become more sophisticated, they become more expensive and specialized, and the technology becomes more capital intensive. Concurrently, the effective management of assets increasingly determines competitive success. Inevitably, management becomes more financially oriented. It is not unusual to have a shakeout in an industry as capital intensity becomes more important, and this heightens the role of capital investment in competitive advantage. Figure 2.3 illustrates this dynamic process.

This work on life cycles has been extended by applying microeconomic analysis of marginal cost and marginal utility.[4] Every technology has a theoretical upper limit of performance imposed by nature. As technological progress accumulates, at some point the marginal cost of developing an additional increment of improvement increases. Hence, a company seeking to retain a competitive advantage from this technology must invest larger and larger sums in R&D. If a new technology is invented that has the potential of replacing that technology, a competing firm will have an economic advantage because the investment required to create an increment of improvement will be less, as shown in Fig. 2.4.

This view, or course, assumes a more or less constant field of applications. As our technology base grows, however, the application of old technology in new fields becomes an important source of innovation and can return the technology to a lower point on the curve. For example, the use of electrical motors—a very old technology—in tape drives imposed extraordinary new demands for acceleration and precision in control, which suddenly made engineering of improved product attributes much more important.

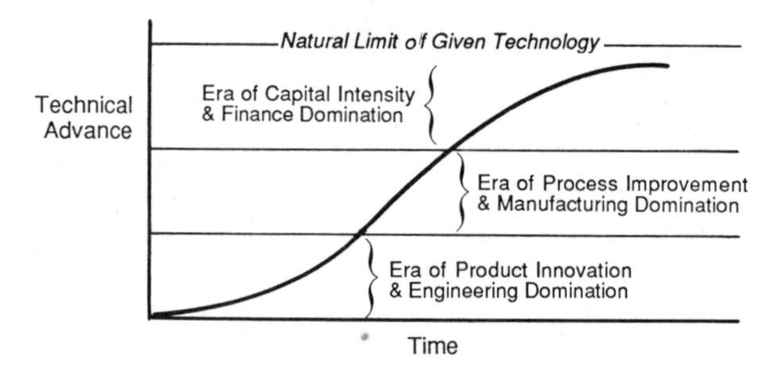

Figure 2.3 Technology maturation.

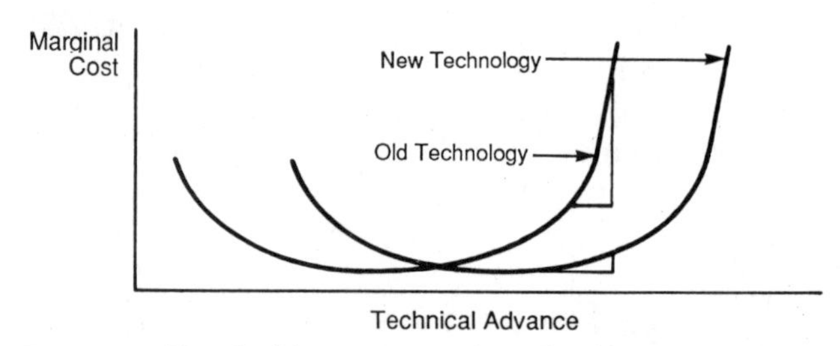

Figure 2.4 Marginal improvement in technology per unit cost.

The slope of the improvement reaches its maximum at some inflection point of the S curve. The inflection point on the curve is difficult to predict in advance. R&D people, deeply involved in developing the technology, tend to become zealots and to anticipate continued advance beyond the point when progress will in fact begin to diminish, as suggested in Fig. 2.5.

When a new technology begins to supplant the present state of the art, we are confronted with a discontinuity that places us at a much lower point on the S curve for the new technology (Fig. 2.6). As we have noted, the new technology is often developed by outsiders who have no commitment to the present state of the art. Since the young new technology yields much larger increments of improvement for a given level of effort, the conventional technology faces a growing and eventually insuperable handicap in attempting to compete. The intrinsic limits imposed by nature, of course, represent the ultimate limit on technological performance—a limit that is approached asymptotically. Obviously, the speed of development varies markedly from one technology to

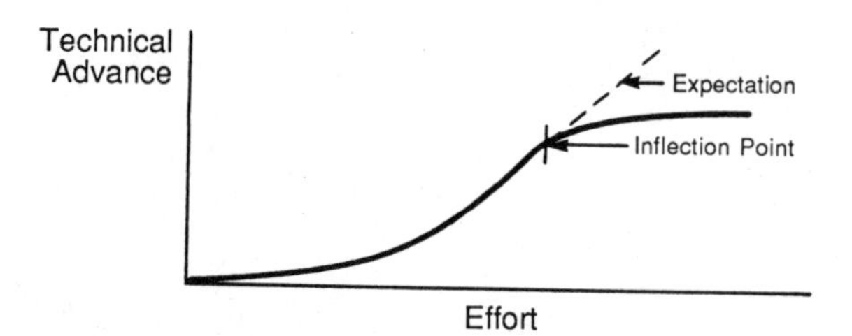

Figure 2.5 Rate of improvement of technology. (*Source: Richard Foster, Presentation at the Industrial Research Institute Senior Management Workshop on Integration of Technology into Corporation Planning, Savannah, Georgia, March 6–8, 1986.*)

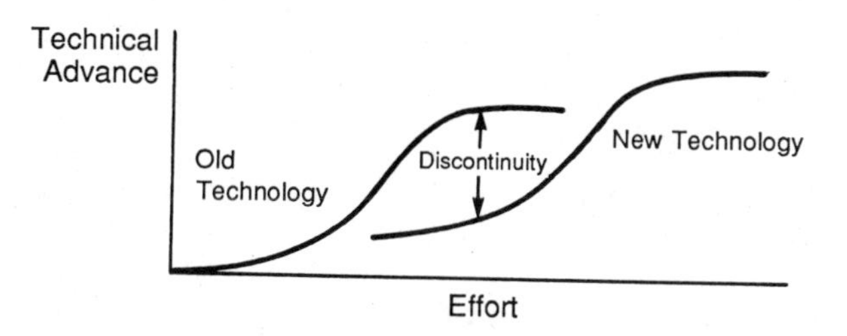

Figure 2.6 Discontinuity created by a new technology. (*Source: Richard Foster, Presentation at the Industrial Research Institute Senior Management Workshop on Integration of Technology into Corporation Planning, Savannah, Georgia, March 6–8, 1986.*)

another, e.g., lasers required years to begin to find uses and thus remained as intriguing technological curiosities, whereas microprocessors moved very rapidly to begin to ascend their maturation curve.

A. D. Little (ADL) also emphasizes the S curve as a representation of the maturation and replacement of technologies and the need to pursue different strategies for technologies at different stages of maturity.[5] ADL divides the S curve into four segments, as shown in Fig. 2.7. ADL divides technologies into three categories:

1. *Base technologies.* Those that are the foundation of the business, but are no longer critical to competitive success because they are widely available to all competitors, e.g., keyboards and power supplies for computers, landing gear for aircraft, shock absorbers and batteries for automobiles, and picture tubes for television.
2. *Key technologies.* Those that currently have the greatest leverage on competitive position, e.g., very large-scale integration (VLSI), operating systems, and applications software in computers; composite materials, propulsion, and electronic controls in aircraft; electronics and aerodynamic design for automobiles; and digital and stereo television.
3. *Pacing technologies.* Those in an early stage of development with a clear potential for changing the competitive equation, e.g., gallium arsenide or vector and parallel processing for computers, the unducted fan engine for aircraft, ceramics for automobile engines, and high definition for television.

ADL asserts that companies tend to overinvest in base technologies relative to the competitive leverage which further progress in those technologies can provide. In addition, companies often are insufficiently aware of new technologies having the potential to replace present key technologies as important competitive weapons.

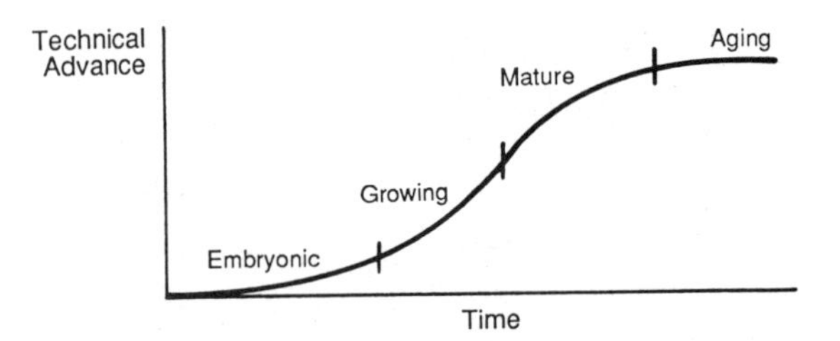

Figure 2.7 Technology maturation sequence. (*Source: The Strategic Management of Technology, Arthur D. Little, Cambridge, Massachusetts, 1981.*)

Technology substitution

Perhaps the earliest rigorous analysis of technological substitution was done by Fisher and Pry.[6] They discovered that the percentage penetration of a new technology (and therefore its level of substitution for present conventional technology) also takes the shape of a classical S curve. They examined many different examples and found that the time scale for substitution varies widely, but the characteristic S shape always holds. They also noted that—at least historically—once the first 10 to 15 percent of substitution had occurred, the process could be regarded as self-sustaining, as indicated in Fig. 2.8.

Retrospective studies help to illuminate the generality of the process, but they may do little to help predict the course of specific technologies. For example, color television appeared well before black-and-white had achieved full saturation. Subsequently, black-and-white sets were sold principally as lower-cost second sets—an application that had not existed before; therefore, sale of one versus the other was not a straightforward substitution.

These various studies demonstrate some general historical principles of technological maturation and substitution and the tension that exists between conventional and emerging technologies. Unfortunately, the future course of events remains murky. At a minimum, awareness that a technology is maturing should alert managers of technology to its growing vulnerability and lead them to an increased monitoring of potential threats from new fields that may supplant conventional technology. The third section of this book includes a critical examination of these concepts and the techniques for applying them.

Clearly, the most important strategic technological judgment is to determine not only the likelihood, but also the timing, of the emergence of a candidate new technology as a genuine threat. A company

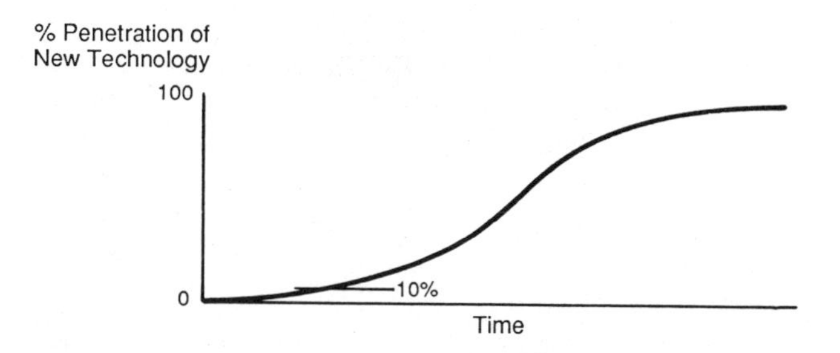

Figure 2.8 Technology substitution. (*Source: J. C. Fisher and R. H. Pry, "A Simple Substitution Model of Technological Change,"* Technology Forecasting and Social Change, *vol. 3. 1972, pp. 75–88.)*

cannot afford to fritter away its resources in pursuing every potential threat that appears—especially considering the poor rate of success for new technologies. However, a company risks extinction or severe trauma if it misses the emergence of a new winner.

Thus, as I discussed in Chap. 1, the most important single task in managing technology is to maintain the proper tension between effort devoted to conventional technology and work that seeks to replace it. Implicit in this tension is the existence of effective programs in each category of work. There can be no tension if all the effort is devoted to conventional technology and one only bemoans the absence of candidate replacements.

Misconceptions About the Introduction of New Technology[7]

The nature of technological progress and the conditions that must be present to foster innovation are, unfortunately, subject to many misconceptions. A better understanding of some uncommon precepts can at least provide guidance in what to look for in evaluating an emerging technological opportunity or threat.

Scientists and engineers are not immune to misconceptions about these precepts—in fact, they sometimes lend credence to them. Few of them, no matter how competent, have the opportunity to see the entire process of creating and applying new technical capability in operation. Consequently, it is not surprising that they do not fully perceive the characteristics of the entire process.

Every misconception does not apply all the time, but in the aggregate they lead to profound misunderstanding of the actual rate, direction, and character of technological progress. Evaluating their relevance in individual cases can markedly improve our skill in anticipating the impact of technological innovation. Judging correctly when a major technical discontinuity is indeed impending or when extensions of conventional technology will continue to prevail is one of the most crucial and difficult decisions a business executive must make.

The drama of innovation is played out over an extended period of time, and frequently some of the participants are barely aware of the other participants or even of some of the important parts of the process itself. Furthermore, individual cases vary greatly, and there is a built-in bias to be more aware of the successes than the failures. What follows could be termed common misconceptions about technology.

MISCONCEPTION NUMBER 1: *"Best possible" determines the choice of technology.*

REALITY: *"Good enough" is the basis for choice.*

The term "good enough" is not very elegant, but it connotes quite accurately the elusive criterion needed for success. Technology is not an autonomous system that determines its own priorities and sets its own standards of performance, subject to physical limits. Quite the contrary, social and economic considerations determine the priorities and set the level of performance for application. Said another way, unwanted technology has little value—therefore, it is not very attractive to create. Similarly, technical performance that is better than customers want nearly always incurs a cost penalty and is rarely viable. Few customers will pay for a racing car when what they really want is a family sedan. A Xerox senior product manager, unhappily recalling a ten-year hiatus in introducing new products successfully, said it well: "We were looking for the champagne, candlelight, and nice steak dinner and what we really needed was a hamburger. We needed a good, solid, meat-and-potatoes product in that market segment. We were screwing around trying for perfection and we weren't happy because we couldn't get it...."[8] The technologist is forced to conclude that "good enough" is the proper target, but this is a decision the customer, not he, makes. In the eyes of the user, of course, "good enough" includes cost. Cost constraints are the constant companion of the technologists. Technology devoid of economic criteria is meaningless. Therefore, a cost barrier should be regarded as a technical barrier.

Almost every product involves a series of technical compromises that determine cost, performance, and life. These compromises must mirror the balance of values that customers want—or the products do not sell. When society was uninterested in energy efficiency or materials recycling or environmental degradation, technologists did not work at improvement. Society was not paying anything for such accomplishments, and technologists got very little in the way of personal rewards for working on them. Instead, much effort went into achieving the lower costs which customers wanted. This lower cost was often achieved as a trade-off against energy efficiency, which was of relatively little interest to customers at the time. For example, the technology existed to make more energy-efficient electric motors and homes. But with energy costs very low and consumers indifferent to efficiency, more efficient and therefore more expensive motors and homes would not sell.

Attempts to "sell" customers on a higher level of technical performance are problematical. Ford attempted to make safety a new feature in the fifties with unsatisfactory results. Building contractors are notorious for installing only basic models of appliances. DuPont's efforts to sell Tedlar, a surface coating with twenty-five-year life, to the construction industry were not rewarding.

When societal values change, technology responds. Motors and houses become more efficient, and processes that reduce environmental degradation are adopted. Furthermore, more effort goes into extending the limits of technical capability. Unfortunately, we often overestimate the speed with which a response can be implemented or the level of improvement that can be achieved in the short run. Later, I will describe some of the reasons for these unrealistic expectations.

Even more unfortunately, some people in science and technology lend credence to the overexpectation. In part, this results from the natural enthusiasm of those deeply committed to a project or an objective. However, those who have not personally sweated out the excruciatingly detailed task of trying to make something work right *all* the time at a cost-effective price are likely to oversimplify the task. The sheer quantity of information that is required to reach such an objective is huge.

This naiveté was evident in proposals by dedicated electronics advocates to move rapidly in replacing electrical and electromagnetic controls for appliances with electronic ones. One manager of manufacturing for appliances objected strenuously. "In order to do that," he said, "within five years I must build a new plant that will employ 5000 people. I don't know how or where to build the plant. I don't know what kind of people to employ or how to train them. I don't know how to select the equipment for the plant or how to lay it out. I don't know where to order materials, how to set specifications for them, or how to evaluate incoming material. I don't know how to establish test and inspection methods for the factory. The whole thing scares me to death!"

Technologists working in the field, however, also sometimes underestimate the magnitude of improvement they will be able to achieve over the long term. It is their reputation that is on the line, and given the inherent uncertainty in their work, it is not surprising that they give conservative estimates. Nickel-based alloys have been the critical high-temperature material for jet engines since the very early days. The 2500°F melting point of nickel was easily established as the ultimate limit. The more important question was how close to that it would be possible to come. In the midfifties the operating limit was 1650°F. Expert wisdom projected improvements to possibly 1850°F, but nobody really knew what was attainable. In fact, the current operating limit is about 2050°F.

MISCONCEPTION NUMBER 2: *Choice of technology results from rational analysis.*

REALITY: *Choice is strongly influenced by convention and past practice.*

The technology chosen is rarely as good as our present level of knowledge permits, as advanced as is physically possible, or even all that society would expect if some series of events led to a different mindset. Herbert Simon's principle of "satisficing" would seem to apply in determining the level of consistency and intensity of societal pressure for progress.[9]

For example, in the area of health care, societal pressure is insatiable and relentless. Science and technology have responded accordingly with nonstop efforts for improvement, and our political system has provided funds. In contrast, environmental deterioration was largely ignored for many years and then was subject to a period of intensifying concern, which led to a significant shift in technical priorities. Likewise, for the many years when energy efficiency had little perceived value, it received scant attention. Within the last fifteen years the intensity of societal interest has varied, depending on whether people view energy as an immediate problem or not. Our energy consciousness has obviously been raised and the efficiency with which we use energy significantly increased, but the intensity of effort on long-term programs to create new energy sources varies widely.

Sometimes "good enough" is better than is actually needed to achieve a goal. With modern detergents, cleansing ability is relatively insensitive to water temperature. Yet belief in the virtue of hot water in assuring cleanliness has been slow to change in the United States and even slower in Europe, despite the savings in energy that would result from using cold water. Lightweight irons have encountered similar resistance.

In the United States small household appliances have come to be thought of primarily as gifts or impulse purchases. This perception affects the level of performance and life that are set as design specifications, not to mention the price and even the method of merchandising. Thus "good enough" must yield a cost that appeals to mass markets as a gift item. In Canada and in Europe such appliances are regarded as important additions to household capital. Accordingly, customers expect and are offered products that have longer life, higher performance standards, and higher prices.

Sometimes "good enough" is based on an erroneous assumption with respect to costs. Targets for quality conformance have traditionally been based on these assumptions: higher volume is always associated with some sacrifice in quality, and beyond some point, the extra cost required to improve quality is greater than the savings generated in assembly, avoiding rework, testing, and so forth. The Japanese, by questioning these assumptions, insisting on higher standards of quality, and then taking the necessary steps to achieve them, have demonstrated that our standards were really conventions that did not yield the lowest total cost.

Herbert Simon's principle of satisficing[10] also applies to the work of technologists. In almost all fields of technology there are areas where improvements *could* be made if technologists saw a sufficient reason for doing so. But opportunity cost is also a factor. Progress results from choosing objectives and then focusing effort. The effort of a technologist, already working on a problem, will be seriously diluted if he attempts to solve an additional, different problem at the same time. Naturally, most choose to work in areas that appear to have a high potential for progress and, not so incidentally, for furthering their own careers. When some major societal change in values and priorities occurs, technologists usually discover opportunities for improvement in areas they previously might casually have said were limited. I recall a manager of manufacturing saying that he had established a special account for complying with environmental regulations, but that the careful analysis which the regulations triggered demonstrated that most of the corrective actions actually led to cost reductions.

The improvements that have been achieved to date in energy utilization in industrial operations have been more or less "off the shelf"—things that have become viable because of the change in energy prices. We have barely started to create new, more energy-conscious technology for the energy-intensive operations associated with processing materials, such as glass, steel, and nonferrous metals, not to mention the selection of materials and the redesign of systems based on greater consciousness of energy costs. No matter what happens to the price of oil, technologists are now much more aware of the cost of energy and will pay greater heed to its cost in future product and process development and design.

MISCONCEPTION NUMBER 3: *Technological advances or discoveries usually are adopted eventually.*

REALITY: *Most don't succeed—and shouldn't.*

Failure to understand this likelihood is both widespread and unfortunate because it leads to unrealistic expectations on the part of technical proponents and laypersons alike. The misunderstanding undoubtedly arises to some extent from the natural tendency to publicize successes but allow failures to die quietly. Even students of innovation can be led to the wrong perception of reality because most companies are reluctant to admit, much less analyze, their failures for public consumption. Generally speaking, the data for analyzing failures are less extensive and more difficult to assemble even for an internal study.

The basis for the very low rate of technological success is twofold. First, our sociotechnical system is immense, intricate, highly interdependent, and exceedingly demanding. Any attempt to introduce a new technical capability must run a gauntlet of barriers—most of which help to ensure that the new technology really *does* offer enough advantages to risk incorporating it into the system. Second, most of the time the new technology is just not enough better to warrant all the actions needed to incorporate it, or it has problems or deficiencies that were not apparent initially.

Thus in assessing the probabilities of success for a potential new technology, perhaps the first question to ask is, how significant a conceptual advance is it?[11] Does it truly offer the promise of providing a new order of capability? If it does not, the likelihood of its displacing existing technology is slim. Without a significant advance in some important parameters of performance or the potential for a significantly lower cost, users will feel the risk of committing to an unproven technology is not worth taking. It has been said that innovations are adopted through fear or greed, and a modest advance does not appeal strongly to either.

If, however, a conceptual advance is major, one should be wary of overweighting the obstacles to its effective deployment. The jet engine offered dramatic improvements in speed and comfort but also presented challenging problems in materials, aerodynamics, and combustion. Nevertheless, improvements in the key problems were rapid. On the other hand, if a conceptual advance is significant, but careful study suggests that the barriers to implementation are truly large and apparently intractable, one can perhaps afford to postpone committing effort and continue surveillance of these barriers.

GE abandoned work on its vacuum interrupter in the 1930s because electrode materials of adequate purity and freedom from gases could not be made. Twenty years later, advances in semiconductor processing provided a solution to the electrode problem, the program was restarted, and a technically superior product was developed.

One must consider not only the magnitude, but also the distribution of barriers. Some barriers are focused and will yield to a single dramatic advance, as in the case of many new drugs or xerography. In such cases, if the invention indeed leads to a new level of capability in an area desired by customers, success is more likely to follow. Other barriers are diffuse or distributed, as is the case with photovoltaic solar energy. The need for lower costs is spread over every element of that system, from high-purity silicon to wafers, to solar cells, to assembly into batteries, to methods of installing on roofs, to control systems for the resulting electrical output. Unfortunately, a dramatic advance at any one phase has only a limited impact on the total system costs, which are spread more or less uniformly.

In assessing the potential of a new technology, one must also take into account its compatibility with related technologies. Consider the interlocking relationships between fabrics, detergents, washers, and dryers. As each of these products has evolved, their interdependence has grown, even though they are produced by three different industries. The advent of wash-and-wear fabrics forced the redesign of dryers in response to the new properties in clothing. If the advantages of the new fabrics had not been so dramatic, it is unlikely they could have induced the needed change in dryers. A change in any element of this interlocking system must take into account its effect on the other elements. For example, the possible introduction of adhesives to assemble garments must accommodate the requirements of detergents, washers, and dryers; similarly, packaged foods and food storage techniques in distribution facilities, stores, and homes are interlocked closely.

The clarity of the apparent use of a new technology must also be considered. If the new discovery or invention creates a "solution looking for a problem," the speed of adoption is likely to be slow even if the discovery is quite dramatic. The laser has faced this problem, as does much of biotechnology at present. The promise of biotechnology is dazzling, but much trial and error will occur before the high-leverage uses that offer value to the user are identified.

Finally, it is too easy to forget that innovation requires not only a creative discovery, but a risk-taking, pioneering user. He needs a powerful motivation to take a chance with an unproven technology—the fear or greed noted earlier. The security of a familiar and frequently underutilized system of available technology and the unknowns yet to be discovered about a new advance create an immense and often justified barrier to the disruption created by innovation. We should be thankful we have barriers; otherwise, we might well be burdened with too many innovations that lack the capacity to survive.

MISCONCEPTION NUMBER 4: *The biggest hurdle is making the original discovery—the downstream development is just a matter of applying the necessary effort.*

REALITY: *Most of what is not yet known about a new discovery is probably bad and requires creativity to overcome.*

The potentially attractive features of a new discovery must become apparent early or it will receive no further attention. As it happens, things are nearly always more complicated than they first appear. Our ignorance is usually greater than we realize, and even exciting new discoveries almost always have some undesirable attributes or limitations.

Sometimes a "birth defect" surfaces quickly and proves fatal. A new fiber was once discovered that looked and behaved more like wool than any synthetic material known. Unfortunately, the fiber disintegrated in dry-cleaning solvents—and its discoverers were unable to solve the problem. Needless to say, this innovation died aborning.

Sometimes a limitation creates immense anxiety, but eventually yields to creative effort and luck. PPO (polyphenylene oxide), the key material in Noryl—one of GE's most important engineering plastics, whose ease of metalizing makes it attractive as a decorative substitute for metals—was almost impossible to mold with technology available at the time of its invention. The discovery of its remarkable alloying properties, namely, miscibility with other plastics, a previously unknown phenomenon, paved the way for solving the problem. Without this discovery, made by a group far removed from the original development, it is problematic whether the innovation would have survived.

Achieving consistent, predictable, cost-effective behavior necessitates great effort to learn more about the phenomena underlying the discovery and to remove or find ways around undesirable features. All this takes time, costs money, and may abort the entire effort. Frequently, the skills needed to overcome barriers are not those of the original inventor. He may be an organic chemist with little knowledge of or interest in plastics molding, or a solid-state physicist with little knowledge of power circuits and applications.

The key to success with new technology is to find pioneering applications where the advantage of its new capability is so high that it is worth the risk. The skills and insights to find these pioneering applications may not reside in the original inventor, yet combining technological adaptation and creative market development is usually the essential ingredient in technological innovation.

Potential users—with very good reason—remain skeptics until they have evidence of successful application. Identifying the leverage point is critical. For example, the little noticed, but excellent dimensional stability of the polycarbonate Lexan (a transparent, high-temperature, high-strength plastic widely used as a replacement for metal and glass) proved to be far more attractive to early users, beset by plastic warpage and swelling, than its dramatic and still exceptional impact resistance.

MISCONCEPTION NUMBER 5: *Technological advances have intrinsic value.*

REALITY: *The customer determines value.*

Unfortunately, the virtues of these pioneering applications cannot simply be demonstrated by careful analysis and paper studies. Actual market test, the analogue of the scientific experiment, must be performed. Sometimes, considerable "experimentation" must be carried out before viable applications are generated.

Originally, General Electric thought of railroad locomotives as the attractive initial opportunity for heavy-duty land gas turbines. This application still has not proved successful. Instead, they first proved viable when used to power remote pumps on natural gas lines. Efforts to use them to turn electric generators were unsuccessful despite strenuous efforts to sell them to utilities. Not until the gas turbine was redesigned for full factory assembly, instead of the on-site assembly that was traditional in the utility industry, could it demonstrate its particular advantage in providing rapid, incremental additions to peaking capacity for generating electricity. The switch to factory assembly, which is inherently more efficient in use of labor than field assembly, also led to lower cost. This change might appear to be relatively trivial compared with the original development of the gas turbine itself—and that is no doubt true—but great skill and ingenuity were required to design a power-packed, high-precision machine that could be shoehorned onto a single railroad car. However, the change had a big impact on perceived value to the customer. Management had to "hang tough" for many years while these lessons were being painfully learned.

As one who has both participated in and observed many similar attempts to match newly discovered technical capability with market pull, I have no patience with those who say, "They should have known better." Much has been made of the value of adopting a market-pull approach to technology development. This approach, if not used carefully, becomes a meaningless tautology—by definition, those innovations that succeed "found" market pull. In prospect that famed "pull" is usually no more than a barely perceptible tug which you struggle with all your wits to sense and interpret.

MISCONCEPTION NUMBER 6: *Radically new advances will win.*
REALITY: *New is not necessarily better.*

This statement is in some respects the obverse of the one above. We are observing this phenomenon today in the attempts to attain higher energy efficiency, reduce environmental damage, and improve productivity. Despite all the talk of dramatic new solutions and technological breakthroughs, the money for development is going, to a great extent, into extensions of conventional technology. These evolutionary advances are less risky, they give promise of more timely application, and in most cases, on balance, they are more cost-effective. They are also less dramatic, but they are "good enough." That discouraging prospect in fact faces most new advances.

The history of the response to the energy crisis is instructive. Much attention and considerable effort were focused on exotic new forms of energy generation and new modes of transportation to increase efficiency. Synfuels, photovoltaic cells, ocean thermal generation, windmills, and fusion—all were advocated as new solutions to the energy problem. Electric cars and more efficient combustion cycles, such as the Stirling engine, were also proposed. The most effective responses proved to be the intensification and extension of oil and gas exploration, learning how to produce more oil in a more hostile environment, enhanced recovery, and cogeneration. The exotic approaches are still waiting in the wings. We consistently underestimate how much improvement is left in a conventional technology until we really turn our attention to it.

The people who make "conservative" decisions to support evolutionary improvement are not necessarily opposed to new technology. They have to bear the odium of being wrong, of seeing an investment prove worthless. They want the application of technology that gives promise of being adequate, which usually also means the technology less likely to fail.

Failure can be expensive. For example, the pioneering application of graphite fibers in the Rolls-Royce RB211 encountered crippling problems in erosion and inadequate impact resistance to birds sucked into the engine. The costly redesign bankrupted Rolls-Royce, and the subsequent delay in delivery and reduction in performance also put Lockheed in dire financial straits.

Much publicity has been given to the promise of alternatives to the present automobile engine. The Stirling engine, electric car, gas turbine, rotary engine, and hybrid electrical–internal combustion engine have been touted. But each of these prime movers has a variety of inherent limitations involving different combinations of operating performance, life, cost, and maintenance that have not yielded to persistent efforts for improvement. In other words, there were sound technical and economic reasons for adopting the present Otto cycle for automotive power. Eventually, some of the problems with these other options may be solved sufficiently to permit their use in general-purpose vehicles. For now, only rotary engines and electric vehicles are viable, even for special-purpose vehicles serving niche markets. But those who have had to respond in the seventies and eighties to demands for better mileage and emission control really have had little choice except to concentrate on improving present engines. To repeat, today's technology system is very exacting and very unforgiving—an "almost" fit isn't good enough.

Obviously, some new technologies prove irresistible, and then a true revolution occurs. That is the pot of gold that energizes the endless search for a major advance. The Pilkington float process for making plate glass is a striking example; the personal computer microprocessor and antidepressant drugs are others. We are seeing the beginnings of a technological competition between electronic photography and silver-halide emulsion. Without knowing the outcome, I would predict that silver emulsion will prove to be formidable to dislodge, but electronic photography will find applications.

Remember the paradox I noted earlier—technologies are fated to be replaced, even though most attempts to do so will fail. Life is not always kind!

MISCONCEPTION NUMBER 7: *The power of a new technology determines its success.*

REALITY: *The infrastructure required to support it is often the determining factor.*

Inventing a new material with outstanding new properties does little good if there are no sources of input materials, no knowledge and facilities to fabricate it, or no engineers who know the design rules for using it.

The history of frozen foods is illuminating. Clarence Birdseye had his flash of insight in 1912. The development of a satisfactory quick-freezing process was only the first step in an excruciatingly drawn-out series of supporting changes. Dietetic information had to be generated comparing the enhanced properties of the frozen foods with canned goods. Experiments were needed to discover which foods tasted better. From there, the problem became more complex and diffuse. New varieties of foods and new methods and arrangements to gather fresh produce for rapid freezing had to be implemented. These changes required new behavior and higher standards of handling on the part of farmers, as well as more demanding criteria for locating processing plants close to sources of supply. New techniques and equipment for transporting and storing frozen foods had to be devised. New equipment had to be developed for both retail storage and display, and merchants had to be convinced to install it. Homeowners had to have available, or be convinced to buy, adequate cold storage at home. Ironically, in the opinion of many, the catalyzing event that led to the widespread use of frozen foods was the decision to discontinue their rationing before canned goods after World War II, because at the time its economic implications were relatively trivial. Over thirty years elapsed before all the pieces were in place for a major new technology to flower!

Obviously, not all inventions call for as complex an accommodation of our socioeconomic structure. Such situations are, however, a powerful argument for the role of large enterprises with extensive financial reserves and, perhaps more important, the ability to marshal the diverse skills that are needed.

In order to succeed, color television needed, more or less simultaneously, action on studio and broadcast equipment, home TV equipment, and then development of programs. It is difficult to imagine a creative inventor or even a well-financed venture capital company assembling the wide variety of resources for accomplishing such a series of tasks.

Those who project the rapid introduction of solar energy or electric automobiles or paperless offices or gallium arsenide are ignoring the constraints imposed by infrastructure. Even industrial and labor practices or the skill and capability of those who apply technology for the customer can prove crucial. The silicone industry developed a new silicone-base roofing material that had excellent properties and long life; it could be applied over old roofs and in very low temperatures. Nevertheless, the material required a level of skill and care in preparation and application that were at variance with the practice of the roofing trade, which reflected the hot, dirty, smelly, low-skill work associated with installing built-up roofing. Territorial boundaries in construction trades precluded the use of painters, whose skills and experience were better matched with the application requirements of the silicone material. The additional time and investment to train contractors, and to verify that they were indeed following prescribed practice, impeded the rapid deployment of an attractive new technology and imposed an additional cost, which had to be factored into its price and its potential promise. Consequently, that affected its cost-benefit ratio. These "additional" cost considerations affected both the size of the potential market and the rate of penetration of the new product.

In order for a company to make the necessary investment in infrastructure, it must be able not only to assemble the resources, but also to perceive a sufficient opportunity or threat to warrant doing so. In the early 1950s scientists at GE invented a new polymer with remarkable high-temperature properties. The equipment and knowhow to fabricate the material did not exist. The company had neither the capability and knowhow for developing the equipment nor the effective relationships for stimulating equipment builders and plastics molders to action. Furthermore, the plastics business was in its infancy; sales were small, and the business appeared to be peripheral to the company's major traditions and thrusts for growth. Reluctantly, work on the new polymer was abandoned.

DuPont, as a major chemical company with a large stake in nylon and much experience in process development and working with plastics molders, saw the opportunity for nylon as a high-performance molded material in a very different light. This combination of a greater centrality of interest, larger in-house capability, and effective relationships with the infrastructure suppliers permitted DuPont to accept the risk of developing the needed improvements in infrastructure.

A few years later when Lexan and Noryl came along, GE faced a very different situation with respect to the adequacy of the infrastructure. Both materials presented challenging problems similar to those of the earlier polymer, but molders by now had experience in working with high-temperature materials, new equipment had been developed, and design experience had been accumulated in designing plastic parts for engineering applications. Each of these plastics became a major business, thanks in part to this infrastructure that had developed in the meantime.

Sometimes an innovation can piggyback on an existing infrastructure—health care delivery is a good example of an infrastructure that despite its regulatory hurdles, eases the way for new technology. Even so, advances in medical technology have had a major impact on its infrastructure, e.g., hospital management, cost structure, and employee skills. If the infrastructure is missing or inadequate, the "barrier" to innovation may be insuperable. This latent problem is often not seen by those who predict the rapid deployment of a new technology. A good practice to follow is to assume complete technical success and then ask oneself, "Now what has to happen in order to get this thing widely adopted?"

MISCONCEPTION NUMBER 8: *Progress in technology comes principally from continuing to improve performance.*

REALITY: *Progress requires establishing standards, imposing constraints, and achieving routine.*

Some would say the course of progress in technology is to make it dull and resistant to change. The early days of any major technological advance, whether it be the jet engine, personal computer, copier, or computerized axial tomography (CAT) scanner, are times of great excitement. People feel they are having the rare experience of writing on a clean slate, or being where no one has been before. The sense of wonder, of so much to learn, of no constraints beyond one's own creativity is enormously exhilarating. After a few years, additional generations of product have been created, basic configurations have become established, the most suitable materials selected, and the key criteria of performance established. Then comes the hard part of tackling the really tough problems of making it better, but in a highly constrained environment. The pioneers will look back with nostalgia and consider themselves privileged to have had an unforgettable experience. The latecomers will find it difficult to imagine those early days.

The engineering group for GE's large steam turbine–generator business believes it has a well-deserved reputation, established painfully over many years, for engineering the finest equipment in the world. The group puts a premium on preserving that reputation. The engineers are not easy to work with; they are hard-nosed, doubting, conservative, unforgiving of mistakes, and hard taskmasters—of themselves and others (all of the characteristics of effective operational management we noted in Chap. 1). Arguably such an environment is resistant to change, and that is true, but resistance to change also minimizes fatally flawed innovation and costly mistakes. Experienced engineers have their own scars from personal encounters with the misconceptions I've been discussing. With today's immensely complex, sophisticated energy equipment and systems, only demanding attention to detail and insistence on routines and standards make engineering design cost effective.

We are observing that phenomenon in software engineering—namely, the development of rules, concepts, and techniques for design, procedures for quality control, and measurements for writing software. Software programmers have been operating with few constraints on their personal idiosyncrasies and sense of elegance. The result has been a vast duplication of effort and the creation of programs that cannot be maintained or modified, except by their creator, without a major investment in unscrambling them. This leads not only to needless cost and time in writing software and debugging it, but also to maddening diversity for the poor user. Anybody who has struggled with the endless shifts in symbology and even in keyboard usage that are required by various software packages for personal computers can attest to the tower of Babel that has existed.

There is universal agreement that we must learn how to standardize software and use modules or packages; though not the most elegant solution, they already exist and users are familiar with them. We must impose rules and standards on the free spirits who have been preparing software. It will, in many respects, make it more frustrating to be forced to use a module that is not ideal for the problem at hand. But the skill and originality of those writing software will undoubtedly be better focused on user-oriented enhancements and improved performance in using the computer itself. This battle to maintain discipline in design is endless. The engineering profession has been trying to sustain it in designing hardware for at least a hundred years, and it still requires constant attention. It should be no surprise to encounter the same problem in the new field of software.

One inevitable consequence of the imposition of structure and order in a technology is that the technology itself becomes increasingly resistant to change. It becomes more specialized for the task at hand; it develops more complex interdependencies with other technologies, but it also becomes more precise, more efficient, and more comprehensive in the solutions it provides. Fitting a new piece into the technology system requires satisfying an increasingly longer, more demanding set of constraints. I am back to my jigsaw analogy.

MISCONCEPTION NUMBER 9: *A new technology can be grafted onto an existing business.*

REALITY: *The new product and the business system developed to produce it should be created together.*

The intimate connection between the technology chosen to satisfy a need and the nature of the business system to implement that technology is too often overlooked. A new technology changes the system requirements. An attempt simply to graft a new technology onto an existing business system, with nothing else being changed, has a high probability of failing. The skills, status relationships, control set points, and modes of communication, for example, of the existing operation are almost certainly poorly matched with the requirements of the new technology. When diesel engines were introduced into rail service, experience soon demonstrated that the old steam-engine repair shops simply could not handle maintenance. An entirely new set of service shops had to be established with new people. The difference was immediately obvious to any visitor because the average age in the diesel shops was much younger.

Conclusion

Where does more sophisticated understanding of these misconceptions lead? It leads to greater conservatism in predicting the success and deployment of new technology. On a probabilistic basis if you bet no, most of the time you will win. Even when an important new capability is discovered, frequently a whole series of ancillary changes must occur—changes involving significant capital investment as well as alteration in behavior, socioeconomic power, and status relationships. It is helpful to ask, even if this is as good as they say, what else has to happen in order for it to be deployed? In our increasingly interrelated world, many advances must be incorporated into a larger system. The value of an advance is dependent on the leverage on the total system that it provides.

Yes, the odds are very high that an attempted innovation will fail.

But I cannot leave it at that. Companies must innovate in order to survive, and nations are crucially dependent on the vitality and effectiveness of efforts to innovate. The benefits of the occasional successes are enormous, not only in terms of direct rewards to innovators and gains to society, but also in the ripple effects generated by the processes themselves. Innovation goads conventional technology to improvement, stimulates adaptability to change, generates greater self-awareness of strengths and weaknesses on the part of the enterprise, and responds to one of the most powerful human drives—to try something new. Participating in or even observing at close hand the sense of excitement and commitment that pervades a group trying to produce an innovation is an unforgettable experience. Yet technology managers and general managers have an inescapable responsibility to understand the realities of the process. We are left with our paradox.

Notes and References

1. John H. Dessauer (in *My Years with Xerox,* Doubleday, Garden City, N.Y., 1978, p. 30) notes the same phenomenon even among the most enthusiastic Xerox pioneers.
2. Steven Greenhouse, "An Innovator Gets Down to Business," *New York Times,* Business Section, October 12, 1986, pp. 1, 8.
3. There is an extensive body of literature on this subject. An excellent review article by James M. Utterback provides an authoritative overview and bibliography ("Innovation and Industrial Evolution in Manufacturing Industries," *Technology and Global Industry,* National Academy Press, Washington, D.C., 1987, pp. 16–48).
4. Richard Foster, *Innovation: The Attacker's Advantage,* Summit Books, New York, 1986.
5. *The Strategic Management of Technology,* Arthur D. Little, Cambridge, Mass.
6. J. C. Fisher and R. H. Pry, "A Simple Substitution Model of Technological Change," *Technology Forecasting and Social Change,* Vol. 3, 1972, pp. 75–88.
7. The material that follows is an expansion of an article in the *Harvard Business Review,* November–December 1985, pp. 133–140.
8. Gary Jacobsen and John Hillkirk, *Xerox—American Samurai,* Macmillan, New York, 1986, p. 76.
9. Herbert A. Simon, *Administrative Behavior,* 3d ed., The Free Press, New York, 1976, pp. xxviii–xxxi.
10. Ibid.
11. George R. White, "Management Criteria of Effective Innovation," *Technology Review,* February 1978, pp. 15–23; George R. White and Margaret B. W. Graham, "How To Spot a Technological Winner," *Harvard Business Review,* March–April 1978, pp. 146–152.

Chapter

3

Management Conventions—The Ties That Guide and Bind

Our brief description of the nature of management focused on the ever present necessity to balance two incompatible requirements: In order to survive, management must oppose a relentless quest for predictable performance, for established goals, and for continuity in operations against a constant probing for change—change in the business portfolio, in the mode of competing, and in internal behavior. This latter search for change, when coupled with the decisions and actions necessary to implement that change, constitutes *strategic management.* It implicitly demands a high degree of self-knowledge, of insight into the distinctive features of the enterprise.

The Concept of the Enterprise

Every business is based ultimately on a few simple ideas, principles, or even assumptions. They address the fundamentals of the business: What products or services do we provide? Who are our customers? How do we compete? How do we define success? How do we behave toward each other? In the aggregate these fundamental features could be termed *the concept of the enterprise.* These features are often embellished with such an overlay of subthemes, counterthemes, and harmonics that the unifying structure of basic precepts is obscured. Furthermore, they are often so deeply internalized that they become invisible. More than anything else, they lead to predictability of behavior, the crucial lubricant that enables an organization to function.

A later section on strategic management will examine some of the issues in changing the concept of the enterprise—the fundamental questions that strategic management must address. In this chapter, where we are still laying some groundwork, we will examine some of the aspects of management that give it a local, as opposed to a universal, character. One of the continuing arguments among managers is over how much substantive knowledge about a given business a competent professional manager needs in order to be effective. As management concepts have developed, I think the answer now would be that at the operating level, specific working knowledge of the business is almost mandatory. At the strategic level, a manager who is very competent, has a sufficient breadth of experience, and is adequately flexible in understanding the differences among businesses need not have detailed operational knowledge of a business. The distinction between operational and strategic management is much more difficult to maintain in a single business. The distinction is more apparent at the corporate level of a multidivisional enterprise. Local concepts, local mores, and local practices are critical to a smoothly working management team, and it is important for us to understand them and give them due weight in examining the management of technology.

Although much attention is devoted to corporate culture, the concept of an enterprise is broader than this. It can be thought of as a kind of internal guidance system that keeps a business on trajectory. Whether or not that trajectory is desirable is another question, but the concept of an enterprise does more or less automatically trigger management actions that keep a business on a given course. It determines the modes of response to any particular new business opportunities: attractive-unattractive, familiar-uncomfortable, safe-risky.... The concept also determines how a business is likely to respond to changes in competition, to changes in the environment, or to traumatic disruption.

The concept of an enterprise is the subject of countless lunch-table conversations: why we succeed at some things and not others; why we made certain mistakes; what our strengths and weaknesses are; why particular management techniques or organizational structures could or could not be used. For example, an executive of a major company, hearing how the corporate management of another company intrudes into operating decisions at will, said such behavior is unthinkable in his company; custom dictates that corporate management must not second-guess operating management. I encounter such diametrically opposed perceptions quite frequently in working with different companies.

Shared Beliefs and Conventions

How does a management—not an individual manager, but a management team—manage? A business is not managed by a collection of managers; it is managed by a team that functions in an incredibly complex and yet highly interdependent and predictable way.

To a remarkable extent a management team manages from a base of shared beliefs and conventions. These beliefs and conventions are not so much taught or inculcated as they are absorbed. Many of them are so deep in the bones that they are not even evident to those who live by them. They may persist for decades and literally go back to the foundation of a company. Historians of technology are discovering that a management's behavior and values regarding the role of technology as well as other aspects of management can be traced to the early days of a company.

The first comment I ever heard about Bell Laboratories management came from Ivan Getting when he was vice president of engineering at Raytheon in the early fifties. He said, "Bell Labs is the only laboratory in the country that runs by the book and makes it work." In many subsequent visits and discussions with Bell Labs managers, I became conscious of how sophisticated the "book" is. I also came to feel that a critical ingredient in the success of a manager in Bell Labs was his ability to understand the book well enough to get around it when he needed to. Managerial success requires using the local management system, changing it, and on occasion subverting it.

What are these shared beliefs and conventions, and why are they so important? I am not referring to the values and practices that characterize leadership, interpersonal behavior, or organizational development, important though they are. I am talking about the way you run a business and determine what it will become in the future. These beliefs and conventions can undoubtedly be classified in different ways, but the following five categories include many of the most important ones:

1. Common understandings and even rationalizations about what business you are in.
2. Shared assumptions about the way you gain competitive advantage.
3. A joint sense of the manner in which the company grew, how it got to where it is.
4. Shared conventions about the quality and extent of information needed for decision making.
5. Shared or perceived conventions regarding the guidance and operational control of the enterprise and the way it will evaluate its performance.

Technological considerations are involved in the solutions developed for every one of these subjects. These considerations form the ambiance within which all technical work occurs. Yet they are rarely addressed by managers of technology, who continue to focus attention on specific products and processes. If these fundamental and often hidden technological considerations become mismatched with the requirements for competitive advantage, then heroic efforts to improve product-specific or process-specific aspects of technology are unlikely to succeed. Many of these conventions were adopted, at least in part, because of what was technologically practical at an earlier point in time. But if they have not been revisited to assay their utility in the light of present technology, they may be dysfunctional.

What business are we in?

The business you are in *does* make a difference. The extractive industries *are* different from manufacturing—they give greater weight to sources of supply and the relative quality of those sources, and they are more sensitive to the logistics of handling huge amounts of relatively low-value material. They sometimes give the impression of looking backward toward their "hole in the ground," because that is their raison d'être rather than looking forward to the market. The merger of Utah International (a major international mining company) and GE obviously brought together two disparate enterprises. At a general managers' meeting shortly after the merger, when the sensitive issues of plant closings and transfer of operations were being discussed, the chairman of Utah International highlighted the differences by noting that "you can't move a mine." I could sense BASF's roots in coal chemistry when I first talked to its management. Its *Weltanschauung*—conception of the world, if you will—is very different from DuPont's or Monsanto's.

Even within manufacturing, businesses are not the same. Suppliers to industry are different from suppliers to consumers: their sense of who their customer is tends to be more specific and detailed, and their approach to attracting his attention is more likely to emphasize economic benefits, breadth of product line, short but certain delivery, and follow-up service. Mass production is a totally different challenge from manufacturing one-of-a-kind or a few very high-value products. This difference was brought home forcefully to one senior executive who came from a background in custom equipment. He had difficulty in accepting the explanation that a warehouse full of lamp bulb blanks represented the minimum economic manufacturing quantity for a *low-volume* item. Process industries are different from discrete-parts manufacturing—even though the more the latter emulates the former, the better. In the process industries, in-process inventory, material flow, and production scheduling are largely predetermined by the original process design, but they are matters of continuing concern for managers in discrete-parts manufacturing.

Ignoring these features inherent in the nature of a business is dangerous, because they determine many of the constraints under which a business must operate, they set priorities, they affect the time horizon and the pace of the business, and they strongly influence the modes of competition chosen. They certainly affect the choice of technology and how it is managed.

I have always been struck by the time and energy people in management spend in reassuring themselves about the rationale that glues their business together. People in GE talked of the "electrical ring," the self-reinforcing sequence of generating electricity, transmitting and distributing it, and then using and controlling it that has traditionally provided the unifying themes of the company. Even the apparent departures from electrical power were rationalized: engineered materials grew out of a need for insulating materials with unusual properties; GE Credit grew out of service to dealers; Apparatus Service grew out of service to industrial customers; jet engines and gas turbines had their roots in high-speed rotating machinery, and Medical Systems, in high-voltage engineering and electronics.

The value that these common themes serve is best illustrated when they are lost or become too diffuse. Many years before RCA's recent merger, a senior staff member in the RCA Laboratories commented with exasperation, "When we were the Radio Corporation of America I sort of knew what to do to try to help it. But what in the world should I do for the RCA company? I don't even know what it is!" He no doubt applauded the actions of management in reestablishing a central theme for the company.

The common theme of "what you are" helps to focus effort in an individual business. It makes it easy to subconsciously exclude a huge array of possibilities, whether they be work in a particular organizational function or exploration of new business opportunities. 3M finds strength and purpose in concentrating on the development and application of coatings and the discovery of new ways to use coated materials, even though it occasionally makes forays into new fields. Hewlett-Packard provides sophisticated measurement and computational tools for scientists and engineers. Its feel for the computational needs of technical people has been a great strength in dominating the market for sophisticated hand calculators. This focus is much more difficult in a multidivisional company and is undoubtedly a factor in the evidence that companies which "stick to their knitting" do better.

As is true with so much of management, there is a negative aspect to this strong sense of focus as well, however. The common beliefs that help marshal and guide effort and foster commitment also constrain.

A former 3M employee commented recently that it is tough to work there if you want to do something besides coatings. What some regarded as wisdom, he regarded as timidity. Managers need to be alert to the importance of the distinction, lest "what you are" become an unquestioned shibboleth. It can become a barrier to hide behind rather than to see beyond.

The constraints imposed by shared beliefs and conventions become more significant if management concludes that future opportunities for growth in present markets are not adequate or that traditional modes of operating and of competing have become dysfunctional. Most of the thrust of strategic planning reflects management's attempts to come to grips with these problems: How do you identify and acquire new sources of growth? How do you change the deeply internalized practices and automatic responses of hundreds of people, across several layers of organization, when you cannot even describe the present constellation of conventions very well, much less specify how they should be changed?

The constraint of "who you are" becomes most serious when a business confronts the possibility of extinction because the purpose it serves may no longer be wanted or viable in the future. The oil industry has been struggling with this situation explicitly for over twenty years—witness the many attempts of the oil companies to broaden their base. As their efforts have shown, it isn't easy. Royal Dutch Shell concluded that its unifying principle was expertise in process technology and purchased a metals company. When that route did not produce returns up to its expectations, Shell decided its unifying theme was energy and became a partner in Gulf General Atomic's high-temperature, gas-cooled reactor—another unrewarding move. Standard Oil (New Jersey) concluded energy was its unifying theme, started a nuclear fuels company, and acquired coal companies. Subsequently, it broadened its horizons to smaller ventures in a wide range of industries. In both cases the diversifications represented tiny attachments to huge enterprises. In Shell they talked of the difficulty of countering the "flywheel effect," the enormous momentum of the oil business that dwarfed the significance of all diversifications they contemplated. Even a dramatic advance in technology would appear insignificant compared with the discovery of oil on the North Slope, in the North Sea, or in Mexico.

One option for a company facing the possibility of extinction receives little attention: simply to keep going in the business it is in as long as possible, rather than change in ways that might not succeed. Based on the record of both approaches, this may warrant more consideration. Some experienced managers would argue that the probability of changing an existing business successfully is so small that it is not worth trying.[1] For example, a business that had great strength in designing and manufacturing small electromechanical components for appliances concluded (rightly, I believe) that its skills were a poor match with the requirements of electronic replacements. Its value-added would diminish drastically if it simply sourced and assembled these components. Its management decided to act like a polar bear on a cake of ice floating south and look for new areas of application for its considerable expertise in electromechanical design and manufacture. Interestingly enough, its perception of what "it really was" now clarified, the business went on to discover markets and applications it had never even thought of before.

How do we compete?

Even with a common understanding of what business they are in, different firms, ostensibly in the same business, can adopt different ways of competing. Every company looks for some way of establishing a competitive advantage. It elevates the importance and power of its chosen approach, compared with alternatives, and develops internal modes of behavior that facilitate its use. Choosing a mode of competing is more than simply a matter of strategy. It is more deeply rooted than that in the traditions, values, and attitudes of an organization. It becomes a kind of security blanket: If we do these things well, we will succeed. Managers and employees throughout the enterprise learn to accept as given a set of approaches to competition and find it difficult to imagine any other way to compete.

Consider DuPont, Dow, and BASF—three giants in the chemical industry. DuPont has tended to conceive of a business opportunity as resting on a powerful proprietary technological position, preferably based on a composition-of-matter patent. Dow places great emphasis on sophisticated process technology and aggressive investment in capacity, which can lead to a dominant share in the market. BASF's roots in coal chemistry tend to be reflected in its approach to competition—i.e., it has been a raw materials–based company stressing efficiency in conversion and logistics in moving quantities of material. An awareness of marketing has been a late addition to its competitive arsenal.

In industrial electronics, Intel emphasizes pioneering technological advances and sales to sophisticated users. Its goal is to obtain a premium price for leading-edge products and avoid experience curve–driven price competition. Texas Instruments is renowned for its aggressive pricing to obtain high market share. Motorola has concentrated more on manufacturing efficiency and being a reliable, low-cost, second source.

Every firm develops, either implicitly or explicitly, a competitive posture—a characteristic mode of competing. Michael Porter has called it competitive strategy.[2] Some companies, such as Intel, Tektronix, Perkin Elmer, and Sony, compete through technological excellence and leadership in innovation. They tend to spend above industry averages on R&D and to invite the risk associated with pioneering introductions of both technology and products. They also sometimes assume that if they but maintain raw technological leadership, success will follow more or less automatically.

Other companies seek to provide value based on a portfolio of attributes deemed important to the customer—parity in product performance and life, a broad product line, competitive pricing, and excellent service. Sears has taken this route with its major appliances.

Cost leadership is one widely used mode of competitive posture. The Japanese have used this technique with devastating effect, as have Texas Instruments and Dow Chemical. It is frequently combined with using technology to be a "quick follower," where competition rests on time to market and product variety. Thus a company avoids the expense and risk of pioneering technology, but maintains both the capability and the mindset to respond rapidly. Motorola has used this method in integrated circuits, and it is endemic in Japan.

Marketing represents another mode of competing, whether it be astute segmentation or sheer power through broad product-line coverage, blanketing distribution, aggressive selling, and promotional pull-through. IBM has long been regarded as the model for this way of succeeding. Proctor and Gamble is renowned for the power and effectiveness of its promotion and distribution, as well as its innovations in new products.

Each of these approaches to competition leaves its distinctive mark on a corporation. It affects the kinds of people who are attracted to the company, the relative status of its various functions, the choice of people who become senior executives, career paths more generally, and the values and traditions that people internalize. A change in competitive posture sends shock waves throughout the enterprise in ways that are very difficult to anticipate.

Do we supply tools or solutions? One of the most striking differences in modes of competing is between components-oriented companies (suppliers of "tools") and systems-oriented companies (suppliers of solutions). Figure 3.1 suggests the basic building blocks of the competitive equation. Not every business engages in every element of competition, e.g., the food companies in general do not engage in agriculture. Nevertheless, since every business has to have inputs of

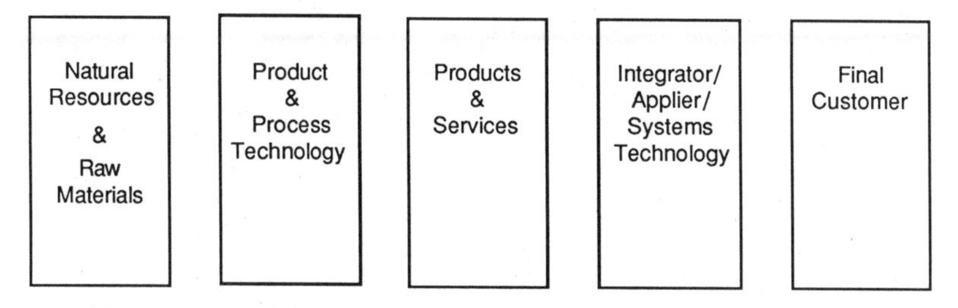

Figure 3.1 Basic building blocks of competitive equation.

materials, it must develop a posture toward the competitive significance of raw material. And if a particular element is to be emphasized, some other organization has to perform the other steps: technology has to be available for development, design, and manufacture; products and services have to be offered; and products have to be selected, maybe joined into systems, and put into use. The attitude toward the basis for achieving competitive advantage and how one offers value to a customer, even how clearly one sees customers, is strongly influenced by the building blocks used by a particular business.

In this schema, the role of the integrator-applier warrants particular comment. This function assembles and integrates whatever products and services are needed to supply a customer with the solution to his need. A turnkey plant in which the builder constructs a factory, installs machinery, trains workers, debugs its operation, and turns over a facility with a guaranteed performance is the ultimate step in this mode of competition. However, the final customer may do this for himself; he may share the role with one or more vendors; an original equipment manufacturer (OEM) may perform the work; it may be performed by a third party, such as a construction firm; or the producer of the product or service may fill the need, whether alone or jointly with the customer or others. This function, sometimes focused and sometimes distributed, sometimes obvious and sometimes invisible, is often critical to finding exact solutions to specific customer requirements.

Figure 3.2 details the competitive outlook of companies whose principal business is locating or producing raw material. As you would expect, their interest in technology centers in exploration and discovery. The eventual user is so remote, he is practically invisible. The United States steel industry traditionally placed enormous emphasis on the competitive value of the sources of iron ore and coking coal that were readily available in the country for decades. One of the reasons for the success of the Japanese steel industry in the sixties and seventies was that since it was essentially precluded from (1) effectiveness at discovery and (2) quality and magnitude of resource base, it had no choice except to emphasize conversion efficiency in pursuit of capacity utili-

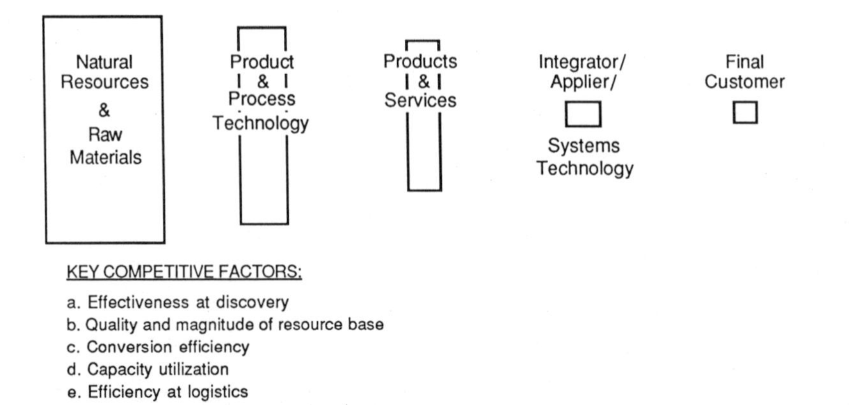

Figure 3.2 Competitive outlook of resource companies.

zation. This different competitive equation proved powerful in attacking companies accustomed to security in (1) and (2).

Figure 3.3 shows the outlook of components-oriented companies. These companies emphasize the value of the product attributes they offer a customer, attributes primarily based on special technology or low-cost manufacturing; but they stop short of assuming responsibility for the utility of their product in serving a customer's needs. In other words, they provide tools and building blocks from which a customer or some intermediary can fashion solutions. Although they warrant that their products will meet their promised attributes of performance, price, and life, for example, and they argue that those

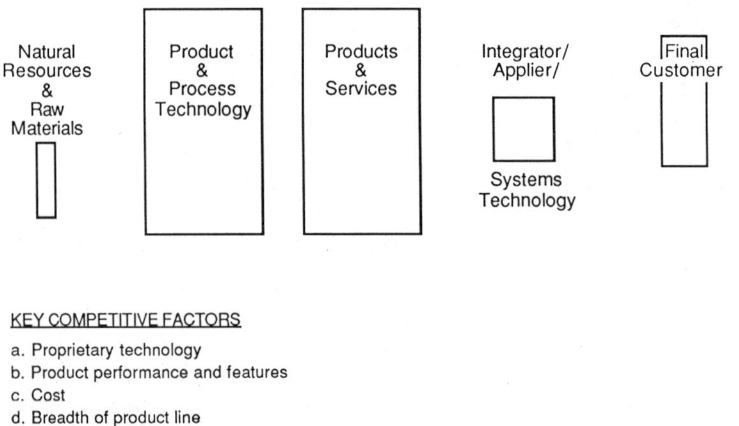

Figure 3.3 Competitive outlook of components-oriented companies.

qualities offer unique value, they do not promise that their products will meet a customer's needs. Nor do they undertake the task of assembling a variety of dissimilar parts into an integrated system. They leave that to the customer or some other agent. In fact, they sometimes do not know who the end user is or how the product is used, even though such information could be very useful in guiding product planning and development. They believe that competitive success derives from making products with more desirable attributes or lower costs than those of competitors. They are aware of the role of the integrator, even though they do not attempt to fill it. On occasion they will, however, share the role of the integrator with a customer or some third party. IBM has tended to do this with large customers, such as banks or insurance companies. That situation also exists, for example, in the relationships among utilities, electrical equipment suppliers, and architect-engineers.

Systems-oriented companies (Fig. 3.4), in contrast, focus on serving the customer's needs. Their approach is to offer solutions, not just capability. They say, in effect: Let us worry about putting all the pieces together. You just want something that meets your need, right? We understand your need and we'll put together a system that is tailored to meet *your* requirements; you can concentrate on using it effectively. The differences between these latter two approaches are profound. Components-oriented companies are much more likely to seek a security blanket in technology. They believe that technology offers the prospect of providing unique product attributes that will overpower competition. They believe that establishing a strong technical position

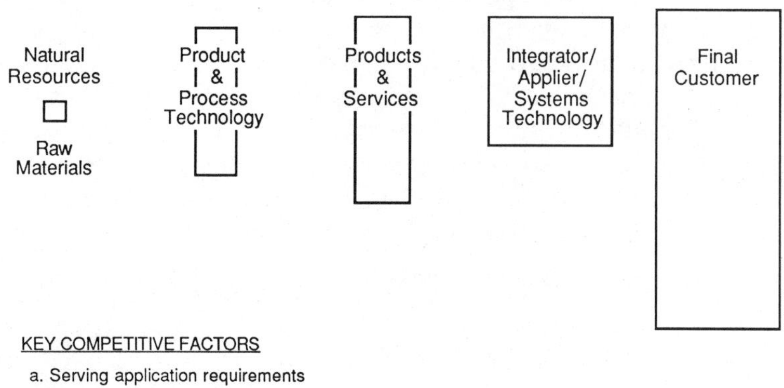

Figure 3.4 Competitive outlook of systems-oriented companies.

is the most difficult, time-consuming, and expensive aspect of running a business and that doing so successfully represents the best "keep out" protection possible. They are more likely to attempt to increase the value-added by providing a higher degree of vertical integration and to create barriers to entry by high levels of capital investment.

Systems-oriented companies place the highest value on knowing a customer and establishing credibility with him. They believe that kind of knowledge and that kind of status are the most difficult to acquire and the most effective "keep out" competitive weapons. They may try to establish a competitive advantage in selected key technologies, but in general they regard technology as something you can acquire as need be. The really critical technology is knowledge of the system. They view vertical integration as something that limits flexibility and makes one vulnerable to shifts in the business environment or a customer's needs.

Components-oriented companies tend to have autonomous operating units, each facing its own market (including internal markets). Internal transfer prices reflect this external market orientation, as do measures of managerial performance and promotion. Such companies find it very hard to assemble, and especially to price, a systems offering. Often, if all internal markups are incorporated in the final price, the figure is not competitive. But ways of allocating the diminished returns necessary to produce a competitive price are inadequate or nonexistent. Managers are deeply suspicious of entreaties to give for the common good. They say: I am judged by the performance of my division, and how do I know that special allowance will be made for the lower profit I'm accepting on this big order—especially if the system package includes components not under my boss's control? Components-oriented companies feel strong pressure to use their own products in the system because to do otherwise calls into question the value of their own products. GE managers are always uncomfortable and defensive about any non-GE electrical equipment included in their products or their factories. In contrast, systems designers in General Electric Company of Great Britain, an unrelated but systems-oriented company, are free to select the most cost-effective component, irrespective of its source.

Systems companies give highest priority to assembling a system and pricing it competitively. In this environment, internal components producers see their job as providing pieces to the system. Their customer is the internal systems assembler, as it were, rather than the eventual user. The internal measurements and rewards provide effective techniques for ensuring that contributions to the system are compensated appropriately. Systems companies are less self-conscious about using technology or components supplied by others, irrespective of internal availability. To a manager schooled in either of these different ways of doing things, the other one can seem totally foreign and ineffective.

These differences are very important because it is rare to encounter companies that are both systems- and components-oriented at the same time, although the distinctions are seldom completely sharp. General Electric and Philips are both mostly components-oriented. Conversely, Siemens and General Electric Company of Great Britain are systems-oriented. Yet all are generally in the electrical-electronic industries. Components-oriented companies depreciate the importance of a systems orientation—"You don't get paid a nickel for providing the systems engineering" is a common assertion. On the other hand, systems-oriented companies belittle the value of unique technology as a key to success. They assert that knowing the customer and being able to serve him better than anybody else is the critical ingredient, and that capability very rarely rests on unique technology. It is the total package that counts.

Why spend so much time on this distinction? The self-image and bedrock of a company is in its perception of its customers and the way to serve them. These features determine the kind of technology it will choose to pursue and the relative priority that will be assigned to different technologies. Proposals to change a components-oriented configuration by moving closer to the customer and adopting a greater systems orientation are one frequent alternative considered in strategic planning. A move like this is one of the most appealing avenues for broadening a potential market and generating additional growth by building on technology. Hewlett-Packard and Xerox are both moving in that direction. Texas Instruments has made some forays. Even materials companies talk about materials systems. These moves are generally made to capitalize on internal technical strengths, without recognizing the strong impact they may have on transfer pricing, measurements of performance and promotion, and sourcing of components. Even more, they will affect the internal power structure. Systems-oriented companies tend to give the highest status to their systems designers and salespeople. In components-oriented companies like GE, systems people (e.g., in power systems) struggle for resources from the components businesses, such as turbine-generator, switchgear, and transformers, because the latter control the pursestrings and the power. Years ago a senior manager in Bell Labs told me that despite Bell Labs' vaunted world-renowned strength and success in scientific research, which has led to remarkable advances in components such as the transistor, the really key people were its systems engineers. Their output provided long-term guidance to the total effort of the laboratories, and their discipline assured the continuing integrity of the entire Bell system.

One important difference between the competitive outlooks is that components-oriented companies are more likely to have adopted a strongly decentralized mode of organization with each business general manager having great autonomy. The accommodation needed to assemble a systems package may require the active participation of higher management. Thus, modifying a venerable and fiercely defended tradition of decentralization can loom as a major roadblock to systems businesses that cut across group or sector boundaries.

Despite the obvious attraction of significantly enlarging a potential market by moving forward into systems while building on a strong base in components, any company considering this transformation should be forewarned: it is embarking on a course that will profoundly alter the entire company. Furthermore, it is likely to generate turmoil within its management.

How you got where you are makes a difference

Clearly, a company's past history strongly influences its character. This is particulary apparent in companies that have played a powerful role in the growth of their industries. Such world-class companies exhibit common features, even though they operate in quite disparate fields, e.g., Goodyear, DuPont, General Electric, General Motors, Caterpillar, and IBM. Each of these companies played a leading role in creating its industry. Their sense of the threat of competition, their sense of the pace and predictability of external events is unlike that of lesser companies, because to a considerable extent what *they* did determined the pace and course of external events. Obviously, these companies have been very skillful at what they chose to do and how they implemented it, or they would not have succeeded as they did. Not surprisingly, that history of success also contains the seeds of trouble.

Companies which have dominated their industries often become intolerant of what is perceived as error, because in the relatively certain world in which they operate, failure to achieve objectives is regarded as evidence of incompetence. Managerial success is based on "track record," which means producing consistently what was promised, in terms of growth and profitability—no disappointments and especially no surprises.

Many of the managers of these enterprises have had limited experience in dealing with the levels of uncertainty that are common in companies occupying a less powerful competitive position. Most of them have worked for only one company and have limited awareness of its salient features or of how those features differ in other companies. They find it difficult to achieve a timely response to a pace of events that moves faster than they are prepared to cope with. The world seems out of control if they are so unfortunate as to be placed in a situation where their actions do not largely determine the pace of events. They are uncomfortable, if not hostile, to the confrontational dialogue that is frequently generated, and may even be necessary, in trying to sort out a very ambiguous, uncertain situation that is in a state of rapid change. These industry leaders, for all their vaunted effectiveness when on familiar turf, may be ineffective when operating in unfamiliar territory.

The priority given to avoiding mistakes can influence a company's effectiveness not only in pursuing businesses in volatile markets but also in investing in rapidly changing technologies, even for internal use. One common characteristic of these leading companies is their powerful and rigorous financial reviews of expenditures. They pride themselves on achieving a high return on investment (ROI) and on obtaining predicted returns on new investments. Consequently, one company found itself woefully out of date in computer systems because its process for reviewing appropriations, which required a rigorous demonstration of benefits in terms of savings achieved, could not keep up with the pace of technological progress.

Failure to recognize these characteristics can lead to traumatic problems if a management decides to participate in markets or in a competitive environment with which its traditions and practices would be badly matched. The merger and subsequent breakup of Dart Industries and Kraft Foods suggests this type of situation.

Companies that have been assembled largely by acquisition, for example, Polysar or Combustion Engineering, present markedly different characteristics. The question, just what are we, anyway? is not taken for granted. Many of their managers have worked in other companies for substantial parts of their careers. They are a hodgepodge of cultures, managerial styles, and experience. These organizations encounter far fewer taboos, but have difficulty achieving consensus and ownership on common purpose and priorities. These problems loom large in mergers and takeovers and, if not handled adroitly (and they rarely are), can wreck the result.

A management that is not sensitive to its own roots, traditions, and unquestioned assumptions is vulnerable. It may make catastrophic errors in changing its portfolio, or it may badly misread the implications of changes in its external environment. The common experience of conglomerates that have grown rapidly by acquisition, which divest the acquired companies and seek to return to their roots, is evidence of the power of these considerations. General Motors' long focus on a single component (i.e., the automobile) in the private transportation system, leaving the overall system for others to worry about, did not prepare it well to recognize the significance, for its product, of changing attitudes toward environmental degradation, safety, highway design, etc.

How much information do you need for a decision?

The extent and rigor of the information expected before making a decision represent another characteristic that guides management. A newly acquired company or a newly hired manager, when first exposed to a business review in General Electric, is likely to feel crucified. Instead, they are simply being subjected to the standard treatment (and I know other companies have the same attitude). "Doing your homework" is critical to establishing credibility with higher management, and that entails extensive effort. The work begins with assembling all the available quantified information concerning markets, costs, competition, technology, capital requirements, cash flow, etc., and even sponsoring special studies, if necessary. It includes a rigorous statement of assumptions, projections of financial performance, identification of possible alternatives, and criteria for selecting among alternatives. An investment in a business proposal is expected to analyze sensitivity to untoward events, develop contingency plans, and identify trigger points for corrective action. Furthermore, the management system requires the systematic use of the staff resources of the company. This rigorous discipline leads to the high predictability of earnings, which is so valued on Wall Street, and the universal image of GE as a well-managed company. Many other companies are more casual or less searching in their scrutiny.

People who move on to other companies comment that their new management associates are willing (and often have no other choice than) to act with less information and to rely more on intuition and qualitative judgment. A manager schooled in the GE tradition feels vulnerable when he is expected to make decisions on the basis of limited data and limited time for deliberation. A manager from another tradition may regard this style as needlessly burdensome and incapable of responding in a timely fashion. These differing conventions affect the dynamics of decision making, the effort devoted to staff work, and the kind of market and competitive environment in which a management can be effective.

The careful attention given to preparing and reviewing plans also means that once they are approved, they are not changed lightly. Achieving agreed-upon targets is a very high-priority objective. One general manager of a rapidly growing business, which was still in the "prove yourself" category, stated that he thought it was mandatory to achieve quarterly goals for earnings in order to continue receiving the additional resources his business needed. The track record is the ultimate arbiter of managerial performance.

This rigorous commitment to agreed-upon objectives provides a very valuable discipline for a business. One of the most pernicious pathologies a management can develop is to begin to establish objectives it suspects (or perhaps even knows) are unattainable. With that uncertainty looming in the background, the management team does not make the necessary psychological commitment to attaining them. Then it begins to lose touch with reality. In its most virulent pathological form, a management can no longer distinguish between what is attainable and what is not—or no longer cares. Such sick businesses are extremely difficult to restore, and they can be sick even though for the moment they are still reaping the benefits of past successes. I recall the astonishment and irritation of a manager transferred from a highly disciplined business, struggling to survive in a very competitive environment, to a business with a long record of high profitability. He said, "These people don't think schedules mean anything; they take slippage for granted and don't even feel bad about it!"

The sanctity accorded operating plans, however, can also be an impediment in accommodating to external events. When electric power output consistently began to undershoot projections, management in the related equipment businesses in GE faced a double task: trying to develop more realistic projections and demonstrating to top management that the lower target was not just a sophisticated ploy to provide a goal that was less of a stretch.

Business operations are controlled by convention

The set of conventions concerns the operation of the business itself: shared understandings regarding how long various business operations should take, in what sequence they should occur, what variances are acceptable, what resources are required, how much they should cost, and what outputs should result. To some extent, these can be regarded as operating conventions that are particular to an individual firm, but they are also characteristic of specific industries and they can differ markedly among countries. There has been a convention in the United States that lower quality accompanies mass production—a convention, as we pointed out earlier, the Japanese have not accepted.

Nearly all the parameters of a business—in-process inventory, accounts receivable, bad debts, both direct and unapplied labor costs, quality costs, meeting delivery schedules, as well as many key operating ratios—are controlled fundamentally by conventions. In reviewing the technological dimension of strategic plans for operating components in GE over a period of years, it became apparent to me that they followed one of two conventions: that engineering effort was unrelated to the volume of a business and therefore should maintain a constant level of technical effort unless special new initiatives were introduced *or* that engineering should represent a constant percentage of revenue. No explicit rationale was presented for either convention, but examination of plans over a period of years made it obvious that some businesses used one and some the other. The relative priority assigned to different conventions can vary from one management to another, even in the same company. As another example, in one division of a company, every operating general manager automatically presented his batting average in meeting delivery promises when he was describing the performance of his business, because the division general manager viewed it as a crucial measure of their performance. Although the information presumably was known in other divisions, it was rarely volunteered.

These conventions usually result in control limits—set points to a process engineer—that establish the acceptable variance in operating parameters. Just as the operators in a chemical plant systematically keep activities within prescribed limits, all the members of management and supervision in every function, as well as the supporting staff, attempt to keep operations within prescribed limits—e.g., quality costs should be no more than 5 percent of shop costs; all overtime should be avoided because it is a sign of poor planning; in-process inventory should be less than 43 days' production, but falling below 35 days' risks a shutdown. An expert in credit and collections lamented that his financial people at headquarters, conditioned to receive payment in 30 to 60 days, just could not accept the nine-month lag in receivables which was standard in a market where the big suppliers were expected to bankroll small, marginal customers. If he did not carry them, others got the business.

The penalties for exceeding an allowable variance can be asymmetrical. While a production superintendent may be admonished about excess inventory, he will be in serious difficulty if his plant shuts down for lack of parts. The same is true for output that does not meet specifications—a plant shutdown has rarely been an acceptable response. In the aggregate, these conventions determine the quality, the cost structure, the response time, and the adaptability of an enterprise. Stated in value-laden terms, they help to define what is virtuous behavior and what is sinful.

These conventions are not just figments of some manager's imagination—most of them are empirically derived. Many are based on the physical limitations of what it is possible to do—or more realistically, on what it *was* possible to do at an earlier state of technological development. But they also include a sequence of safety factors to cover a variety of contingencies. One of the most important sources of uncertainty comes from limitations on what is, or was, knowable about the state of a business at any point in time. This underscores the heavy emphasis on information processing in this book. Almost all of our conventions were established in a world of relatively primitive information processing. Weeks, sometimes months, elapsed before management received information on business transactions. It once took approximately nine months for GE's lamp business to ascertain that homeowners had indeed changed their buying habits—they were resorting to more "bulb snatching," replacing bulbs from other sockets in the house.

The delays introduced to provide safety margins influenced the conventions associated with purchasing, manufacturing, finished-goods inventory, and a host of other parameters. They also affected capital investment in plant and equipment and plans for product development. The time delay and limitations on available information required safety margins to ensure uninterrupted operation. Additional allowances also had to be made for imperfections in the system—imbalances in operations that existed simply because not enough was known about how the system operated to anticipate or identify and correct the imbalances. System simulation of factories can now provide these insights. Today, even in the most sophisticated companies, it is almost a certainty that operating conventions have not yet been adjusted to take into account the present level of capability in information technology, much less what will be possible in the future.

Conventions also make allowances for protection against perturbations—breakdowns in equipment and delays in delivery, for example. Finally, they just plain provide insurance against the unknown and reduce the anxiety of management.

I believe no one would argue that present conventions represent optimum performance. In fact, we have no idea what optimum performance truly is. Since the safety factors are cumulative—and sometimes multiplicative—the reduction in the effective use of resources can be major. A division of one high-technology company, making complex test and measurement equipment, reduced manufacturing cycle time for its most complex product from 75 days to 18 days. If asked in advance of any careful consideration of the problem, they almost certainly would have said: Impossible! There are even conventions about the conventions. For example, supervisors learn that they should triple the amount of inventory the standards call for.

Perhaps the most important lesson the Japanese have been teaching us is that our conventions—our set points—have not been optimal. Our conventions on inventory, which we thought ensured uninterrupted output, also obscured imbalances in the flow of manufacture. They hid avoidable irregularities in delivery and unacceptable quality of parts from vendors and also protected incompetence in production management. Too often they provided wasteful insurance to keep out of trouble. In similar fashion, our conventions on quality encouraged product design that needlessly introduced production difficulties. They accepted inadequate inspection and shop supervision that fostered worker indifference, even resentment. Unfortunately, they also fostered attitudes toward service that alienated customers.

One of the most pernicious conventions involves the use—or more properly, misuse—of standard cost data. Standard costs are developed primarily for purposes of control or to help value inventories when accounting for profit. They are not an adequate guide for decision making. They help to ensure that all of the costs of a business will end up being allocated to products. Everybody recognizes that these costs represent averages and that they do not apply strictly to any given product. A difficulty comes when they are used to project the cost of a product under development, especially in their use of overhead rates based on historical data. In this guise, they strongly influence the development itself. Since direct costs, and particularly the cost of materials, are the most visible cost, they become the target in product development and in cost reduction programs. The problem arises because these costs reflect the cost structure of the business at a particular point in time. When that cost structure changes, as it inevitably does, the costs are no longer applicable. Every product development, and every product modification program, has an impact on the cost structure itself, and that impact is not evaluated. For example, an electric motor business gave high priority to reducing the amount of copper used in a motor, because that was the largest single element of cost. Eventually, the business ended up with 6000 wire sizes (a manufacturer's nightmare), most of which were introduced as cost reductions.

A general manager of a very successful business once said to me, "When I look at the individual products in our line, they make good sense. We identified attractive market opportunities, and we did a good job of developing products to serve them. But when I look across our line, we have an awful mess. We have too many products with too many different components. We've created terrible problems in production planning, in inventory, in materials management, in employee training, and in field service—because we have too many parts and too many models for our service people to handle effectively."

Information technology has the potential to provide much more realistic projections of true costs—the costs of a complete business, not just isolated product costs constructed from standard cost data.

Changing the set points, and the conventions that underlie them, is an enormously complex task. Efforts to recast them require great skill and persistence and will change the character of an enterprise in dramatic ways—just ask the managements of General Motors, Caterpillar, or Bethlehem Steel.

It takes no great perspicacity to realize that the common beliefs and shared conventions that enable a management team to work together effectively can also contain the seeds of its failure. Any combination of shifts in market requirements, advances in technology, alterations in the nature and extent of competition, or evolution of the company itself can call for a change the management may not even perceive the need for, much less be able to implement.

Technology and Management Conventions

The technical dimensions of management conventions play the dominant role in determining both the character of the technical effort and the technical strategy. DuPont, with a concept founded on proprietary composition of matter, gives high priority to product innovation. Its R&D effort tends to be higher than industry norms. Because of the heavy investment in technology, it establishes high standards for margins and return on investment. Since it is often in the position of introducing new materials, it faces the task of discovering applications in which the material offers sufficient value to risk its early use. Consequently, DuPont also places great emphasis on application engineering and technical service for its customers.

An examination of differences among companies in the electronics and computer industries is illuminating. Part of the concept of the enterprise in GE is a requirement for growth with stability in earnings. The company also expects above-average ROI. It sets demanding performance standards and is intolerant of failure to achieve them. Consequently, it has rigorous procedures for justifying investment. In this atmosphere, the volatility and cyclicality of electronics, especially integrated circuits, do not generate enthusiasm. GE has participated in consumer electronics, but has not been a leader. It has been most comfortable and successful in developing proprietary technology for industrial applications. It eventually overcame its ambivalence toward integrated circuits and reentered the business, but only after escalating investment requirements increased the financial barriers to entry and the pace of progress became more predictable. In this more "mature" environment, the company's financial strength and ability to apply rational management to a less volatile situation looked like a better fit.

I suppose that some would argue that a truly competent management should be able to operate effectively in any environment. Bosh! The beginning of wisdom is to know your own strengths and weaknesses. Centuries ago Sophocles said, "Know thyself"—good advice.

IBM has a powerful marketing orientation. Its technical orientation has been toward aggressive application of the state of the art or "proven" art, combined with extensive process development. Its preference has been for larger systems, with the customer assuming responsibility for applications. DEC has segmented its market. Its technology has focused on machines that are designed for technical applications and for easy upgrading to more powerful equipment.

Intel starts from the premise that it will succeed by pioneering new technology-based products. Consequently, it focuses its technology on that objective. It eschews commodity products where success depends on process refinement and where margins are dependent on being ahead in the learning curve. The character of its technology in turn affects its marketing because Intel's products appeal to sophisticated customers; therefore, it must be involved in application engineering.

Technical strategy also in turn strongly influences the selection of the operating characteristics and competitive strategy of a business. GE's Medium Steam Turbine and Gear (marine drives) business established a renowned, worldwide reputation for the efficiency and reliability of its product. Its engineering effort emphasized carefully tailored custom designs to wring the last ounce of performance out of a given installation. In consequence, engineering expense was heavy and manufacture was high-cost because of so much variation in design. Less visible, but also important, logistics and field service were slow because of customized parts. With the advent of diesels as serious competition in marine drives, it eventually became apparent that the concept of the business had to be rethought. The shift called for a limited product line that would be less well matched to individual applications, but that would be inherently less expensive to manufacture and could provide much faster field service. The change proved wrenching for marketing, engineering, and manufacturing, and in fact took years to accomplish. It altered not only the way engineering designed products, but the sequence for releasing drawings to manufacturing and thus the way it scheduled its own work. The new emphasis on cost of manufacture elevated the importance of the manufacturing function and therefore its relative status compared with engineering.

Changes in technology can force other changes in the concept of the enterprise. Although the U.S. tire companies are large and powerful businesses, their industry evolved a tradition—similar to that of most vendors to large original equipment manufacturers (OEMs)—that the auto companies would largely determine product characteristics. The ride and handling of a car are determined by the design of its suspension and the characteristics of its tire. Traditionally, the auto companies integrated the two to establish the eventual performance of a given car—and in the process specified what kind of tire they required. Because of that tradition, U.S. tire manufacturers asserted that radial tires would never penetrate the U.S. market—car manufacturers, they said, would not redesign cars to capitalize on the different characteristics of radials and would not tolerate attempts to influence car design by urging their adoption. Sears and Michelin, which did not participate in the OEM market, saw radials as a means of offering something different in the replacement market. When the superiority of radials eventually induced car manufacturers to use them, a massive technological effort and investment by the tire manufacturers became necessary.

Businesses that stress technological leadership tend to place their primary emphasis on product performance (in the chemical process industries, product cost and process yield are also important). Although it is characteristic of industries in early stages of technological development, Hewlett-Packard, Intel, Digital Equipment, Alza, and Tektronix still exhibit this same emphasis, even though they are now two or more decades old. On the other hand, Caterpillar—at a much later stage of maturity—also places a high value on offering the finest equipment in the world in terms of performance, ruggedness, and life. In general, organizations like these give less weight to cost or time—until competitive pressure forces them to. In fact, since they usually push themselves to the limit on achieving technical goals, they tend to develop a forgiving attitude toward failing to meet time schedules: That's in the nature of technical work, they say. As one senior development engineer commented to me, "Hell, if I thought I could do it, it wouldn't be any fun!" A change in the competitive climate, which forces a rebalancing of cost, time, and performance, becomes traumatic. A whole complex of attitudes and values must be altered, and even status relationships changed, e.g., the "bookkeepers" become more important and engineers face scheduling constraints; this in turn necessitates rethinking technical goals and questions of manufacturability. It also generates fierce debates over whether creativity is stifled by such constraints.

Since technology, and especially past deficiencies in technology, has played a major role in the growth of conventions, managers of technology have a special obligation for leadership in recognizing the need to question conventions and in prodding the organization to a critical self-examination of the ways it functions. This requires broadening their purview from the traditional focus on products and processes to consider the functioning of the total business and ways in which technology could become the basis for improvement.

Examining Conventions—Key to Improved Efficiency and Effectiveness

U.S. industry faces the challenge of achieving dramatic improvements in the efficient use of assets and resources. The line of competitors who want its markets is as long as the number of countries seeking economic growth and higher employment. The sources of competitive advantage lie not just in product attributes, product cost, or market segmentation. Increasingly, major sources of cost lie outside the production cycle. These costs are partly rooted in conventions regarding product planning, product development, product line structuring, and many similar activities. Our accounting procedures measure expenditures of money, but they do not measure expenditures of time. Yet time represents a cost not only in interest charges, but also in excessively long development cycles, market windows that cannot be exploited, and competitive reactions that take too long. Too often the engineering function has labored mightily to reduce the cost of labor, material, and capital required to manufacture products, but has been comparatively oblivious to delays created by its engineering design changes, to excess inventory resulting from a lack of discipline in choosing components, and to product lines with an unwarranted variety in features or attributes.

Many of the inefficiencies in the use of assets and resources result from outmoded conventions. Some come from a failure to incorporate new technology—especially information technology. Many others, however, result from a failure to ask, does it have to be this way? Effective management requires both doing the right things and doing things right.

Thus the third principal theme of this book is that effective management of technology requires addressing these conventions which affect the total business. Technology can be a powerful force for change, and failure to include conventions in the candidates for change is a failure in managerial vision.

Conclusion

We have now laid out the basic characteristics of the territory encompassed in managing technology. Technology includes a diverse range of related activities that need to be considered holistically. The fragmented view fostered by organizational structure, educational focus, and career patterns must be replaced by an integrated view across the spectrum from basic research to product service and embracing product technology, process technology, and information technology. We must recognize and indeed nurture the inherent contradiction in technology—the need for rigor, discipline, and conservation in opposition to the need for innovation, change, and risk taking.

In Chap. 2, I pointed out that the process of technological progress responds to an internal dynamism in which the life cycle of a technology moves from the development or invention of a new capability, through a gradual transition of emphasizing process development and manufacturing efficiency, and finally, to capital intensity and emphasis on scale and timing of investments. Although this cycle is inexorable, its progress in any given case is subtle and full of uncertainties. Predicting the timing of transitions is fraught with difficulty. If, however, one examines a large number of cases, there are consistencies that can provide some guidance for those who must manage the process.

In Chap. 3, I placed technology in the context of a business. Management is a team effort, and a management team is guided by shared conventions that guide decision making and foster predictability in behavior among team members. These conventions include what business one is in, whether its focus is on providing the customer with tools or solutions, how it competes, and how it behaves. Technology helps establish these conventions, is in turn guided and constrained by them, and can be a powerful force for change.

Notes and References

1. Donald W. Collier, "How Management Should Use R&D to Set New Directions," *The Journal of Business Strategy*, Fall 1986, pp. 25–28.
2. Michael E. Porter, *Competitive Advantage: Creating and Sustaining Superior Performance,* The Free Press, New York, 1985.

Part 2

Operational Management

This part focuses on operational management. I do not intend to emphasize "how to" prescriptions and answers, for I believe strongly that home-grown solutions are most effective. A search for easy answers by buying or borrowing a solution from others is a tempting but ineffective approach; ultimately it is an evasion of managerial responsibility. This is not to say that an informed external view cannot be helpful in changing perceptions and clarifying a problem. However, the fit between external solutions and internal requirements is rarely adequate. Even worse, if a management avoids the tough task of immersing itself in a problem from the inside, it is unlikely to understand the ramifications of the problem or to recognize the limitations of the solutions being adopted.

My focus will be on illuminating the forces at work, on characterizing the conditions and processes which enable effective solutions, and on outlining the important elements of any workable solution. The part addresses in detail one major source of the tension in technology management: the demanding, intolerant-of-error, resistant-to-change posture of operational management that defines success as technology that performs as promised, all of the time.

This part consists of four chapters. Chapter 4 examines the task of selecting programs—more or less the first step in managing technology (assuming that one has already selected the basic fields of technology that will be exploited—a topic covered in Part 3). This chapter evaluates the strengths and weaknesses of various approaches, emphasizes that different techniques are appropriate in different phases of a program, and elaborates on the process considerations that strongly influence the success of program selection. Program selection cannot be done in isolation by technology management, and orchestrating the involvement of needed participants from other areas is vital.

Chapter 5 takes up the subject of risk. It argues for the value of a more careful delineation of the difference between risk and uncertainty. It develops the concept of "risk space" and points to the danger of the cascading effect of assuming increased risk in more than one technological dimension at the same time, i.e., in product and process technology, product development, and markets. If risk is approached objectively and rigorously and disaggregated according to levels of uncertainty, then options are available for reducing risks, options that too often are unused through ignorance or false pride.

The chapter then examines the need for more discriminating techniques to evaluate risk using methods that discount the future. It points out that the relationship of risk to reward and punishment must be perceived as an incentive to risk taking if management wishes to encourage more risky endeavors.

Chapter 6 addresses project management. It points out the difference between the serial functional handoffs that have been traditional in project management and the full-team approach that evolved during World War II. It examines the strengths and weaknesses of the matrix approach. The principal theme of the chapter is that most project management techniques were devised to deal with complexity, to delineate the intricate sequencing needed to coordinate a large-scale, multidimensional task. They were never intended to deal with uncertainty, where unknown unknowns (unk-unks) can require changing objectives as well as redefining program tasks. The discussion strongly emphasizes the need to identify the degree of certainty associated with every objective, to identify the additional information that is required to reduce uncertainty, and to assign responsibility and schedule downstream inputs to reaffirm or modify the initial objectives and assumptions.

The part concludes with Chap. 7 examining the problems of technology transfer—of moving new technology along from a laboratory or development setting to successful commercialization. It first explores the cognitive aspects of transmitting information from a source to a recipient—identifying potential users, determining what information needs to be provided, alternative mechanisms for transferring the information, etc. It then examines the affective features of information transfer—the differing points of view of the inventor and the applier, problems in eliciting enthusiasm and commitment, and understanding the psychology of change. Finally, it explores the problem of transferring technology over a diffuse network where one needs to foster continuing interchange of information rather than move a particular development program into operations.

Chapter

4

Selecting Programs

The notion of selecting technical programs can be somewhat misleading. It is sometimes interpreted as an occasional process of choosing, on a once-through basis, those programs that will be pursued. Actually, the process is much more like the funnel in Fig. 4.1, with a series of screens that programs must pass, over time, in order to continue. The screening criteria and the screening process become more extensive and rigorous as the funnel narrows, and the resources committed to a given program tend to increase. A snapshot of the funnel at any point in time will show programs being actively pursued at all locations from entry to exit. The entire process can be thought of as gestation, with genetic defects and adverse accommodations to the environment causing abortions or miscarriages all along the route.

The stage at which the selection process starts varies depending on how far upstream into R&D a company is active. Most smaller companies or companies that engage in little or no R&D typically begin

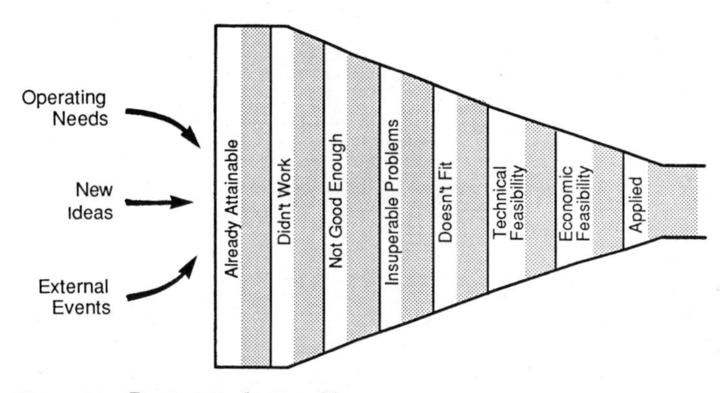

Figure 4.1 Program selection filters.

with technical feasibility. Others may begin with original ideas where it is unclear whether nature will cooperate.

The screening process itself represents a shifting equilibrium of contradictory influences. A selection process that is too demanding will serve to deter presentation of new ideas, yet one that is too relaxed is ineffective. Nascent ideas are typically associated with a paucity of information and great uncertainty, which make impractical a screening system that demands extensive, detailed inputs. Yet, obviously some threshold must exist, for not every idea is worth pursuing.

Nature of the Process

At the entry level, two criteria are dominant—technical merit and relevance to the business—as suggested in Fig. 4.2. This screening is almost exclusively judgmental and is rendered by peers and by managers. The latter cannot avoid this responsibility, but it must be exercised with care. Two of the most effective managers I have ever known were careful not to criticize ideas themselves. One said, "If I have some doubts about an idea, I suggest, 'Why don't you talk to so-and-so about it?' knowing full well that if the proposer survives that ordeal, the idea is worth pursuing." The other waged a constant battle to get people to relax their intellectual censor and think more imaginatively (he continually proposed wild schemes himself to challenge people), but he *never* criticized anyone for trying something that did not work out. In fact, nearly all R&D managers bemoan the dearth of new ideas. They are more concerned about stimulating creativity than screening out poor bets.

The criteria of technical merit and business relevance are somewhat interrelated. The greater the business relevance, the less rigorous the evaluation of technical merit. In other words, in self-defense a busi-

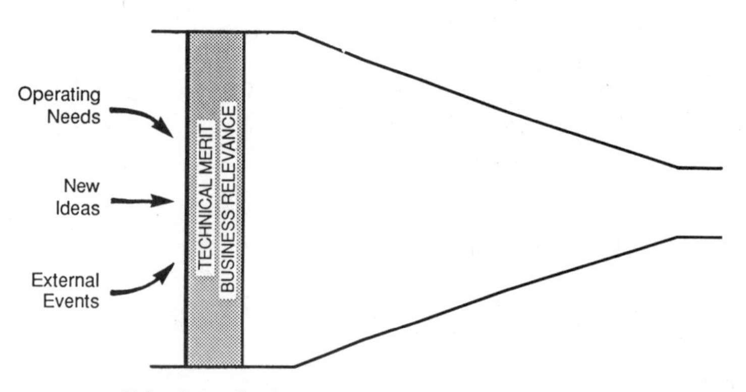

Figure 4.2 Selection criteria.

ness can justify, indeed often dares not ignore, working on ideas that bear directly on its activities, even if they seem somewhat improbable. Conversely, managers of an established business should feel that they need a strong technical argument or see a very large potential to invest in ideas far removed from the business.

At this early stage, nature itself is the best screen. Most bright ideas turn out not to work or to produce uninteresting results or to exhibit other undesirable results that make further effort unattractive. Fortunately, activities at this exploratory level are inherently small in scale and require a very modest commitment of resources. A more difficult situation arises when definitive results are slow to come, often because of difficulties in developing experimental or analytical techniques that yield meaningful, reproducible results. In one program that was eventually successful, it took a full year just to learn how to conduct the first valid experiment. In the case of man-made diamond compacts, one of the researchers commented, "We knew what we had to do; it just took us seven years to find out how to do it."

So, a third criterion in the early stage is evidence of progress that suggests the goal is still attainable, even though it may still turn out not to be worth the effort. Figure 4.3 suggests the natural attrition in programs that occurs and the concurrent increasingly rigorous evaluation of progress that must be made as effort continues. The level of effort per program during these early phases may increase gradually, but growth is small. For those programs that survive, the resource commitments then grow rapidly, as shown in Fig. 4.4.

Up to this point programs tend to be relatively large aggregations of small-scale efforts. The objective is to discover and lay the foundation for a new technical capability. But the balance shifts as the demands of programs grow. As programs progress, they move from exploring phenomena or evaluating the potential of new discoveries to technol-

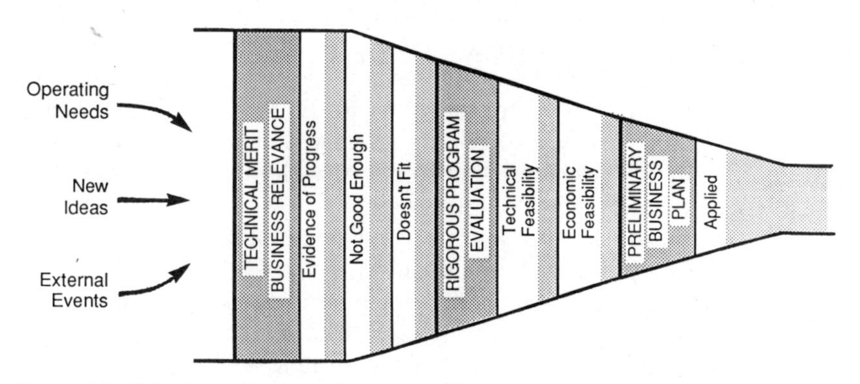

Figure 4.3 Selection criteria and program filters.

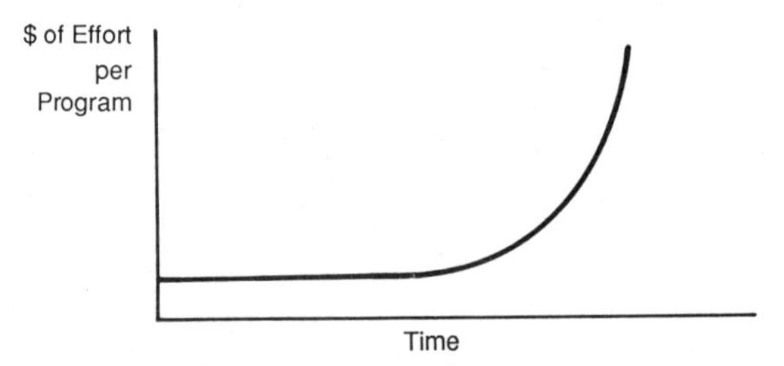

Figure 4.4 Growth in resource requirements.

ogy or product development. They require rapidly escalating resources, which tends to inhibit other exploratory effort. As they are completed, effort usually shifts back to a more exploratory character. In an engineering operation, work of this more exploratory nature might vary from almost zero to perhaps 10 percent of the total. Of course, some companies may engage in no exploratory effort of this sort.

The programs that survive this initial period (and as the funnel diagram suggests, they become progressively fewer in number and larger in size) face a gauntlet of ever more stringent scrutiny in which the nature of the questions changes. Initially the focus is on "can we" questions: Will nature permit us to do this, and if so, are the results potentially interesting? Once these fundamental questions have been resolved, the questions take on much more of a "should we" character: As work progresses, are new deterrents appearing? Is the promise of the original advance shrinking to the point where it is no longer attractive? Do economic evaluations look promising? Do we have the resources to do this? and so on.

These "should we" questions begin to involve other parts of the business. Although the program is, and will be, in R&D for much additional time, inputs from operations become increasingly important. Operating people generally have better access to relevant data, but even when they do not, they are much more likely to question the validity of data generated by "amateurs" in R&D. Perhaps most important, seeking the assistance of people in operations to evaluate programs is a useful way of beginning to nurture their involvement and eventually their commitment. Developing a sense of ownership in an innovation is critical for its success, and ownership can only arise from a sense of having contributed to shaping the work. Ownership, as the term connotes, implies developing a personal sense of commitment to the importance of the work and a resolve to see that it succeeds.

In considering more rigorous and demanding screening techniques, it is important to distinguish between the cognitive or analytical activity that addresses the substance of the issue and the screening *process* that takes into account how information will be generated, who should be involved in the process, and what sequence of events needs to be orchestrated.

Technical managers tend to feel ambivalent about rigorous analytical screening techniques. On the one hand, their training and intellectual style lead them to value rigor and quantitative procedures. They genuinely want to use procedures that are objective and that reflect the most relevant and valid inputs possible. On the other hand, they also recognize the large elements of uncertainty and gaps in information that are inherent in research and development. They are wary of heavy-handed or mechanical use of techniques that are all too fallible. Applying space-age methodology to stone-age inputs is a dubious exercise.

Formal Program Evaluation

All screening techniques that attempt to apply more rigorous analytical approaches begin from the same set of principles. They seek to maximize the return from investing in technology. That return is defined as a function of the benefit the technology produces, the probability of success, and the investment required.

$$V = \frac{RP}{I}$$

where V = index of value
R = net return
P = probability of success
I = total investment required

An immediate problem arises in attempting to apply this technique because the relationship noted above ignores the contribution needed from all other functions in the business (i.e., market development, process development, etc.) in order for the innovation to succeed. If each function went through this same exercise, the aggregate projected return to the enterprise would far exceed the actual return. Consequently, in using any of these procedures, it must be clear that they are being used to select programs, *not* to provide a quantified measure of the return on investment in technology.

Every attempt I am aware of to determine the true return from investing in technology founders on the difficulty of allocating the benefits of the undertaking among the various contributors to its success.

Pushing the technique too aggressively simply arouses resentment in other parts of the organization. For example, the manager of marketing for man-made diamonds always asserted that the truly critical contribution was figuring out how to price them to penetrate the market without setting off a destructive retaliatory action by the diamond syndicate. He seemed to take the original technical achievement for granted. Fortunately, the overall project was so successful that there was nothing to be gained by arguing over the credit for it—there was enough for all.

A variety of analytical techniques exists for applying the basic relationship noted above. They vary in their degree of mathematical rigor, their requirements for input data, the complexity they introduce, and the extent to which they include nonfinancial as well as financial considerations.

Scoring methods

Scoring methods of varying degrees of complexity are possibly the most widely used formal methods. Each constructs a figure of merit for every project by applying a set of criteria. The number of criteria can be as large as the user wishes. The criteria typically include financial considerations, such as estimates of cost and return. They usually include an estimate of probability of success as well as a variety of nonfinancial considerations: time to completion, market size and rate of growth, competitive status, fit with the business, and so on. The criteria are often weighted, and the results can be aggregated either additively or multiplicatively to a single figure of merit.

Much of the discussion of program selection in the literature focuses on the intrinsic merit of particular techniques or the comparative value of one versus another. Most of the time, there appears to be a sort of subliminal assumption that if a sufficiently powerful *technique* can be developed, its effectiveness will lead to its adoption. In practice the problem is quite different. The application of *any* technique requires the manipulation of information, information that typically is nonexistent or exists only in the most rudimentary form. Generating information takes time and costs money. Perhaps more important, it frequently can be done only by the responsible manager. Consequently, he must continually weigh the diversion of his own limited time and intellectual energy to generate the information he needs against the improvement in decision making it provides.

One might ask, why not turn the task over to a group of specialists? Such a group could undoubtedly develop greater expertise in applying sophisticated techniques, it could build up a data base and sources of information that would strengthen and facilitate the process, and it could also attain greater consistency than is likely from a collection of managers who have little incentive to devote effort to achieving consistency.

The proposal founders on two difficulties. First, a separate group that is charged with program selection, or even with program evaluation and recommendations for line management decision making, will almost inevitably be in an adversarial role vis-à-vis line managers. Line managers will resent and seek to subvert the work of a group charged with determining what will be worked on, but that is not accountable for accomplishing the task. They will feel impelled to develop countervailing activities, will demonstrate great vigor and ingenuity in doing so, and will probably win the battle of wits. The other difficulty is more subtle, but equally compelling. Selection of technical programs is a critical function of technical management. Demonstrated skill in doing so should be a major consideration in career development and promotion. To deny line managers the central role in one of their most important responsibilities is to stand management responsibility on its head.

Despite years of effort to develop more rigorous program selection techniques, their most noteworthy feature is limited use. For example, a survey by the Industrial Research Institute of its own membership found that one-third or less used formal methods of any sort.[1] Another report in the literature noted 90 percent agreement between less demanding scoring techniques and much more sophisticated but costly approaches—a finding that corroborates the judgment of managers in rejecting the adoption of the latter approaches.[2]

Besides heightened effort and cost, another frequent problem with more elaborate methods is a mismatch between the fit of the proposed technique and the thought processes and data sources of the user. A typical pattern is for some interested person to develop a technique and sell it to (or impose it on) a management team. The technique is used as long as its advocate is around, but falls into disuse after he or she leaves. For example, many managers are unfamiliar with linear programming as a technique for selecting programs. Consequently, they will be reluctant to divert the necessary time and energy to learn unless they see high value in the results it promises.

A formal selection procedure has an important additional virtue: it leaves tracks—and as programs increase in size, the more needed and valuable such tracks are. Memories fade, and reconstruction of the bases for informal, mostly verbal decision making is frequently virtually impossible. This virtue provides a clue to how formal a program selection procedure should be at this stage of understanding. The objective is not necessarily more formal and rigorous program selection; the objective is more consistent program selection when many people are involved in the process. What criteria should be used? What objectives should be sought in attempting to develop a better system? How could it be done better next time?

First, managers must ask the right questions to ensure that they are applying the correct criteria in evaluating programs. As we have noted, technologists tend to focus on "can we" questions; indeed they tend to push program objectives to the point that "can we" questions become dominant. Studies consistently show, and my experience certainly confirms, that programs which reach the stage of being put on a development time schedule rarely fail because the technical objectives are unattainable. Rather, other factors, such as time to completion, size of market, customer acceptance, cost, fit with the resources of the company, or manufacturability, prove decisive. The program selection system should encourage addressing these other issues with increasing rigor as a program continues. Unfortunately, projects sometimes fail because the drama of attacking spectacular technical problems which prove solvable drowns out adequate attention to other data which could have demonstrated that the project should not be pursued.

Second, the system should foster the cumulative development of a data base and sources of information that will lead to improved decision making. Much of the information on market potential and competitive status exists outside the technical organization, and managers must be motivated to develop networks of sources that will enable them to make more informed decisions.

Third, the real focus should be on improving management skill in program selection. Consequently, the emphasis must be on ways to get line managers to become active participants in and advocates of the process. If they feel the process requires more work than it produces in benefit or if they feel the criteria and values reflected in the system are not consistent with those the organization should be seeking, they will certainly find ways to subvert the system. This means the program selection process must come to be viewed as an aid to managerial judgment, not its replacement.

An example of a scoring method. No single system is likely to become universal. In fact, given the need to elicit managerial participation, an element of home-grownness is probably an important, if not a mandatory, ingredient. Furthermore, an effective system is likely to be fluid and change over time as managerial sophistication grows. One system with which I am familiar underwent substantial revision over a period of seven to eight years. It became a tool of managerial and institutional learning as much as an aid to program selection.

The first step in implementing the system was to recognize that attempts to measure return on investment at the early stages of a focused R&D program are generally misleading and unrealistic. The uncertainties are simply too great. Rather, the focus was to be on the size and attractiveness of the potential market or alternatively the size of the potential cost reduction. The system focused on two sets of factors: leverage or attractiveness and probability of success. These leverage factors were elaborated as follows:

1. Size of the potential market or of the cost reduction. The size of market was estimated by analogy to existing businesses in the company, for which a typical dollar range could be determined. This proved helpful because managers clearly felt more comfortable in predicting—for example, this would be no more than a product section or this could be a new company division or this could revolutionize an existing division—than they did in forecasting an abstract magnitude.
2. Market rate of growth (expressed in relation to the company average).
3. Likely market share. (This takes account of the competitive situation.)
4. Sensitivity of the market to technical considerations. (Technology is dominant in some fields and relatively peripheral in others.)

The probability factors were

1. Nature of the technical barrier(s) to be overcome.
2. Technical competition.
3. Fit with organizational skills and resources.
4. Difficulty of transitioning to operations.

Except for the market size and rate of growth, all of the factors simply called for systematic exercise of managerial judgment in carefully constrained categories. Even market inputs could be based on guesstimates, if no better data were available, but managers were asked to indicate the basis for the estimate to give an indication of how solidly it was supported. They were encouraged to rely on inputs from marketing people in operations. Major industry studies from external sources, such as SRI International, A. D. Little, Data Quest, and many others, were also valuable. One ancillary value of this approach was that it sensitized R&D managers to the need for including market potential in their evaluation, and it increased their competence in locating useful sources of information.

Mechanically, the system worked as follows. (Illustrative labels are used to characterize more extended definitions.)

1. Impact

M = Market size in millions of $ (10–30, 30–100, 100–300, 300–1B)

G = Market growth
well above company average
above company average
equal to company average
below company average

K = Market share
dominant position
one of two or three leaders
one of the pack
a minor player

S = Sensitivity to technology
dominant factor in success
one of several important considerations
other factors more important

$$\text{Impact} = M \cdot G \cdot K \cdot S$$

2. Probability of success

B = Technical barriers
no problem, just a matter of applying resources
some uncertainty over the path, but little doubt of goal
a tough challenge, but we believe we can do it
may be impossible

C = Technical competition
we lead the world
one of two or three leaders
one of the pack—no credible advantage
a dark horse

F = Fit with resources
we have the skills and facilities
stretching, but we believe we can handle it
calls for new skills outside our experience

T = Transition difficulty
an operating component wants it and can handle it
not clear who should get it

faces hostility in operations
no apparent home—don't know what we'll do with it

$$\text{Probability of success} = B \cdot C \cdot F \cdot T$$

3. Step definitions for all factors except market size were on a semilog scale from 1.0 to 0.01 (or less), but no arbitrary number of steps was imposed.

As one can see, this system applied all factors in impact as potential discounts against the key variable of market size. The intent was to establish a generally optimistic estimate of the likely impact on revenue if the program were commercialized successfully. Revenue could result from new products or from present products that were protected by the incorporation of improved technology. The system could be applied visually on a matrix (Fig. 4.5) by keeping the impact and probability factors separate, because it often was helpful to consider the two dimensions separately. In particular, some of the elements in probability could be influenced by management actions, e.g., transitions to operations. Obviously, programs with low potential impact and low probability faced tough sledding.

In its original implementation, programs focusing on cost reduction were not easily incorporated. Subsequently, an alternative was devised that applied an average return on sales (ROS) of 15 percent (figured pretax to make comparable with cost reduction), and cost reductions could then be used as an alternative. The 15 percent ROS was somewhat above the company's average, but it was applied uniformly because more "precise" calculations were unrealistic. Experience in using the system demonstrated that it was not applicable to exploratory programs where the uncertainties are inherently very high or to internal service activities, such as analytical chemistry or mechanical testing.

The reaction of local management varied from initial resentment at additional paper work of dubious value, to a recognition that the system forced attention to critical issues. The screening system was

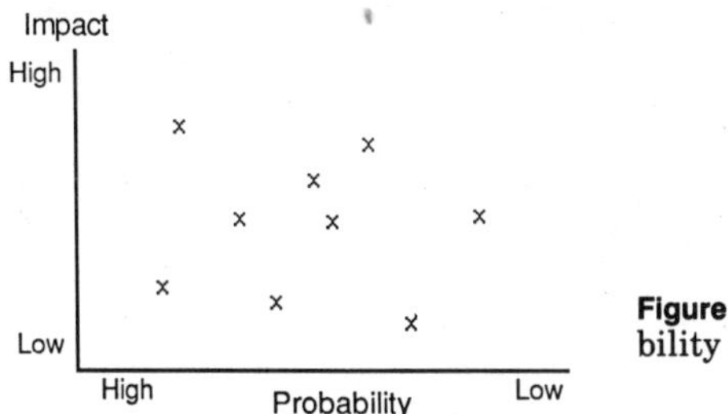

Figure 4.5 Impact versus probability matrix.

viewed as a useful tool in encouraging consistency among a group of approximately forty managers. It forced recognition of all factors rather than the special interests of individual managers. It also served to guide and focus managerial discussion when there was controversy over program value. The program value calculation was never permitted to override managerial judgment. Its consistent application over a period of years did serve to raise managerial consciousness on important issues beyond technical merit that should be incorporated into program selection. Interestingly enough, the system also served to improve the credibility of R&D management with corporate management. The latter saw it as evidence of a commitment to improve the effectiveness of R&D management and to incorporate business considerations at the early stages of technical programs.

This program value system illustrates the use of scoring methods—probably the most widely used formal techniques for program selection. The criteria can vary, as can the weighting and the method of aggregating. Scoring techniques have several advantages. First, they can easily accommodate both financial and nonfinancial considerations. Second, they can include both judgment factors and harder data. Third, they lend themselves to being home-grown. Consequently, they can be as elaborate and rigorous as local management deems fit.

These systems are clearly not sufficiently rigorous to screen programs as they begin to approach business commitment rather than just R&D. Furthermore, they cannot easily deal with the influence that different levels of expenditure and different completion dates may have on the range of outcomes or probabilities. This limitation, however, is not intrinsic. In principle, different versions of a program could be analyzed, but the managerial load to undertake such a step for a large number of programs would overburden the system and probably kill it.

Cost-benefit analysis

As noted, when programs move further through the funnel and closer to business commitment, the screening process must become more rigorous and demanding, and in general scoring techniques will suffice less well. Progress on a program will slowly answer technical questions, and issues increasingly become matters of finance and timing. Conventional cost-benefit analysis—with suitable incorporation of probabilities—becomes more appropriate and more illuminating at these later stages. Treating future program costs as a stream of future expenditures and subsequent return as a stream of future income permits the application of standard discounting techniques to establish equivalent present values. Even so, many aspects of cost-benefit analysis, especially in marketing, are difficult to quantify, e.g., to fill out a product line, to prevent domination, or to shut out competition.

There is an insidious danger in applying this technique in the appraisal of a new technology. One may unconsciously assume that the option of using conventional technology will continue. In fact, if a new technology comes to be accepted, the conventional technology may simply no longer be viable. For example, as CAT scanners gained acceptance, the market for conventional medical X-ray systems shrank catastrophically.

Managers of technology are wise to adopt this or perhaps more powerful techniques for the small number of programs that reach the stage where a substantial commitment by the company begins to loom. These techniques are widely used and readily understood by nontechnical, financially oriented executives. Furthermore, they are now being applied to programs whose information base can justify more rigorous treatment. Cost-benefit analyses do focus on financial considerations. However, other factors, such as human resources, availability of space, and fit with the company, in most cases should already have been addressed as the program moves to the stage of warranting cost-benefit analysis. One should not make the mistake, however, of assuming that just because an apparently hard number has been generated for a program's value, the results can be accepted unquestioningly.

Linear programming

Even more sophisticated approaches, such as linear programming and dynamic programming, have been developed to take explicit account of varying time frames, varying levels of resources, varying rates of program completion, and so on. The information requirements for these approaches severely restrict their usefulness in an industrial setting. They may well have greater utility in an institution that selects and funds programs, but does not manage them. In principle, they can provide guidance not only on program selection, but resource allocation as well.

Need for Increasingly Rigorous Selectivity

One general limitation of program *selection* techniques is that they assume that one is beginning with a clean slate of candidate programs. In fact, a typical portfolio contains only 15 percent new programs. Perhaps 85 percent are ongoing programs, where the primary question is the level of expenditure, not whether or not to fund. Program selection techniques, especially scoring procedures, do not address this question. The requirements for resources needed to carry the programs forward become dominant at this point. In theory, managers could analyze each program under a range of levels of effort. In fact, the work required to flesh out such alternatives severely restricts their utility.

The final phases of development programs are in fact the region in which consideration of alternative levels of funding and comparison with available resources are most badly needed. A basic objective of applying rigorous selection techniques is to foster greater selectivity at the point when expenditures begin to escalate. One of the endemic problems of R&D organizations is a tendency to spread themselves too thinly. Programs that have survived to these later stages have developed a constituency. They employ a number of people who have made by this time a considerable investment of their personal lives in the effort. Quite naturally, they feel strongly about the promise inherent in the program. This group of advocates includes both managers and technical contributors. The tilt toward continuing the programs that these advocates represent must be counterbalanced. The critical question is not, is the program worth pursuing? (if it has survived to this point, it probably is—at least in the sense of having a minimum acceptable projected return); but rather, in the array of programs that must have increased effort in order to reach commercialization, should this program be allocated additional funds from our limited resources?

The tendency to spread resources too thinly is reinforced by a parallel tendency to underestimate the amount of effort and the time required to carry a technical development to successful application and to achieve commercial acceptance. Many managers—even experienced technical managers—have had relatively little hands-on experience in these final stages. The general manager of a business with a decades-long record of successful innovation commented ruefully about a new product on which he was still losing money: "We never seem to learn! We always underestimate the problems we'll have with these new products—and are too optimistic about sales and profits in the early stages."

Programs that fall behind schedule may miss their market window or require greatly elevated effort—even beyond what would have been needed had they been budgeted realistically—to get them back on track.[3] An effective selection process at this stage should force consideration of contingencies, such as a significant (50 percent) increase in time to completion, a reduction (25 percent) in projected benefits, or a significant (100 percent) increase in cost. Real changes of this magnitude are not unusual, and unless increased effort and funds can be scheduled to meet the time target, frequently the project may well not be worth pursuing.

As J. H. Arnold pointed out in the *Harvard Business Review,* even worst-case scenarios for capital expenditure (where the technical risk normally has already been substantially quantified) often fail to consider drastic enough contingencies. He states: "A RAND Corporation study found that the first estimate of construction cost for process plants involving new technology was usually less than half of the final cost, and many projects experience even worse performance." Arnold asserts that more than 50 percent of projects will encounter major problems.[4]

Despite the need to consider contingencies, threshold criteria may at times be dangerous. One common criterion is a minimum size of projected sales that an opportunity must promise before it is regarded as sufficiently attractive to warrant pursuing. At a workshop on innovation this criterion was identified repeatedly as the major barrier to innovation. In principle, its validity is unarguable—any business person should prefer to invest in opportunities that promise large increments of growth. The difficulty is with the validity of the forecasts. Philips developed an early electronic hand calculator but abandoned the product because the projected market was too small—a few scientists and engineers. Obviously, the potential of evolutionary advances is easier to predict than that of major advances.

In our discussion of the dynamics of technology we pointed out that once a technology is accepted, many new converts begin to look for uses. Examples of notoriously low estimates of market size are common. My own experience indicates that the one thing that can be said with certainty is that all forecasts will be wrong. I know of several businesses in which the dreams of the most committed apostles were for a doubling or tripling of the present size. If one had told them that, well before they retired, the businesses would grow to 10 to 20 times their present size, they would have said, "You're crazy!" And yet, that is what happened. When dealing with a truly new product, it is exceedingly difficult to estimate the size of the eventual market. I know of one case where even after more than ten years of attractive growth and profitability, enthusiastic advocates dreamed of doubling or tripling their sales, when in actuality sales increased by an additional order of magnitude over the next fifteen years. The question of size is certainly legitimate, but it must be used with flexible discretion and not applied as a rigid barrier that must be hurdled.

The term *hurdle rate* is conventionally used to describe the minimum return on investment that a project must promise before it will be approved. The fluidity of the concept is reflected in the comment of a top corporate officer regarding a proposal toward which he was favorably disposed: "You find a finance guy who will show 15 percent ROI and I'll approve the proposal." Return on investment is clearly a legitimate criterion to apply. But again, it must be applied with discretion. We will examine this subject more fully in Chap. 5 on managing risk.

Project selection starts out as the exclusive domain of the technical community. It moves rapidly to a phase where business inputs are important and progresses to a stage where financial considerations are dominant. Successful managers of technology must understand this sequence and anticipate the changing nature of information needed for decision making.

Building Commitment

Because of the intense commitment engendered by projects that have a considerable investment, it often is helpful to ask for an independent appraisal by experts outside the component or even outside the company.

We noted earlier the overriding significance and the constraining influence of the information base required for program selection. It is important to recognize that program selection is not only a cognitive exercise, it is an exercise in consensus building. A number of different constituencies with somewhat differing objectives must feel that they are parties to the process. Inputs from a variety of sources must be stimulated and orchestrated—members of the technical staff, technical managers, both functional and general managers from operations, corporate executives, those indispensable market development and sales people, and eventually, external people and institutions either as customers or vendors. This is particularly true with respect to data on costs, market size, and pricing. People in development often feel that marketing people lack the vision to provide credible figures. The wrong solution is for technical people to generate their own marketing inputs. They are not experts at it, and their efforts will be resented and all too easily derided. A much more effective approach is to begin involving marketing people early in the project and to develop a relationship that enables one to participate in developing marketing data, but that also incorporates market feedback into final product specifications. Success requires a tight link between the generators of the technology and those who will use it.

Unless this process of including the inputs of diverse groups and recognizing the values and priorities of various constituencies is reflected in the program selection process, the most sophisticated techniques in the world will be of no avail. Perhaps it is not surprising that effective managers devote more attention to developing, nurturing, and fine-tuning this aspect of program selection than they do to developing more rigorous analytical tools.

Are Screening Techniques Worth Using?

Despite my emphasis on the limitations of formal screening techniques, I believe they should be used. My concern is with slavish or mechanical reliance on them or the imposition of a system that demands more in managerial inputs than it produces in better decisions and increased confidence. In addition to the advantages noted at various points earlier, there is another important advantage. Effort devoted to developing or improving more rigorous techniques for selecting programs is tangible evidence to higher management that managers of technology are striving to be good stewards of the resources entrusted to them, are striving to be better managers according to criteria that are applicable to all managers, i.e., are using the most rigorous techniques and quantified data practicable in making decisions.

Notes and References

1. Robert H. Becker, "Project Selection Checklists for Research, Product Development, Process Development," *Research Management,* September 1980, pp. 34–36.
2. Frank Krawiec, "Evaluating and Selecting Research Projects by Scoring," *Research Management,* September–October 1983, p. 22.
3. Donald G. Reinertsen, "Whodunit? The Search for New-Product Killers," *Electronic Business,* July 1983, pp. 62–66.
4. J. H. Arnold III, "Assessing Capital Risk: You Can't Be Too Conservative," *Harvard Business Review,* September–October 1986, pp. 113–121.

Chapter

5

Managing Risk

Innovation involves risk. That assertion is certainly common wisdom. Unfortunately, risk pervades virtually all of the application of technology, not just innovation—witness the admonitions in engineering design to design fail-safe systems or, short of that, to build in precautionary redundancy. Even in apparently mature technologies, knowledge is always less than total, and operating conditions cannot be comprehensively specified in advance.

Risk looms as a pervasive, amorphous threat, which takes on different meanings, depending on the viewpoint of the risk taker. To the technologist, it is interpreted as a technical goal not achieved or a device, component, or system that fails. Although risk is often interpreted as implying opportunity for greater return, that is, strictly speaking, not a valid inference. Potential return is an independent analysis associated with benefit foreseen from taking risk. To a business person, risk is often seen as investment that does not earn its projected return or, worse, that becomes worthless. Risk is also a psychological hazard that creates anxiety. Paradoxically, it is also a stimulant that generates excitement. Thus once again, we see the need to maintain an uneasy balance: in this case, between the comfort and security of low risk, with the danger of being lulled into vulnerable self-satisfaction, and the excitement of high risk, with the danger of tumbling to failure.

The total thrust of the effort to come to terms with risk, to treat it rationally, should be aimed at putting it under a microscope, dissecting it so that we understand its key features, and imposing structure on it. When risk can be treated in a more discriminating manner, it becomes less threatening and more amenable to management, if not to elimination.

Risk vs. Uncertainty

The first step in treating risk rationally is to distinguish between uncertainty and risk, because failure to do so is cause for much confusion.[1] If we define risk as having two components, a probability of occurrence and a magnitude of consequences, the distinction between risk and uncertainty becomes easier to understand. We often hear technical programs with very low odds of success described as high risk even though the expenditures required may be quite small, and therefore the true financial risk is inherently modest. Of course, if I commit a chunk of my life to an endeavor, the risk may be large for me, although in this case "consequences" would be measured in percentage of my professional career rather than expenditure.

Risk (R) in a technical program, as I will be using it, is defined as the probability of failure or partial success (P) times the magnitude of consequences, usually expressed as expenditure on the program (E).

$$R = P \cdot E$$

Risk implies structure, an ability to perceive and differentiate among the forces that will determine the course of events. It also includes a capability to identify the possible courses of events that might transpire and to estimate, with varying degrees of confidence, the probability of their occurrence. One way of illuminating the meaning of risk is to say that it involves dealing with "known unknowns."

Because of its psychological impact, risk must be considered in both absolute and relative terms. A potential loss of $100,000 may represent a significant fraction of the resources of a small startup and be invisibly small to an IBM. This factor is important. It is relative risk that must be managed carefully.

In contrast, uncertainty resists analysis. It exists in situations where information is so lacking that one cannot even characterize the possible courses of events, much less estimate the probability of their occurrence. Consequently, the results are entirely problematic. In contrast to the "known unknowns" of risk, uncertainty is concerned with "unknown unknowns."[2]

Unfortunately, the appropriate distinctions between risk and uncertainty are rarely drawn. Consequently, the possible outcomes of a situation about which knowledge is limited loom as an undifferentiated, threatening cloud. There is a perception that little can be done except have the courage to accept the risk or the prudence to avoid it. Unhappily, this dichotomy is often converted into guts versus no guts. Risk becomes a *macho* symbol.

Risk *can* be managed. Proposed courses of action *can* be dissected. The sources of uncertainty *can* be identified and management can then determine the information required to convert the uncertainty into risk. In fact, it is helpful to regard the work of R&D as converting uncertainty to manageable risk. In most businesses in which there are many activities under way, with varying degrees of risk for each, it is possible to treat risk actuarially. It is even possible, as we shall see later, to partition the risk, pool it with other businesses, and thus reduce its consequences.

From the foregoing, it is apparent that information is central to evaluating and managing risk. Acquiring information requires time and resources. Hence one of the central issues for managing risk is the need to address specifically the amount of time and effort that should be devoted to gathering or generating information before taking action on the basis of incomplete information. Line managers tend to have a natural inclination toward decisive action. One of the most consistent problems I encounter among operating managers is this "let's get going" syndrome. As a result, program objectives are often developed on the basis of a little information and a lot of guessing, and set as unquestioned targets. Then, unrealistic targets lead to disappointment and concern over "what's wrong with our project system, we consistently miss our targets?"

The Japanese adopt a quite different allocation of effort between preparation and action. Many people have noted the apparent ability of the Japanese to move rapidly. What they fail to perceive is the extended period of time and effort devoted to preparing the organization and gearing up for action. The pattern in the United States is to move as rapidly as possible to decision making, with an extended sequence of implementation and modification to complete the work. The Japanese pattern is the reverse: to devote sustained effort to considering alternatives, to invite inputs and build ownership, and to delay freezing plans until necessary. This extended "winding up" process can then result in what appears to be explosive action when it is finally unleashed. As we shall see, this careful treatment of the information-gathering aspect of addressing uncertainty is the central factor in managing risk.

Treatment of risk as a component of managing technology involves three different perspectives: intrinsic risk—the risk inherent in the program or activity being contemplated; personal risk—the perception that individuals considering participation in the program will have of the likely consequences for *them* should the endeavor either succeed or fail; and actuarial risk, which involves the opportunities for a business to treat a portfolio of risky undertakings on an actuarial basis and thus reduce the overall risk for individual businesses. I will discuss intrinsic and actuarial risk in this chapter and personal risk in Chap. 11.

Intrinsic Risk

As we noted earlier, technological risk is primarily associated with innovation, but it is in fact a component of all technological undertakings. Utilizing technology involves a spectrum of risk, as shown in Fig. 5.1. At one end is application of the state of the art, where the technological risk is quite low. One is undertaking the kind of application any competent professional technologist trained in the relevant art should be able to carry out. Projection of product performance, cost, and time for development can be made with great confidence. Next is extension of the state of the art—termed "microinvention" by one of my associates. This inching ahead with improvements in efficiency, speed, life, power, etc., constitutes the lifeblood of industrial innovation. In this case, technological risk is somewhat higher, and therefore projections of cost, for example, are less certain. Nevertheless, in my experience a situation in this category in which a technological barrier wrecks a project is rare. The principal technical challenge is to determine which of the options available—and there nearly always are technical alternatives—is optimal.

Next in level of risk is application of technology developed externally. One might ask why adaptation of technology developed externally is deemed a higher risk than improving the state of the art. Application of technology is a highly local and unforgiving phenomenon. Science may be universal, but application of technology involves complex interactions with the work force, sources of supply, capital equipment, plant layout, and so on. One common mistake is to assume that a major component of risk can be eliminated simply by acquiring technology whose performance has already been demonstrated. In these cases, contingencies for the unexpected are all too often not taken into account. Unfortunately, some of the critical characteristics that make a technology effective in one locale may be invisible even to its developers. Furthermore, competent technologists who contemplate licensing the technology often do not

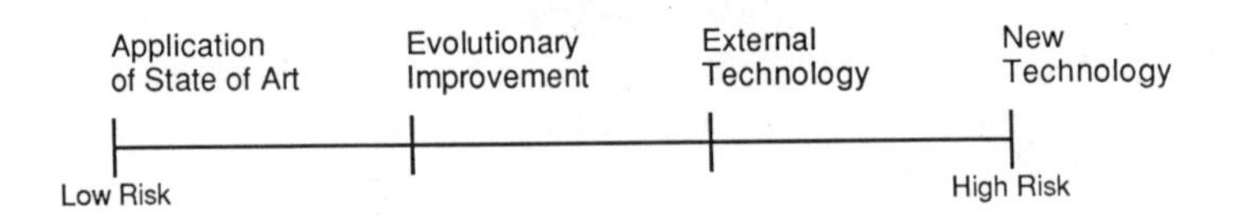

Figure 5.1 Spectrum of technological risk.

know the critical questions to ask that will uncover any limitations which may affect its performance in their environment.

I know of a company that once built a plant based on licensed technology already debugged and working successfully in another plant. As a further precaution to ensure that the licensed technology would be applied properly, the company hired a senior technical person from the original plant already operating to act as a consultant. It also asked the highly competent internal professionals who had some experience with the technology to review the plan for the installation. Even with this careful attention, the new plant failed by a significant margin to achieve its design yield. After urgent effort the problem was alleviated empirically, but the fundamental cause for the difference in performance never was fully understood.

At the highest-risk end of the spectrum is the development and introduction of new technology. This activity generates the greatest uncertainty. Projections on product attributes, cost, and time have low probabilities of validity; they are virtually educated guesses. The level of uncertainty perceived under these circumstances is strongly influenced by one's familiarity with them. Thus the psychological dimension of risk is most important in efforts to develop new technology. The human tendency is to make perception of risk inversely proportional to proximity to the phenomenon. Those whose exposure to major innovation is remote and infrequent tend to see it as a threatening cloud likely to rain disaster. Those whose involvement is more immediate and frequent see the varying degrees of uncertainty that exist in the fine structure of the activity. They can say: these things we are quite certain of, these have some alternative paths that we can evaluate, and these we can only make guesses about at this point. But they do perceive a way to proceed rationally even in the face of great uncertainty. A major requirement in managing risk is a willingness to commit the necessary time and effort to illuminate the fine structure and the alternatives available. Ignorance tends to breed anxiety and to lead to scattershot decisions rather than carefully targeted ones.

Obviously, knowing where a particular development is located on the spectrum of technological risk is important in evaluating total risk. Technological innovation is never solely a matter of technological risk. Even in cases where a proposed innovation is based on application of the state of the art, with almost no technical risk, there may be a substantial commercial risk. Lest readers regard technological innovation based entirely on conventional technology as a logical contradiction, let me remind them of the snowmobile and the Sony Walkman—new products that required no new technology. The creative act was in conceiving the product in the first place. The technology for implementation was already in existence.

In evaluating risk, one dimension that is too often ignored is the risk of no action. In a completely static world, that may be an acceptable posture. In the real world, inaction involves risk—that a competitor may act, that customer preferences may change, that a new technology may appear from the shadows, or that factor costs may change. The vulnerability inherent in doing nothing must always be included as an explicit component of managing risk. This can be accomplished by constructing an "inertial forecast": If one simply lets the forces already known to be operative continue to work, if trends already visible continue on their course, what future configuration of circumstances is likely to exist? What are the implications of such a scenario for the business? If this is done carefully, with attention to all the relevant forces in action, the future posited often becomes unattractive.

Risk Space

Risk can be thought of as resulting from movement away from the existing state of affairs in any of three dimensions: technology, product, and market, with the first being separated into product and process and the last being subdivided into customers and channels of distribution. The distinction between product and technology is important, though at first glance it may appear redundant. One may introduce new technology into a well-established product, e.g., the substitution of plastic for metal housings in power tools or the introduction of electronic controls into automobiles. Conversely, one can develop a completely new product based largely or exclusively on conventional technology, e.g., the food processor. In the latter case, the originality lies in the concept, not in the technology chosen to implement it. The distinction between product and market is also important. When Proctor and Gamble introduced Pampers disposable diapers, it was clearly a new product, but for Proctor and Gamble it served a market that was much less new than if 3M had tried to introduce the same product. P&G had extensive experience in evaluating consumer preference, powerful techniques for promoting new products with such customers, and a massive sales and distribution capability to gain wide availability. These dimensions create a kind of risk space, shown in Fig. 5.2.

The product innovation spectrum can vary from an incremental improvement in performance, cost, or features (e.g., successive models of the Boeing 727) to a major advance in those same parameters (e.g., the Boeing 747) to a totally new product providing a new function (e.g., instant photography). In similar fashion, a product innovation may in-

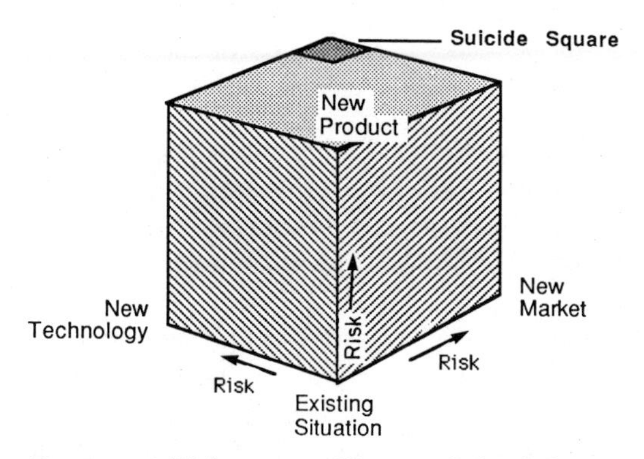

Figure 5.2 Risk space. (*Source: Lyle Ochs of Tektronix suggested this helpful graphical representation of risk.*)

volve dealing only with present customers and present channels of distribution (as in the case of VCRs), using present distribution to reach new customers (as Tandy did in introducing its personal computer), or having to establish new channels to reach new customers (as IBM had to do when it first introduced the PC).

The farther along each dimension that a development is situated, the greater the risk. Movement out on two dimensions at once *greatly* increases the risk. The locus of greatest risk is movement along all three. The point at greatest distance from the existing situation along all three dimensions has sometimes been called "suicide square."

Use of this space to characterize a venture can quickly illuminate the critical issues. For example, when Steve Jobs and Steve Wozniak developed the Apple computer, it was clearly a new product and, once past the dedicated hackers, addressed a new market. On the other hand, while I am not deprecating the creative wizardry of Wozniak, neither the product nor process technology was new. In fact, purchased components were used extensively. There was little question that the product would work and was manufacturable. The critical questions were, would anybody want it and how would you reach them? Thus the venture was far out in the risk scale on product innovation and in customers and channels of distribution, but not in technology.

Successive generations of Apples have had comparatively lower risk in terms of product and market, and still have not moved to higher-risk technology. The Lisa, on the other hand, was a new product aimed at a different market and with technology not yet well characterized and demonstrated; however, the Lisa did use conventional process technology. The combination of a new market, which Apple was not well equipped to serve, and a high price, resulting from use of labor-intensive conventional technology, proved to be an insuperable problem.

The Macintosh built on and extended the Lisa product technology, but it introduced new design features and highly automated process technology, which permitted a price that greatly broadened the potential market. Thus the product-technology risk was less, but the process-technology risk was greater.

The contrast between EMI and GE in the case of the CAT scanner is illuminating. EMI, a musical instrument and record company, was clearly in suicide square. It was introducing a totally new product, based on new technology that was totally outside its tradition, into a market for which it had no distribution or service and to customers with whom it had had no prior contact.

General Electric was also introducing a new product based on new technology (but the product was not as new as when EMI started—it was now clear that customers wanted it). However, GE was addressing customers with whom it already had a strong franchise, in a market for which it had an extensive and effective service organization. This difference proved to be crucial. GE was able to use its access to customers to make effective use of their inputs in designing a product with advantages over EMI's CAT scanner. Furthermore, GE was able to use its demonstrated resources and responsiveness in service as a powerful sales tool for selling an expensive, complex piece of apparatus.

Clear-headed appraisal of where a proposed development fits in the risk space is both very important and exceedingly difficult. Tales of woe about innovations gone awry often begin with the plaintive comment "we thought..." or "we didn't realize...." GE's X-ray business *thought* its market was medical equipment and its customers were doctors when it attempted to broaden its line to include patient monitoring equipment and other diagnostic devices. It soon discovered that in reality its products were diagnostic equipment and its customers were primarily radiologists and surgeons. It rapidly discovered that a franchise with these customers had little leverage on other medical specialists. Consequently, a presumed advantage in entering other segments of the medical services business was largely ephemeral.

Options in Managing Risk

Back off on the technology

Hardheaded consideration of where a project is located in risk space can lead to ways of reducing the risk even before the project is initiated and of phasing the risks assumed. One option is to back off on the technology. For example, when GE was developing a high-bypass fan jet for commercial aviation, the potential weight savings from using graphite-composite material in the fan blades were very enticing. Its aircraft engine people concluded, however, that not enough was known about the performance of these new materials under flight conditions. Consequently, they decided to stick with titanium despite the weight penalty. The disastrous consequence of Rolls-Royce's use of the graphite-composite material has been cited earlier.

Back off on the application

Another option is to back off on the targeted application. When GE was planning the development of a whole-body CAT scanner in response to the dramatic head scanner introduced by EMI, it decided to develop first a breast scanner for earlier detection of cancer. This development was less demanding technically and permitted accumulating some clinical experience before the effort was escalated to the whole-body product. The breast scanner was never offered commercially, yet it was a useful test vehicle on the way to the more demanding, but commercially much more attractive, whole-body machine.

Perhaps the most complex option for reducing risk is to take the time and expend the necessary funds to obtain additional information before making a large commitment. Here, the risk of hurrying introduction and getting into trouble with an inadequately debugged product must be balanced against possibly losing a market window to a competitor. Fear of the latter too often has stampeded a company into premature action. The detailed and extended process of establishing and fine-tuning product specifications and features that Ford went through in introducing the Taurus is a classic study in generating information to reduce risk.

Resist panicky market response

With the exception of "keep out" patent positions that can be obtained in pharmaceuticals and materials, one should not leap to the conclusion that being first is a requirement for succeeding. No doubt, to be first is very often rewarding, but realistically it cannot happen all the time. It is a mistake to conclude that just because a competitor has introduced a new product the battle is over and there is no chance of competing. General Electric did not invent and introduce the CAT scanner, but it managed to turn a superior product and marketing strength into a resounding success. The IBM PC was introduced well after Apple and Tandy had become major players in personal computers. Panasonic was not stampeded by Betamax and turned its VHS, introduced eighteen months later, into a winner. Hewlett-Packard beat Apple in introducing a laser printer as a lower-cost printer, but the Apple LaserWriter has been a great success. Philips and Sony developed compact disk technology, but by no means dominated the field. The Japanese have repeatedly demonstrated the power of bringing out better products, even if not necessarily being first to introduce them.

Require field testing

I learned this lesson early in my career from an ex-general manager, who lost his job because he hurried the introduction of a product. He lost no opportunity to issue the admonition: *Never* introduce a product without adequate field testing.

In short, locating a program accurately in risk space is to no avail if one misreads the probabilities of success. The market is always the eventual arbiter of success. A macho attitude of "we'll make it, because we won't make those mistakes other people make" needs to be tempered by a discipline which penalizes those who misconstrue the probabilities. This may seem an unfair burden on innovation, but a system of all rewards and no penalties fosters extremes of unrealistic behavior.

A large variety of avenues exists to obtain more information before making final decisions: for example, test marketing such as Proctor and Gamble is famous for; construction of special test facilities, such as wind tunnels or "shake" tables, to evaluate key components; or small-scale applications to demonstrate feasibility, such as Shippingport for nuclear power. In fact, an important argument for undertaking fundamental research is to acquire a more comprehensive understanding of the physical phenomena that underlie a technology so that one can predict behavior with greater confidence and over a wider range of operating conditions. GE introduced a new refrigerator compressor that eventually began to fail within the five-year warranty. This led to a substantial effort in sophisticated chemistry to understand the very slow, subtle chemical interactions that occurred among the Freon refrigerant, the insulating enamel used in the motor, and the lubricants. The company also hired a statistician to develop advanced methods for predicting with greater confidence the failure rate of a very large population based on a very small sample. Although this investment to develop better basic understanding and to apply sophisticated statistics was made after one costly experience, it served to reduce risk in future developments.

There are also useful alternatives to "going it alone" in order to reduce risk. All of these options require a departure from an attitude insisting on total control and reaping all the rewards. They require accepting the possibility that a share of success is worth more than sole ownership of failure.

Get a partner

An increasingly common option is to look for a partner who can both share the risk and provide expertise that reduces it. Partners can provide complementary skills, market knowledge, and market entree, and also provide funds that reduce financial exposure. Corning Glass has a remarkable record of forming joint ventures with partners having technical or marketing resources—for example, Owens Corning Fiberglas, Dow-Corning (silicones), and Siecor (optical wave-guides). Notably, these joint ventures have also enabled Corning to capitalize on inventions whose exploitation internally would have significantly diverted Corning from its central theme: to be a sophisticated supplier of specialty glass and ceramics. In the mideighties, Corning's joint ventures were growing more rapidly than the parent company, had net sales larger than those of Corning, and on an equity interest basis represented almost half of total net after-tax income.

GE formed a production partnership with SNECMA (Société Nationale d'Etude et Construction de Moteurs d'Aviation), the French jet engine manufacturer, first as a means of gaining access to the European market. Subsequently, a pioneering joint venture was created to develop a new jet engine for new applications. This partnership continued the advantage of market access, but also shared investment requirements and broadened the base of technical resources by enabling GE to concentrate on the high-temperature engine core while the French developed the fan and compressor.

Purchase a company

A variant of a partnership is establishing an ownership bridge. DuPont purchased Endo Laboratories, a small pharmaceutical company, to provide marketing and product development knowhow for DuPont's thrust into pharmaceuticals. IBM's purchase of ROLM, and equity position in MCI, provided telecommunications skills lacking in IBM.

One step removed from a partnership is heavy reliance on, and unusually close ties with, selected vendors. When GE was developing an urgent response to the CAT scanner, it relied heavily on Floating Point, a small, sophisticated supplier of high-speed array processors, for an absolutely critical component in the system. High-powered "number crunching" was central to the reduced scan time that was the target, and developing such a computer internally would have required too much time and overtaxed an already tightly stretched technical staff. The symbiotic relationship benefited both companies. Airbus Industrie made extensive use of U.S. suppliers for everything from jet engines, to aluminum alloys for the skin, to electronic controls because the proven capability of U.S. vendors reduced both real and perceived risk. Polaroid purchased from archcompetitor Kodak the final development and manufacture of its instant color film because Kodak's skill and resources in sophisticated process technology far exceeded Polaroid's.

Use consultants

Consultants with expert knowledge in specialized fields can also provide advice and counsel that reduce risk. Kodak used a world-renowned retired executive from the pharmaceutical industry as a consultant for its thrust into biotechnology.

Use sophisticated customers as a test bed

A final option is to rely heavily on selected customers. Eric von Hippel[3] has demonstrated that in some industries—notably instruments—customers are very important developers of new or improved instruments. He has also called attention to the value of working with a lead customer. Thirty-five years ago Philip Sporn, president of American Electric Power, was widely known for performing this role. He was deliberately sought as a purchaser of new generating or transmission equipment that advanced the state of the art. He was regarded as able to appreciate the technical risk inherent in the new equipment and therefore to be a demanding but sympathetic customer if trouble arose.

In summary, a variety of options is available if the level of risk or uncertainty in a program seems too great to be borne alone. Obviously, the key step is to look at the situation in a very clear-headed manner early on, to sort out the regions and sources of uncertainty, to evaluate one's own skill and resources in addressing them, to look for alternatives to plunging boldly ahead, and to accept the possibility that traumatizing trouble may appear.

Risk Varies at Different Stages

So far we have been considering a program as though it were a single, unchanging entity from beginning to end, with a single set of probabilities for various elements of the program. Typically, programs consist of a series of phases, and since substantial additional information is generated at each phase, the residual uncertainty at each phase diminishes. Moreover, the usual pattern is for expenditures to be relatively smaller in the earlier phases when uncertainty is greatest. Since risk includes magnitude of the consequences as well as probability of their occurrence, the actual risk to the enterprise during the earlier stage is not necessarily large. Hodder and Riggs[4] have pointed out that conventional discounting techniques, which are used to place all projects on an equal footing, nearly always apply a single discount value (hurdle rate) for the entire time period. This approach heavily discounts the remote future on the grounds that it is inherently less certain. However, the fact is, in many—maybe even most—cases, those larger funds for the remote future are not yet truly committed, and when the time for decision comes, much more information will be available and the risk associated with committing those funds much less. Consequently, a more discriminating and credible evaluation of risk would require using three different components for the discount factor: a risk-free time value of money, a premium for expected inflation, and a series of premiums for program risk, which diminish for later stages when more is known.

For example, a risk-free rate of 3 percent with 10 percent expected inflation and 6 percent risk premium would give a nominal discount rate of approximately 20 percent. However, the risk premium in the early R&D phase should be higher than for the product development, market development, and early market growth phase. Each of these phases requires an outlay of funds for a time interval that can be estimated. Because the outlays will tend to grow in each phase but the probability of failure will decrease, a decreasing risk premium should be applied to each phase in calculating either discounted cash flow or internal rate of return, with very little risk premium assigned to the final phase.

In some industries, such as pharmaceuticals, where a large number of development efforts must follow exactly the same obstacle course to commercialization, one can build up a reasonable statistical base of the survival rate of projects at each milestone. This provides the basis for realistic estimates of probability of success at each stage. If discounting is applied in this more discriminating manner, the frequent criticism that it tends to favor near-term contributions over longer-term, but presumably larger, advances would be substantially muted.

After all the careful appraisal and actions to reduce risk and uncertainty are made, a residuum remains for all projects that are undertaken. Is there anything else that management can do to ensure the success of a program or an expeditious termination if its projections turn sour? Fortunately, the answer is yes; there are many things that can be done both in the initial program planning and in execution to improve the probabilities of success, and we will explore them in the next chapter on project management.

Treating Risk Actuarially

Managing risk can sometimes be thought of as an actuarial problem. A startup venture has no choice except to have all its eggs in one basket. An established enterprise has the option of treating risk as a portfolio of ventures. It also can evaluate risk relative to the total resources of the enterprise. As noted earlier, a risk that is unacceptable to a $100-million enterprise may be invisibly small to a multibillion-dollar corporation. One inherent advantage of a multidivisional enterprise is its greater freedom to make risk commensurate with resources. By elevating or sharing responsibility for risk at higher levels of the organization, where it is pooled and arrayed against a larger business base, management can encourage risk taking at lower levels.

Much attention has been focused on the vital role of champions in successful innovations.[5] It has been my experience that champions are needed at two different levels to perform two different roles. One might be termed "the ground commander," who risks his career in a substantial way by undertaking a given program. He may or may not have originated the idea, but he makes the commitment of resources needed to exploit it. The other champion provides "air cover." He is nearly always a respected member of senior management who provides assurances that the work is being performed competently and that the program continues to hold promise, despite the inevitable vicissitudes along the way. He shares the personal risk, and since he typically has responsibility for several businesses, he can treat the program as one of a portfolio of ventures and relate the financial exposure to a larger business base. His reputation is at stake, but he is not betting his career in the same way as the ground commander. Furthermore, he is expected to be less personally involved in the project and therefore in a position to call a halt if he believes the promise of the venture is evaporating. His support must be firm, but conditioned on continuing progress. In a sense, his role is analogous to that of the venture capitalist.

Involving Higher Management

One of the delicate transitions in risk management is the elevation of risk responsibility to include higher echelons of management. Corporate management, eager to encourage entrepreneurship, sometimes establishes corporate venture programs to ensure that ventures receive the continuity and support they need. This elevation is nearly always viewed as a mixed blessing by operating management and sometimes as the kiss of death. Elevating a program almost always exposes it to review that in itself is disruptive; the program can easily be "stared to death." When DuPont initiated a corporate ventures program, operating managers quickly concluded they would prefer to "bootleg" programs and avoid all the "help" that headquarters provided.

Higher management, used to dealing with large investment decisions and large-scale business decisions, and always eager to discover promising new sources of growth, all too often escalates expectations for a program and attempts to accelerate development by providing larger resources. The fledgling business then develops an infrastructure its sales cannot support, and large losses ensue. Eventually, top management becomes disillusioned and support turns into criticism. Rare skill and wisdom are required to provide support and encouragement while at the same time continuing the pressure to achieve profitable operations and realistic rates of growth.

I used to return from visits to fledgling businesses railing at higher levels of management so insistent on demonstrating profitable growth that they were, I asserted, guaranteeing that these businesses would fail to achieve their potential. Experience taught me better. Having seen businesses handed a new factory and a full staff before sales appeared and then being buried in criticism when overhead delayed profit projections until far into the future, I am convinced that in general, fledgling businesses must feel relentless pressure to achieve profits early. The operation must be kept in scale with the level of sales activity it is able to generate; otherwise, it develops a cost structure that is not viable. Even more pernicious, a new business protected from such pressure develops an unrealistic perception of the rigorous discipline that is required to operate profitably in an unforgiving and demanding environment. GE's experience with factory automation is a case history of such a situation.[6] The company laid out a master strategy for a comprehensive attack on the problem of manufacturing productivity. It sought to put in place the capability to offer a full system solution to the problem by investing in product and software development, licensing technology, and acquiring companies that provided missing pieces in the system offering. These ambitious moves were accompanied by projections of dramatic increases in sales and earnings and by the scale-up in human resources and facilities needed to carry out the plan. When the market developed less slowly and when full system solutions did not appear to be what customers were willing to buy, it became necessary to scale back both projections and resources to a level that could indeed be operated profitably.

Another subtle problem in treating risk actuarially is the difficulty of recognizing the changing magnitude and character of risk as a program advances and of elevating the source of sponsorship when need be.[7] A typical scenario is as follows: A promising idea is developed and sold to an operating executive. As progress is made, additional tasks need to be undertaken and a widening variety of information must be developed. These successes, of course, require more people and larger expenditures. From the point of view of the aspiring innovator, the risk (he really should be thinking of it as uncertainty, because he tends to exclude the financial aspects in his appraisal) is growing smaller. Progress is being made. Potential barriers that could have killed the endeavor have either been overcome or proved unreal. The goal seems palpably more attainable. The innovator becomes more enthusiastic. The sponsor, however, may well be growing more anxious. As resource requirements escalate, the program begins to represent a disturbing fraction of his available resources. The innovator can thus encounter the puzzling phenomenon of success appearing to discourage his sponsor and stimulating reluctance to provide larger funds. The sponsor is beginning to think, "Boy, I'm not sure if I can afford this. If things don't pan out, I'll *really* be in trouble." Before this scenario develops, the innovator should anticipate the requirement to enlist additional support from somebody with deeper pockets.

How can the management process foster elevation of sponsorship without reducing the sense of ownership? I readily admit it is difficult and often done poorly. A mechanism that automatically triggers a review when a program reaches a set percentage of total programmatic effort (the latter being defined as those efforts not associated with the production and sale of output within the budget period) can be helpful. The automatic trigger removes the threat that a review has been initiated by questions about the project manager's judgment. Furthermore, it avoids the subjective problem of deciding case by case when a review should be conducted. Such automatic triggers are routinely applied to capital expenditures, and development programs are similar to investments in nature. Even more important are the mood and tone of the review process. The tone must be: "How can we help? We need to be well enough educated to be informed supporters rather than blind (and possibly nervous) advocates." To be able to question carefully and still be supportive is an art of skilled managers. Members of the Knolls Atomic Power Laboratory, the naval nuclear propulsion facility managed by GE, have told me that for all of Admiral Rickover's renowned irascibility, he managed to convey a tone in his reviews of, What are your problems? How can I help? rather than, Why haven't *you* already taken care of this?

Conclusion

Some have suggested that a formal champion of innovation be designated within senior management. (Edwin Gee is reputed to have performed this role for DuPont during his time on the executive committee, as did William Nicholson at Union Carbide.) The success of such an assignment is obviously crucially dependent on the style and skill of the incumbent and on his personal commitment to the task. This is not an assignment that can simply be handed to somebody just because he reports to you. There are very few role models to follow and developing successors is not easy.

What can be said, however, is that if a management team wishes to address in comprehensive fashion the challenge of both encouraging risk taking (an almost universal aspiration) and avoiding disaster, it must develop responses to all three elements of risk: intrinsic, personal, and actuarial. To address intrinsic risk, management must recognize the central role of information in reducing risk, explicitly include an evaluation of confidence limits for objectives, and then assign responsibility for generating needed information and scheduling downstream review of objectives and plans. It must accept realistically the pyramiding character of risk in products, technology, and markets and not overestimate its ability to beat the odds. First-rate technical outfits can and should be pretty cocky. Unless they are remorseless in condemning failure, they can degenerate into unrealistic risk takers who see a challenge in accepting risk rather than in demanding success. Effective management of risk makes use of a variety of options to reduce risk even at the price of sharing benefits.

Finally, effective risk management requires an automatic process for elevating risk taking to higher levels so that it remains commensurate with available resources, pooled in a portfolio of undertakings with varying levels of probability, and arrayed against a larger business base.

Notes and References

1. Donald A. Schon, *Technology and Change,* Delacorte Press, New York, 1967, pp. 19–42.
2. Norman R. Augustine, *Augustine's Laws,* Penguin Books, New York, 1986, p. 66.
3. Eric von Hippel, "Users as Innovators," *Technology Review,* June 1978, pp. 30–39.
4. James Hodder and Henry Riggs, "Pitfalls in Evaluating Risky Projects," *Harvard Business Review,* January–February, 1985, pp. 128–135.
5. Donald A. Schon, "Champions for Radical New Inventions," *Harvard Business Review,* March–April 1963, pp. 77–85, first called attention to the value of this role. Others have also noted its importance, for example, A. K. Chakrabarti in "The Role of Champions in Product Innovation," *California Management Review,* Vol. 17, Winter 1974, pp. 58–62.
6. "How GE Bobbled the Factory of the Future," *Fortune,* November 11, 1985, pp. 52–63.
7. Tait Elder, retired 3M executive, emphasizes this point.

Chapter

6

Project Management

Project management is a specialized subset of the field of management. Its most distinctive feature is that it is temporary. A group is assembled to accomplish a specific, carefully constrained task. Once that task is completed, the raison d'être for the group disappears and it is disbanded. Project management is a central feature of operational management for technology because most technical work is performed under the aegis of specific programs with a defined target and a date for completion. Project management applies to all phases of technology from exploratory research to the development of a major new system. In the former case, the requirements are less stringent, emphasizing disciplined thought and study. As work moves through the technology spectrum, project management becomes more rigorous and more quantitative in specifying resource requirements, time intervals for completing work, and convergence of separate chains of activity. The terms *project* and *program* are often used interchangeably. I have chosen "project" because I believe it more unambiguously applies to work that is discrete with a clear terminus, whereas "program" is sometimes applied to work of a more ongoing character.

Functional Project Management

Project management has two different traditions, one serial and one parallel, and each leads to a different approach. The first, and still quite probably the most widely practiced in U.S. industry, arises from the necessity for an ongoing business to conduct two different kinds of technical activities concurrently: the daily activities needed to supply customers with products in order to meet the current year's sales target (sometimes called "running expense"); and a different, but also more or less continual activity that could be termed "programmatic effort," consisting of those programmed activities carried on in each function to enlarge capacity, improve effectiveness and productivity, and strengthen and extend the product line. These programmatic activities constitute a drain on current profits, and in that sense are temporarily expendable if operating results are disappointing, but they are vital to the continuing competitive health and growth of an enterprise. In the realm of technology they include redesign to reduce costs, to eliminate problems that have become apparent, and to provide evolutionary improvements in performance. They also include extensions of the product line, to serve new market segments, for example. While some of these programs may involve innovation or entrepreneurial endeavors, those labels are not descriptive of all the programs.

Typically, each function develops a portfolio of these programs, and the general manager decides which are most urgent and how much the business can afford to undertake in a given year. The financial criteria for investing in most such programs are usually quite stringent. Paybacks will rarely exceed two years and will often be much less than that.

In this realm of programmatic improvement, product development programs generally begin in engineering, even though the basic design objectives, cost targets, and product attributes have presumably been worked out in conjunction with product planning in marketing, with inputs from manufacturing on producibility and finance on costs. They then move sequentially to manufacturing, and back to marketing for promotion and sale. This carrying out of work in series, with a handoff from one function to the next, characterizes a significant fraction of all project management. In this context, the functional manager is responsible for the project work in his function. It is not uncommon for a functional project manager to be responsible for more than one project and for functional specialists also to be working actively on more than one such endeavor. In many, many companies this is the only kind of project management practiced. For projects aimed at the evolutionary improvements that maintain the continuing vitality of a product line, this is probably the most effective approach. In the quest for useful incremental change, a sense of continuity, of past decisions and trade-offs, can be very useful. The most severe problems of cross-functional constraints have already been confronted.

Multifunctional Project Management

The other tradition adopts a quite different approach. It probably originated in the industrial response to wartime pressures in World War II. It involves attempting to identify at the very beginning all of the skills and resources that will be required to carry a project to completion. These needed skills are assembled, or at least committed, at the outset, and the project manager's responsibility is to complete all phases of the project—not to hand it off to some other group at some predetermined point. Thus he must act as a generalist. In fact, experience in managing such projects is regarded as excellent preparation for becoming a general manager.

This mode of operation responded to the overriding wartime imperative for speed. It recognized the time that was required to assimilate new skills and new team members. It also recognized that when diverse, interacting skills were required to complete a program, it was more effective to have them assembled and interacting from the beginning, so that problems in differing objectives and priorities and differences in terminology could be addressed early on. This approach of assembling a team to carry out the complete project also recognized the value of allowing time for complex human relationships to be developed and for a sense of common purpose and commitment to the program to evolve.

Students of organization identify a spectrum of forms from the purely functional to product or project teams, to varieties of matrices. It seems to me that the critical distinction, expressed in computer idiom, is between serial and parallel processing. The functional form, with its handoff between functions, represents serial processing, and all the other varieties are different ways of addressing parallel processing.

Need for Parallel Action

Irrespective of which tradition is dominant in most organizations—serial development with a functional handoff or parallel development under a single project manager—events are forcing movement toward development projects to be done in parallel rather than in series, with all functions being involved from the beginning. Although the term "matrix management" may not be used and no formal parallelism be maintained between functional managers and project managers, special multifunctional teams are becoming the conventional approach. Serial development invites problems in assigning responsibility for difficulties that emerge in manufacturability, quality, meeting schedules, satisfying market requirements, etc. Because of their interrelatedness, all functional perspectives must be present from the very beginning. Establishing product attributes is not independent of cost and quality considerations. Problems of manufacturability must be addressed from the beginning of a design. The tendency toward shorter product life cycles and the need to shorten response times no longer permit the recycling of product development projects until "you get it right." Parallel development in multifunctional development projects represents the most effective approach to shortening time cycles.

All of these attempts at parallel action require organizational learning to become effective. One participant characterized them as antithetical to the western tradition of individualism. A period of three to five years may be needed before the people who must work in matrix teams and those who manage them acquire the requisite skills and behavioral responses. Even learning what functional skills are necessary for a successful project team requires trial and error. An auto executive once told me that his company had learned only from downstream difficulties in manufacture that the makeup of the initial team had been incomplete: manufacturing people representing fabrication and processing had participated, but those representing assembly had not.

Dealing with Uncertainty

Every technological development begins with a large number of imponderables, areas where information is insufficient to support decisions with low risk or small error. The most basic objective of a development project is to generate information to reduce or eliminate these areas of uncertainty. Surprisingly, perhaps the most common problem is failure to delineate these areas of uncertainty adequately. Rather, the emphasis is on delineating and then spotlighting the specific objectives that have been set. A typical project plan contains a series of objectives for product or process configuration; for product attributes, such as performance, life, and availability; and also for cost constraints and time-to-completion goals. The common wisdom holds that it is desirable to establish these project objectives firmly at the beginning; otherwise, it will be difficult to focus attention on specific goals. Just as important, it will be difficult to encourage commitment and to measure progress, to ascertain that participants really are moving toward their target instead of just enjoying themselves learning interesting things.

Although these justifications for establishing firm objectives at the beginning are sound, they also involve a risk. They submerge the varying degrees of uncertainty associated with the objectives; they may even suppress vigorous examination of uncertainty. The process is necessarily tilted toward narrowing focus and motivating action. Activists—and by definition, managers tend to be activists—would argue that one never has enough information to eliminate uncertainty; therefore, the role of the manager is to make decisions in the face of uncertainty. However, it is arguable whether making a commitment to a course of action with insufficient information is always the sound approach. The line between indecisiveness and temerity in making decisions is influenced by the number of scars one has accumulated from past mistakes.

Although a truly effective project planning and managing process must indeed make provision for freezing objectives, it also must include two additional elements. For every project objective or specification established, the underlying information base must be examined and then a confidence level assigned to it. Second, having made the level of uncertainty explicit, the process then must ask the question, does this objective or specification have to be frozen at this point in time or can it be delayed until later in the sequence? The guiding principle should be that no objective or specification should be frozen until a decision on that particular element is required in order to proceed with the work. That change in perspective can shift significantly the attitude toward project planning.

Typically, in planning a technical project, the technical organization paints itself into a corner by advocating technical, cost, and timing targets that stretch its capabilities to the limit. Consequently, all questions become "can we" questions rather than "should we" ones. Then the total resources of the project team must be devoted to achieving these demanding goals. Frequently, as the project nears completion, it is discovered that some of the objectives are wrong or at best questionable. Everything stops while the work to refine objectives is performed, which often leads to a recycling of the project to achieve new targets, with resulting delay and additional expense.

An approach that focuses on uncertainty should seek to identify the level and source of uncertainty associated with key project objectives and specifications. It should determine the information needed to reduce the uncertainty and assign responsibility for gathering it to the function best equipped to generate that information. Finally, it should identify the latest time at which a decision must be made and schedule that action as part of the project plan. The review process that monitors progress on the project should continually examine the current level of confidence associated with key objectives. This approach helps to ensure that the needed focus on "should we" questions is not lost in the crunch to keep the project on schedule.

A conventional project plan, and indeed the principle focus of such planning techniques as PERT and Critical Path, carefully articulates the timing, sequence, and relationships among all of the work elements of a project. In other words, the planning process assumes that the objectives and specifications are both valid and attainable. It addresses rigorously and in detail the questions of what has to be done, in what sequence, and with what requirement of time and resources in order to accomplish each objective. With those inputs, it can lay out a master plan that completes the project with a minimum of slippage from failure to fit all the pieces together as tightly as possible. These planning systems are valuable because they anticipate that not all of the initial projections on time and resource requirements will be valid and they can show very rapidly the downstream effects of failure to achieve objectives at any stage of the project. The project manager can then decide whether to increase resources to get back on track, to change the overall schedule, or to adjust the objectives. Once the decision is made, the necessary realignment of all the effort can be carried out quickly.

The value of this sort of rigorous, detailed, step-by-step planning is unquestionable. Complex projects are possible only if such an approach is used. Combining many tasks in a complicated sequence is a very complex challenge, and failure to do it well can lead to disaster. The problem arises when the limitations of PERT-type planning techniques are not recognized and they are used in circumstances for which they were never intended, such as in a typical industrial R&D project. Originally, sophisticated project planning and management techniques were devised primarily to deal with complexity and large scale—not with uncertainty. Setting objectives that are either wrong or unattainable also leads to disaster, but conventional project planning and management will not uncover these latent problems. Furthermore, for the smaller project usually found in R&D, the complexity of PERT-type planning represents overkill.

An effective project planning and management system for technology must retain a focus on the role of uncertainty throughout an endeavor. It must never submerge the ever present relevance of "should we" questions under the pressure of "can we" questions. The central task of a technology development project is to generate information rather than to complete certain tangible structures or systems. Consequently, the initial phases of planning must highlight the level of uncertainty inherent in each objective and assumption, the source of the uncertainty, the kind of work needed to reduce it, and the most logical people to do the work. The process should allow explicitly for an open-endedness that would be heretical in conventional project planning, i.e., it should address directly questions such as: Do we have to decide that at this stage? Should we make explicit provision for reexamining the objectives and assumptions on which we are proceeding, and if so, when should they be reexamined? Who should generate the information needed for that reexamination? Table 6.1 indicates the sequence of steps required for effective project planning, the foundation for effective project management.

Every prospective project manager, whether he will be managing a

TABLE 6.1 Steps in Program Planning

Step 1

a. Lay out the information or assumptions that stimulated the initiation of the proposal.
b. Evaluate the confidence level that is associated with the information and assumptions.
c. Identify the information needed to improve the confidence level and determine how critical reducing the uncertainty is to the success of the program.

Step 2

a. Lay out the program objectives.
b. Evaluate the level of confidence associated with the importance, the validity, and the attainability of each objective.
c. Identify the source of the uncertainty, i.e., the missing information that creates it.
d. Determine the work needed to produce the information necessary to reduce the uncertainty.
e. Ask, in the light of this analysis, whether it is necessary or wise to freeze the objective at this point.
f. Lay out a schedule or work plan that includes not only all the work needed to accomplish the objectives, but also the delayed decision points, where they have been created, and the periodic downstream reviews of the continuing validity of the objectives and assumptions on guiding the entire effort.

Step 3

a. After the complete program plan is established, conduct a final review that addresses the question: Have we now made provision, within the limits of our resources and the time available, for generating all of the information needed to answer both "can we" and "should we" questions? In other words, under what circumstances could we approach the completion of this effort only to discover that it was misguided in some fundamental way?

technology development project or one for product development, must identify the key questions necessary for success and illuminate the levels of uncertainty associated with each question. Since product development projects nearly always incorporate more diverse activities and require larger resources, Table 6.2 indicates the kinds of questions that should be raised as part of developing the project plan.

The questions listed in Table 6.2 are the sorts of queries that any effective project plan should address for market-oriented, rather than cost-reduction, programs; obviously, some of the issues are not applicable for the latter. In projects that focus on product development, the critical questions tend to revolve around product-attribute–price–volume relationships. Continuing close interaction among product engineering, manufacturing, and marketing is required both to question and to help resolve the "should we" issues.

As we noted earlier, the natural inclination of technical people is to

TABLE 6.2 Questions for Project Management of Product Development

1. Do I have all the needed views represented on the team?

2. Do I know the customer?
   a. How completely has he been identified?
   b. How well are his needs and preferences understood? How stable?
   c. How good is my franchise with him?

2. How well do I know the distribution?
   a. Relationship with suppliers
   b. Relationship with customers
   c. My own franchise with them
   d. Preferences and hang-ups
   e. Leverage on promotion and sales

3. Do I understand the application?
   a. Needs to be satisfied; rate and direction of change
   b. Priorities among objectives
   c. How much leeway in serving?

4. How certain are my product attribute specifications?
   a. Have I distinguished between those that must be set now and those that can be delayed pending further information?
   b. Have I assigned responsibility and scheduled work and decision making on specifications kept tentative pending further information?

5. How solid are my price-volume projections?
   a. How sensitive are they to product attribute specifications?

6. How realistic is timing of introduction?
   a. What would be the effect of slippage?
   b. How late can I hold off starting the ball rolling?
   c. Do I have a fallback position?

7. How solid is our projection of competitive reaction?
   a. How carefully have we identified potential competitors?
   b. How well do we know them?
   c. Does our plan adequately anticipate their likely behavior?

8. How well characterized is the technology?
   a. Have we included both product and process technology in our thinking?
   b. Are representatives of every key technology truly on board as members of the team?
   c. Have we ascribed a level of confidence to every critical technical objective?
   d. Have decisions that should be left open because of inadequate information been identified and responsibility assigned for reducing the uncertainty?

9. Is the review process comprehensive?
   a. Have frequent probing reviews of progress against plan with trigger points for corrective action been scheduled?
   b. Have periodic comprehensive reviews of basic objectives (both "can we" and "should we" questions) been scheduled?

push technology to the point where "can we" questions are dominant. However, those issues often are resolved relatively early, and attention shifts to doing the necessary work to meet objectives. Meanwhile, "should we" questions remain as the increasingly important issues that all too often are accepted as given rather than questioned.

Figure 6.1 summarizes the critical steps needed in what could be termed preproject planning to ensure that uncertainty considerations get the attention they warrant in the final plan.

This approach to project planning and management contains an additional advantage. Customarily, the project plan must be presented to and approved by a higher level of management. The approving manager often has not been deeply involved in preparing the plan; consequently, he may have only limited insight into the varying levels of uncertainty associated with key objectives or into the kind of work required to reduce that uncertainty. Furthermore, he has limited opportunity to probe for the areas of uncertainty. He must depend on the judgment and experience of those who prepared the plan. A process that addresses uncertainty directly and openly gives him much more insight into the true risk associated with the project. It also alerts him to decisions that have been held in abeyance and to additional actions needed downstream on his part at scheduled intervals in the project.

Using Review Boards

The concept of a dedicated project team focusing exclusively on achieving its objectives carries an inherent danger—loss of objectivity on the validity of goals and on the status of the effort to achieve them. People inevitably become advocates for a cause and internalize the values and priorities implicit in a project. For this reason, where a project is of great importance to a business, it is desirable to establish a continuing review panel at the very beginning, or failing that, to schedule a rigorous independent review before a project is completed. Such a panel can become familiar with the work of the project, but retain a degree of remoteness from actual participation. In most technical projects, objectives are based to a substantial degree on judgment, and it is important for that judgment to be subjected to sharp, periodic scrutiny. A review board can provide a countervailing force to the commitment and mutual reinforcement that are actually desirable among team members. This is another example of the deliberate ten-

Figure 6.1 Preproject plan.

sion required in effective management of technology: dedicated, action-oriented commitment, tempered by critical "prove it to me" restraint.

Balancing Conflicting Demands

Product development plans are another illustration of my theme of balancing tension. A number of factors must be balanced, and the balance should vary from product to product:[1]

Product cost. Has the lowest-cost design been found?

Product performance. How closely do product attributes match perceived customer desires?

Development speed. How well does development time meet market dynamics?

Development costs. How generous should the budget for development be?

Risk taking. How much effort should go into reducing uncertainty?

In preparing project plans, one of the most common mistakes is to underestimate the dollar value associated with time. A veteran general manager I know, making his first public comments to the staff when taking over a major R&D operation, emphasized that the cost of trying to get work back on schedule was many times what would have been necessary had a more realistic plan been devised in the first place.

A parallel cost is involved in missing a market window because of delayed entry. Some businesses have estimated that 50 percent or more of the available volume had been lost because a development project had to be recycled in order to produce an appropriate product. One should at least examine very carefully the leverage that additional resources would provide in reducing time to completion. The lowest project cost is not necessarily the most cost effective.

Some companies speed the development cycle by beginning test manufacturing before the final design has been frozen in order to provide early feedback to the development team. Shorter development allows less time to make changes in product specifications (often prepared by marketing). Some would assert that this is a benefit.

Parallel Projects

If a business has adequate resources and an enabling culture, another approach to dealing with uncertainty is to run multiple projects in parallel. Some Japanese consumer electronics companies adopt this approach. For example, a visit to Matsushita's product center reveals many products that never make it to market. This approach provides an opportunity for encouraging competition among teams. The parallel-project approach permits management to delay until late in the process the decision on what to take to market. Although this process would appear to be wasteful of resources, it avoids the delays associated with recycling projects until "you get it right" with the attendant risk that the market window is missed. A more subtle benefit is that these multiple projects often provide the development necessary for rapid introduction of second- and third-generation models. The product chosen for initial introduction is not necessarily the most advanced technically nor does it offer all planned features and options. Those come from follow-on products and upgrades, provided that the sequence makes allowance for good product flow and adaptation to market feedback. The initial product may be the one that seems to offer the best balance of acceptable properties and technological certainty at that stage of development. U.S. competitors are often caught off guard with the speedy appearance of new models and wonder how such speed is possible. The multiple-project approach is one important ingredient.

Achieving Cross-Functional Integration

Probably the most critical ingredient in successful technology development projects is achieving effective participation from all functions, but especially engineering, manufacturing, and marketing. What do we mean by effective participation? It is the kind of participation that results from a sense of ownership in a project; it is not just somebody else's project for which you are providing some inputs.

I have learned that organizations which express concern over ineffective project output often treat projects as basically engineering development projects. The differences are apparent when listening to project managers describe their projects. If they are engineering-focused, most attention will be given to the technical options that were available and the reasons for the chosen course. Managers will describe in detail the technical work under way to meet goals and will highlight the problems and difficulties that they foresee. The content smacks more of a design review than a development project review. Little mention will be made of uncertainties about pricing or projected sales volumes; possible mismatches with market requirements may not even be mentioned. Significant participation by people in other functions is not apparent.

Project management becomes even more difficult when the endeavor involves more than one operating component. Fighting over turf and bickering among components about relative contribution and reward can virtually paralyze the effort. In these situations one of two courses of action can be adopted: the project manager can be given sufficient power to overrule any of the warring factions (a circumstance that is difficult to establish for a project which is small compared with the scale of the operating components involved) or an operating manager at a higher level can be designated to ensure adherence to schedule by ready intervention to resolve stalemates. Even so, the project manager is in a delicate position. If he turns to higher authority too frequently, he is weak; if he waits too long to get things moving, he is ineffective.

Building Ownership

How is ownership established? One manager used to distinguish between "responsible" and "irresponsible" inputs. He was not casting aspersions on the integrity or competence of the source of the input. He simply assumed that where one's own welfare is not at stake, the nature of response will be different. Ownership results from having a voice in five critical features of the project planning and management process:

1. Providing opportunities to help shape the project. It is much easier to develop a sense of responsibility and commitment for proposals one has come to think of as one's own. If the process provides no opportunity for all functions to participate in defining objectives, then those frozen out of key decisions will feel like what they are—peripheral—and they will behave accordingly.
2. Allowing sufficient time for team members to build a sense of ownership. The required time cannot be compressed significantly. One of the disadvantages of the functional-handoff approach is the time lost while new team members develop a sense of ownership.
3. Setting priorities that are accepted as reasonable given the nature of the work. If a project is labeled "high priority," but is obviously not receiving adequate support, the disparity will be impossible to hide. One manager with whom I worked routinely established very demanding, and frequently unnecessary, completion dates. I, and others, soon learned to avoid any sense of commitment to his schedules.
4. Defining project objectives that continue to be regarded as attainable as work progresses. One of the most devastating things that can happen to a team is to continue to be held accountable for technical, cost, or schedule objectives that have come to be considered as impossible.

5. Creating appropriate rewards and penalties for contributions to a project. This is especially true in smaller organizations where limited resources require part-time participation in project teams. If people discover that involvement—good or bad—in a project has no perceptible effect on one's compensation or career, the project-team approach faces an almost insuperable handicap. This problem is particularly severe in matrix organizations. Who determines compensation? If the project manager does not control it, he will feel that he lacks adequate command of resources. If the functional manager does not control it, he will feel that he lacks the power to maintain adequate discipline in technical work. Preferably both should be required to concur in a judgment, and the individuals on the project team should know how decisions are made.

Some relatively obvious contributions to team building are often overlooked. Physical contiguity is important. It is difficult to build a sense of common purpose and momentum among people who are physically separate. And the distance does not have to be large. Thomas Allen[2] has demonstrated that the probability of even weekly communication among technical people falls off very rapidly as a function of distance: it is <0.1 after 10 meters and becomes virtually asymptotic after 30 meters. One engineer who moved approximately 100 yards to a pilot facility caught himself saying with some surprise, "I spent the afternoon up at the laboratory" (meaning the building he had just moved from). There is also evidence that vertical distance is a greater psychological barrier than horizontal distance. Nevertheless, the number of times one discovers members of project teams working in different buildings, sometimes miles apart, indicates that the enormously inhibiting influence of physical separation is not yet given adequate weight.

Providing off-hours opportunities for people to share relaxed time can be helpful, especially for people working under severe pressure. Offering psychological reinforcement by openly rewarding progress, in trivial or even zany ways, builds teamwork. Extensive use of publicity to note progress and to highlight people rewarded for special effort or contribution is also important. Generosity in sharing credit for success is especially vital, and it has the interesting property of not diminishing any individual's credit for contribution—in fact, it may well enhance it.

Effective project management must pay attention to the substantive issues of staying goal-oriented and continuing to address "should we" as well as "can we" questions, and at the same time undertake the activities necessary for establishing ownership and sharing credit. If a project team is kept informed of changes in the environment, it can accept changing targets for work with greater equanimity. Even more important, it is less vulnerable to the devastation of finishing the development and discovering that output did not meet the market need.

I have learned, from my own experience and from observing others, that effectiveness as a member of a project team is a skill that can be learned. For example, if you want to control a team meeting, get to the blackboard first: the person who determines what is written down and the words that are chosen controls the meeting. The existence of a pool of people with experience in project teams is a very valuable asset.

The dynamics of working in or managing a temporary organization such as a project are quite different from those of an ongoing organization. A chapter entitled "Managing Temporary Organizations" in *Innovation in Big Business* by the author examines this subject more fully.[3]

Project Control

Efforts devoted to generating a sense of ownership and to ensuring that project objectives match technical and market requirements are useless if the project itself does not succeed in coping with the inherently large uncertainties and unexpected events in technical development. A typical project consists of a large number of highly interrelated tasks, each involving large elements of uncertainty. Maintaining effective control and a semblance of order in a situation that persistently borders on chaos requires exceedingly rapid, effective communication and an almost endless juggling of goals, timetables, workloads, and even resources.

The secret to success is extensive partitioning of the problem into its constituent components and then aggregating each into subtasks that can be tracked. However, a fine balance must be struck between a control system sufficiently detailed for effective monitoring and control and one so demanding that it suffocates the project.

Need for discipline

Technically trained people recognize the uncertainty in a development project. They accept delay and disappointment as inherent in the work. In fact, if they were not teetering on the edge of failure, they probably would feel that the technical goals had not been set high enough. Marketing, quite naturally, is unlikely to discourage proposals to achieve demanding performance levels. Unfortunately, this acceptance of inherent uncertainty can easily become an excuse for failing to meet deadlines: everybody *knows* you can't schedule inventions, so people can't be faulted for missing milestones. That situation can in turn deteriorate into a persistent formulation of unachievable technical goals, because nobody feels guilty or is penalized for missing targets. Thus the situation becomes a self-fulfilling prophecy. In its pathological form, an organization literally loses the ability to judge whether project objectives, resources, and timing are realistic.

The kind of discipline that is imposed on a technical project depends on the relative priority assigned to the three key parameters: technical performance objectives, cost targets, and schedules. One could think of it as a business resting on a three-legged stool, as shown in Fig. 6.2, in which the legs always touch the ground no matter how tilted the surface. Some businesses having a powerful market position are able to give highest priority to technical performance, with less attention to cost and time. I know of two major corporations that until about 1980 never included a cost target in their product development plans. Their powerful market positions and customer franchises enabled them to set prices at profitable levels, irrespective of costs. Thus, setting an objective to keep the stool level is not necessary. Rather, one should be very aware of the balance that has been struck and of events that might require rebalancing.

Other businesses, facing severe competition or a rapidly changing environment, are time-driven. Companies in industries that are maturing are forced to give greater priority to cost. Managers in commercial businesses often deride the lack of cost consciousness of contractors to the federal government. A senior organization consultant for a company having both kinds of businesses commented that while the government-focused businesses might be less cost conscious, he felt they were much more disciplined with respect to time. He also noted that on balance he was not sure which was more effective. The subtle equilibrium that must be achieved among these three elements is but one more example of the prevalence of balancing tension as a central issue in effective management.

Information requirements

A key ingredient in determining the nature of the control that will be exercised in project management is the kind and timing of informa-

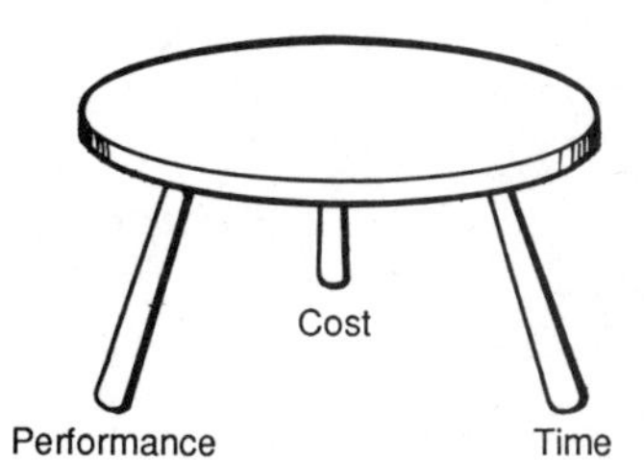

Figure 6.2 Key business parameters.

tion that will be provided the manager in reporting on its status. If cost data are incomplete or delayed, if compliance with schedule is not made evident, it is small wonder that project managers do not give high priority to cost and schedules. Problems in achieving effective project control and installing an effective management information system are more likely in businesses with a tradition of parallel development than of handoffs between functions. The information system commonly in use in such a business is structured around functions. It is designed to tell the functional manager whether his expenses are in line with the budget. Often data are not available until sometime after the end of the reporting period and are classified according to major budget categories, for example, labor, material, administrative and office expenses, telephone, travel and living, utilities, and depreciation. Such a system is not well equipped to accumulate charges across functional lines, to indicate progress according to schedule, or to provide rapid feedback to managers. Perhaps the most critical step in establishing effective project control is to determine what priority will be assigned to the various parameters of the project and then to charge the information systems group with providing information that will illuminate the status of each parameter. While project cost data can usually be extracted from the cost data accumulated for other reasons, the project manager who must take the time and effort to do that for himself has good reason to believe that the business does not take project control very seriously. Increasingly, personal computers have been used by many project managers to enable them to perform their own project cost accounting because the conventional reporting has not done it—or is too delayed to meet their needs. Although this results in an inefficient accumulation of "private" data bases, these managers cannot be faulted for using new technology to overcome deficiencies in their business information system.

Measurement and Reward

The final component of effective project management is the system of measurement and reward used for team members and managers. Perhaps the most critical issue in establishing an effective system of measurement and reward is, what constitutes success? Is a project that meets its objectives for performance, cost, and time, but that fails to gain market acceptance, a success? Conversely, is a project that fails to achieve its technical targets, despite clearly first-rate technical work, a failure? Obviously, there are no unequivocal answers to these questions. The important point is for management to be explicit and forthcoming with team members and project management. One might expect the system to be tilted toward attaining commercial success. Otherwise, technical people could with impunity continue to behave as though the project is a technical development, with commercial success being the responsibility of others. However, people cannot be expected to take risks if they perform competently, even brilliantly, on a technical project that does not succeed commercially, and that therefore is labeled as a failure.

The key requirement is to be aware of the changing nature of the risk profile as a project progresses. When technical uncertainty is high, effective work in reducing that uncertainty must be rewarded. As uncertainty is reduced, a point is reached where the projected benefits of the project begin to be incorporated into the business forecast. As a project reaches that status, success clearly should be measured in commercial terms, not in technical terms. Success should be measured in terms of what the project was expected to accomplish at the various stages. Even some commercial introductions are really tests to reduce uncertainty about the size and nature of the market. In such cases, poor market acceptance is not necessarily failure.

The other important dimension of measurement and reward is who has control over the process. If the project manager cannot determine or at least strongly influence recognition and compensation, he faces an almost impossible task. Control of the pursestrings is the most effective way to get people's attention. Inability to promise and deliver rewards commensurate with contributions is a major barrier even to recruiting capable people to work on projects, much less to motivating them to intense, dedicated effort. Being farmed out to a project, even a high-priority one, can make one an orphan when it comes time to reward performance—and it is a painful experience.

Keeping Resources and Commitments in Balance

So far, we have been examining the issues involved in managing an individual project more effectively. A quite different, but equally important, subject is maintaining control over an entire programmatic endeavor. Two basic requirements must be met. First, the management team must ensure that the total resources devoted to programmatic effort are in fact deployed in a distribution consistent with the basic strategy of the business. Second, in an environment in which the initiation of projects is diffused among many people and distributed over time, managers must install measures that will prevent the aggregate of commitments from exceeding the resources, especially human resources, available.

The easy solution to the first requirement—keeping resource allocation consistent with strategy—is to establish the allocation at the beginning of the budgeted period and freeze it. Unfortunately, that is totally impractical. "Managing by exception," the central theme of operational management since F. W. Taylor, recognizes the inevitability of juggling resources, objectives, and time in the face of perturbations and changing circumstances. The challenge is to do it in such a way that the strategy is not inadvertently distorted or vitiated. How often have managers struggled successfully with this task, one project at a time, only to discover at the end of the period that their business had not achieved what they planned, they did not foresee where it would be positioned strategically, and they did not like the results of their "successful" efforts?

An adequate control system to maintain consistency with strategy for a portfolio of projects must contain the three following features: (1) a set of categories that correspond to the basic strategic thrusts of the business under which the individual projects can be grouped (the value of strategic technology categories will be discussed in detail in the next section on strategic management); (2) project expenditures aggregated into these strategic categories; and (3) provision for periodic (probably quarterly) review of programmatic effort to ensure that changes that are made in response to exigencies in individual projects are not distorting the strategic balance of effort in an undesirable direction.

The second requirement—ensuring that project commitments do not inadvertently exceed resources—requires a more multidimensional response. In principle, simply limiting project initiatives to the beginning of the budget period would at least severely curtail the possibilities of overcommitting resources. In practice, it is impossible to know at the start of the period all of the work that must be activated before the period ends.

The foundation on which effective control must rest is self-discipline. If both individual professional contributors and their managers know that failure to fulfill commitments will result in penalties, they will be more careful in examining their present commitments before agreeing to work on still other projects. Obviously, the introduction of such a project control system will require formal inputs from many people regarding project commitments they have made. This system will almost certainly engender resistance as another burden of generating "needless" data. Again, if all involved recognize that variances from plan that are not noticed and addressed will be penalized, they will be more receptive to a system designed to help keep them out of trouble.

Terminating Projects

No discussion of project management would be complete without a consideration of the unpleasant subject of terminating projects. This eventuality is viewed with about the same enthusiasm as preparing wills or buying cemetery plots. If one has ever witnessed the demoralizing trauma of a team that progressively loses touch with reality because it has lost the capacity to call it quits, one is forced to recognize that terminating projects is an inescapable management responsibility.

The sequence is almost tragic in character. The very ingredients required for success—sharp focus on objectives to the exclusion of all else, total immersion in the problem, intense intellectual and emotional commitment to the work—create a momentum that is difficult to abort. It is no accident that the activity is frequently called "killing a project"—as though it were an animate object.

Unfortunately, projects are rarely patently hopeless. In fact, termination is seldom necessitated by an insuperable barrier. Much more frequently some combination of shrinkage in projected benefits, slippage in schedules, appearance of unexpected problems, and cumulative investment with inadequate prospect of return creates the conditions for termination.

The problem is magnified by the fact that members of the project team have made a deep personal commitment to the project. Even worse, they have invested an irretrievable chunk of their professional careers in the endeavor. They cannot be expected to view the continuing viability of the project dispassionately.

A wise manager will conclude that it is much better to consider the conditions that would warrant termination in advance when alternative probabilities can be weighed more objectively. Some even argue that a contingent termination plan based on milestones should be a routine feature of every project plan. I would not go that far because I do not believe that one can anticipate with adequate precision all of the circumstances that will prevail at the various milestones. I would argue that one should identify the parameters that will determine the continuing viability of the plan and specify the review dates at which continuing viability will be determined. The process should make clear the requirement of accountability for results and the existence of independent review.

Two different aspects of managing project viability should be included: (1) what substantive considerations will be weighed in determining viability and (2) what process will be used to reach expeditious decisions and minimize the trauma that is associated with termination.

Substantive considerations

First, and unfortunately the easiest, is the continuing technical merit of the project. If nature says no, or says yes but yields uninteresting results, the task is easy. Much more common is the circumstance where the perceived range of benefit narrows or the expected benefit lessens, but the opportunity does not disappear altogether. Conversely, the size of the opportunity may remain, but the project effort needed to realize it grows.

Obviously, one cannot specify in advance what decisions will be made. It is important, however, for these various contingencies to be raised in advance and to make clear that the continuation of the project will depend on a rigorous periodic review of its attractiveness at scheduled milestones. The image one wants to convey is that of an enthusiastic, committed, understanding, but nevertheless demanding sponsor.

In weighing technical merit, there are some danger signals to be alert to. If work on some obviously key problems is postponed or given lower priority than effort on more peripheral, but solvable problems—watch out! GE's project on fuel cells was eventually abandoned because extensive investigation revealed the barrier of the platinum catalyst. With adequate platinum to provide acceptable life, the cost of the system was too high; with platinum reduced to acceptable levels, the life of the cells was uneconomic. Those who were not sensitized to the import of the platinum barrier were inclined to fret over the threat of ongoing work elsewhere which implied that an economically viable fuel cell was indeed an attainable goal. I pointed out that none of these projects addressed the platinum problem and that progress on other features had little significance until the platinum problem was solved. The laboratory had taken its best shot with the most competent people it had, who knew of no additional promising avenues to explore. Although they might indeed be proved wrong, the time had come to "take their losses and run."

Another danger signal is a persistent refusal to consider downstream engineering problems early in the endeavor. Both fusion and magneto hydrodynamics (MHD) research suffer from this difficulty. GE abandoned work on closed-cycle MHD after a relatively brief effort because nobody could ever postulate the availability of the kind of materials that would be needed to achieve acceptable life in the incredibly destructive environment of an MHD generator.

The second parameter that must be kept under continuing review is the downstream infrastructure needed to support the endeavor through to successful commercialization. This infrastructure includes capability to manufacture, access to markets, and ability to distribute, sell, and service the new product. Of the various components of infrastructure, probably the need for service is most likely to change as a project moves along. For example, the need for an extensive service network to support use until more extensive field experience had been accumulated became a major barrier to introducing a very high quality projection TV system.

The final parameter is adherence to schedules. Unfortunately, adequate consideration is rarely given, up front, to the economic value of timing in exploiting a market window. Again, it is important for this element to be addressed explicitly early on so that members of the team are forewarned of the importance attached to timing.

Process considerations

What action can be taken to ensure effective and yet sympathetic consideration of project viability. First, it is important to be forthright and explicit up front about the criteria that will be weighed in determining survival. Team members should be under no illusions about accountability for results. They should also be made aware that the criteria involve more than technical progress.

Second, it should be made clear that management will make the decision. In other words, the team will not decide the fate of the project. If the size of the project warrants it, it is sometimes useful to establish an independent review board to monitor progress and recommend courses of action. Such a board must not be regarded as a threat, because it has the obligation and the opportunity to be deeply knowledgeable about the project and can provide protection and support as well as criticism.

Third, the review process should be made as impersonal as possible. This is not easy, but it is important to keep the focus on the work and not on the competence or performance of individual team members. I remember one disastrous example in which a manager voiced problems with his superiors. The response he got was, "My God, you're looking to us for help on that? What do you think we're paying you for?" His reaction was, from now on, I'll simply have no problems!

Fourth, it is important to try to arrange a soft landing for all team members. Even when a team agrees with a decision, aborting a project is traumatic. Lives are suddenly in disarray. It is important to have other tasks to which people can be assigned. But also it is important to be explicit in recognizing the inevitable existence of an interregnum during which people must be supported.

Finally, the people I have known who are most effective are matter-of-fact and low-key. They do not defend their decision extensively—this simply invites reopening the argument. They recognize the traumatizing nature of the event and evidence a desire to be helpful, but they make it clear that it is necessary to get this episode into the past and move on to other things.

Conclusion

Effective project management is at the heart of effective operational management because much of the most important work in a business is performed by project teams. Techniques that were developed primarily to articulate and control complex physical tasks, where the activities of perhaps thousands of people must be coordinated, cannot without modification be applied to problems in which dealing with uncertainty is a major component of the work. It is not enough to make certain that every member of the team is doing what the plan calls for and at the right time; one must return again and again to the issue of whether the initial objectives and assumptions are still sound.

An effective project management system also builds in rewards and penalties that to a great extent make the system self-disciplining in terms of eliciting commitment and evaluating the attainability of commitments. Finally, the system not only encourages successful accomplishment of individual project objectives, but also provides for periodic overview to check consistency with strategic thrusts and available resources.

Notes and References

1. Donald G. Reinertsen, "Whodunit? The Search for New-Product Killers," *Electronic Business,* July 1983, pp. 62–66.
2. Thomas J. Allen, *Managing the Flow of Technology,* MIT Press, Cambridge, Mass., 1977, pp. 234–248.
3. Lowell W. Steele, *Innovation in Big Business,* Elsevier Publishing Company, New York, 1975, pp. 189–205.

Chapter

7

Transferring Technology to Operations

The previous chapter focused on project management, starting at a stage where technological development had already advanced enough that a plan to commercialize it should be prepared and carried out. There is another kind of project management further upstream: that involved in nurturing a technical discovery or invention to the point where it can be turned into a product or process development project. The objective of this kind of project management is not only to generate the additional information for demonstrating technical feasibility, but to do it in such a way that—assuming nature has cooperated—the new technical capability will be picked up and commercialized by operations. Although I do not question the value of effective project management for this earlier R&D phase, it is not as complex and multidimensional a task as that associated with commercial development. Rather, the transfer of technology from creator to applier is frequently the point at which the system breaks down. Whenever R&D managers get together, the problem of getting operations to pick up new technology and carry it forward to active use is raised perhaps more than any other.

The change in perspective from creator to commercial developer is crucial. The former has pride of creation. He quite naturally tends to place a high value on invention or discovery. Consequently, he tends to value an advance in terms of its technical elegance and the new capability it makes possible. He attributes intrinsic value to new capability. The developer, on the other hand, has no pride of creation or discovery. He must address a very prosaic question: Is there sufficiently certain prospect of new capability or more cost-effective performance for me to sell somebody on running the risk of trying something new? Often the prospective user can already do what he needs with other technology. The question then is, does this do it better or at lower cost, or both? Of course, with *new* technology, the underlying questions always are, will it *really* do what is promised? and, will it be available on schedule?

In many respects the term *technology transfer* is a misnomer because it implies that something is moved, more or less untouched, from one place or one organization to another. It connotes interfaces and barriers to be overcome. The process is more complicated than that because the technology itself is changed as a part of its movement from one organization to another. In Corporate R&D (CRD) in GE, this process is called *transition*—and the word is used both as a verb and a noun—because it implies a change of state as well as movement. Others have used *joint commercialization* to emphasize the partnership that is required to achieve success. The creating and commercializing organizations must develop a common vision of the opportunity and integrate activities for pursuing it.

Technology transfer involves two different dimensions. The first, or substantive one, addresses the problem of transferring information about physical phenomena, equipment, analytical and manipulative techniques, terminology, etc., associated with the technology. The second is affective—it concerns the feelings and attitudes required in both organizations in order for two sets of people with different skills, values, and priorities to become successful in passing the baton from one to the other. In this chapter we will examine each of these dimensions and then consider some of the special features of technology transfer in multidivisional companies, and in particular, between R&D and manufacturing.

Transferring the Substance of Technology

The subject of uncertainty has come up repeatedly in our discussion. The uncertainty associated with product development, as we noted, has a large component of "should we" questions. The technical uncertainty is associated principally with time and cost. At this stage, the question, will nature permit it? is rarely critical. In contrast, at the earlier stage of development associated with an R&D effort, the program is loaded with unknowns and unknowables. Planning consists mostly of following intuitive hunches and guesses. Perhaps most disconcerting, there is no perceptible "path" to follow and in fact no certainty that one is even making progress. An analogy to a maze is possibly most illuminating. The steps being taken may lead to a blind alley, they may lead you out of the maze but not to success, they may lead to success, or the maze may be *all* blind alleys.

Obviously, in such a situation the term *planning* is not very apt. One may know what one needs to learn, but have no assurance that the work being undertaken will produce the desired information. A major program with which I am familiar was dropped after several years of effort because the team came to realize that at a much earlier stage it had chosen the wrong material system to work on. The team members will never know whether the choice of a different material would have led to success; they just know that they chose a wrong one. Any group working on a problem at this stage must have great autonomy and flexibility, and the level of interaction among team members must be very high.

One of the endemic problems of working at this stage is that there is no way to evaluate the importance of information being generated. Unless the information being generated is made available to people with different outlooks and experience from those doing the work, its full implications may be missed. Obviously, it is also important to keep detailed records of information produced, irrespective of its apparent significance at the time. DuPont used to require that a highly trained chemist be present in its dye research group whenever a technician was "striking the color," the crucial step in synthesizing a new dye; otherwise, critical features of the experiment might have been missed.

Information requirements change

The nature of the information required undergoes a transformation when a technology development moves from the creation phase to the demonstration and utilization phase. Frequently, problems of scale-up and market development require information quite different from that associated with the work that led to the invention. Rarely is the information transferred what the receiving group needs. Unfortunately, neither party to the transaction knows what is truly needed.

When man-made diamonds were invented, the group in R&D could transfer information about equipment, materials, and process requirements. It also set about producing enough diamonds on laboratory equipment to make one grinding wheel. The group in operations needed to know: How did diamonds work in removing material? Could this process of material removal be improved? Were man-made diamonds identical with natural ones? If not, were there properties that could be exploited? Now that a supply of industrial diamonds was readily and reliably available, were there new applications that could be exploited?

Even with respect to the technology already developed, much more information was needed. How did the process work in the first place?

How could you control particle size? Could you make bigger ones to open up new applications? How did you specify raw material, not only for material to be used in the process, but also in manufacturing the high-pressure cells? All of these questions were far removed from the interests and, to a great extent, even the present skills of the physicists and physical chemists who had made the original discovery. The information needed far exceeded the competence of the fledgling group in operations, which was concentrating on getting into production and developing marketing and pricing strategy.

Close and effective communication and a strong commitment to success are necessary on both sides if technology is to be developed and transitioned successfully. Decisions about who should do what must reflect the needed skills, experience, and equipment rather than conventional views about organizational charters and functional turf. They must also reflect a strong commitment by the originating group to change gears and help develop the new and different information needed.

Even when care and attention have been given to ensuring the flexible, effective communication needed to deal with a very uncertain situation in a state of constant change, the necessity remains to actually transfer the specific information associated with the new technology. This task is often cited as the major barrier to technology transfer. How do you transmit understanding of complex new terminology, new phenomena, new intellectual constructs to the uninitiate? Admittedly, the task is not easy. It requires careful attention to details and persistent follow-up. However, there are literally thousands of examples that demonstrate the task is doable.

In order for technology transfer to occur, a number of requirements must be met. The first may seem ridiculously obvious, but is distressingly uncommon—the receiving organization must be capable of and interested in receiving the information. Where this is lacking, transferring technology has been likened to trying to push a string. A minimum pull must exist from the target organization in the form of a basic receptivity to ideas. But the requirement goes beyond that. The members of the two organizations must, to some extent, have an overlap in their training, skills, and experience. Without some common base, those in the recipient organization will lack assurance that they truly comprehend what is being transmitted. People frequently feel they understand something when in fact their knowledge is superficial. Furthermore, they may lack the self-confidence to ask questions and really probe the limits of the information being transmitted. Finally, the people in the recipient organization may be unable to distinguish between changes in process or technique—and changes are inevitable—that are permissible and those that vitiate the new technology. Unfortunately, some people reinforce the potential breakdowns in communications by hints of arrogance in flaunting their superior knowledge.

An additional reason for needing skilled receptors results from an inherent limitation on the part of the transmitters. The people who create a new technology rarely perceive with sufficient clarity what they in fact really do and do not know. Moreover, they often cannot discern the truly critical information from other details, and therefore are unable to transmit it until they are quizzed. Corporate R&D in GE was able to transfer materials-related technology to mechanical or electronic businesses if there was even one person there who had a materials background. It failed repeatedly if only electrical or mechanical engineers were in the recipient organization—they simply did not know how to ask the right questions. A report on a program, no matter how detailed and how carefully prepared, is woefully inadequate for transmitting the information needed.

The match required between transmitting and receiving groups is somewhat analogous to the matching of impedance in designing electrical or electronic circuits. Dr. Eduard Pannenborg of the Philips Company has argued that technology must be transferred to groups at the same location in the technology spectrum.[1] The wider the gap in language, culture, and technical experience, the more important this becomes; but the problem of impedance matching is always present to some degree and the manager must not ignore it. Thus work still at the research stage in one group cannot effectively be transferred to a development group at another location. Once the information has been transferred successfully to the counterpart group, the development can succeed (Fig. 7.1).

The impedance match can perhaps best be established by transferring people from the original R&D group—"The best way to transfer technology is to transfer people," as the saying goes. Even temporary assignments of people to the other component, from operations to the R&D group or vice versa, can be very useful. Other arrangements will work, however—for example, hiring or transferring the requisite skills from other organizations or arranging for the

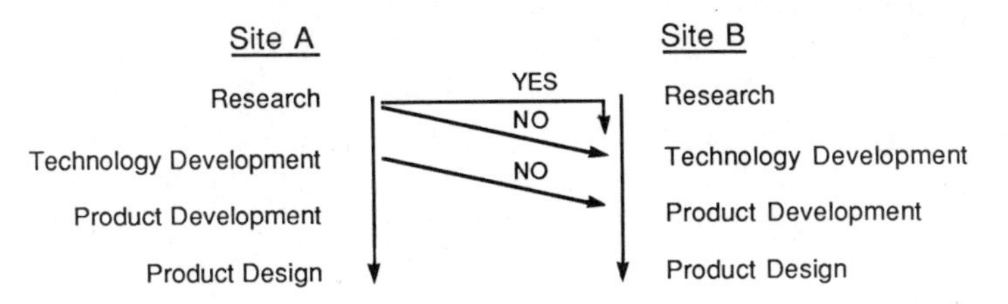

Figure 7.1 Transferring technical information.

receiving component to be located physically (at least temporarily) next to the originating group. The process is an iterative one between originator and developer that requires easy, and frequently intense, interaction.

Transferring people from R&D to the operation undertaking commercialization is especially helpful. The view of problems and priorities is different in R&D. Greater weight is likely to be given to problems that are most interesting technically, not necessarily those that are most important for commercial success. In operations, people will be bombarded by messages from marketing about the problems that are inhibiting customer acceptance. The CAT scanner that GE's Corporate R&D turned over to operations generated specious images called artifacts. The key technical people in CRD were inclined to say the problem was trivial and should be taken care of in operations. They were interested in beginning work on a new, more sensitive X-ray detector. CRD people who transferred to Medical Systems solved the problem by dint of a major and very sophisticated effort. It is arguable whether they would have been sufficiently motivated back in CRD to make that necessary effort. More likely, they would have blamed a lack of competence in operations to solve this "trivial" problem.

Additional considerations

Regardless of how the process starts or how effectively communication proceeds, other general requirements exist. One nearly universal need is to expand the mix of skills available to carry the development program forward. The original skills are still needed, but as we saw in the example of man-made diamonds, a large number of new questions must be addressed. Furthermore, the depth and detail of information required in order for a technology to perform reliably over an extended period is staggering. Thus the scale of effort in development rapidly dwarfs that of the original project.

Another especially important and frequently mishandled requirement is the need to assess the status and pace of development of competing technology. There is an unfortunate tendency to compare an emerging new capability with the present state of the art, rather than what that art will be by the time the new technology becomes available. The continuing improvement in silicon-based solid-state devices has been a shifting barrier to progress with gallium arsenide. Rapid advances in VCRs aborted the thrust of laser video disks for home entertainment. Advances in sulfur hexafluoride circuit breakers proved to be a major barrier to GE's very elegant vacuum interrupter. Development programs aimed at internal usage are particularly vulnerable in this area because competition from vendors is less obvious (except to those trying to beat them) than competition in the market.

The appraisal of competing technology must include not only presently used technologies, but also other candidates waiting in the wings, because their pace of progress may determine how long the window of opportunity may stay open. Thus appraisal of the potential of magnetic bubbles for computer memories had to involve consideration of both magnetic-core and semiconductor memories.

It may be wise to have the competitive comparison made by knowledgeable people not associated with either technology. When the distinguished inventor of EMI's scanner was asked to appraise the potential of a proposed competing technology based on electronic scanning rather than his mechanical scanning, his response was, "It will never work!"

An important component of the information necessary for technology transfer is preliminary data on costs and market prospects. Generating this information presents the development organization with a dilemma. It may feel that such information should be produced by its own people, who best understand the technology and its potential applications and who have a vision of its promise. That may well be true, especially for new technology that departs from past practice, but unfortunately it ignores the problem of credibility. Operating people feel that only they—hardheaded creatures who understand life in the real world—can produce reasonable projections on costs and can assess potential market opportunities. Both sides have much to support their position, but people in operations will be the ultimate judges of validity and one excludes them at one's peril. Probably the most effective approach is to ask operations to take the lead in developing data and then to work very closely with them to ensure that the data reflect the special nature of the technology and are truly representative of projected cost and market factors. Sometimes a skilled consultant can be helpful in generating credible data.

Affective Aspects of Technology Transfer

Earlier we considered the challenge of stimulating commitment and enthusiasm in a project team. In the present situation the challenge is how to transport this commitment and enthusiasm to a different team or to kindle it afresh. The members of an operating component have good reason to be hostile to the disruption and diversion of resources a technology development program represents. In fact, their response has been likened to the body's response to an organ transplant—they try to reject it.

Barriers and bonds concept

The conventional wisdom holds that technology transfer is most effective when people are in close physical proximity and when organizational barriers are minimal. This perception has been systematized into the "barriers and bonds" theory. Close physical proximity is a bond, distance a barrier. Membership in the same organizational unit is a bond, in different components, a barrier. The theory postulates that when both distance and organization operate as barriers, technology transfer is so difficult as to be improbable. If at all possible, arrangements must be made for one or the other to operate as a bond.

This theory originated with Jack Morton of Bell Laboratories, who noted the difficulty that its centralized facilities had in transferring technology to geographically dispersed Western Electric plants.[2] The response was the development of so-called satellite laboratories that were located on site at Western Electric plants, but that were organizationally part of Bell Labs. This arrangement, it should be noted, also meets Dr. Pannenborg's requirement of transferring technology to an organization working in the same segment of the spectrum of technology, in this case, from development group to development group.

Developing a sense of ownership among people in operations is just as crucial as in the project team we examined earlier. Again, ownership takes time. The R&D organization needs to begin early, informing potential recipients among the operating components (if there is more than one potential home) and informing more than one function within each business. Often people in R&D make the mistake of assuming that their sole point of contact should be with the technical function in operations. It is important for R&D to understand the power structure in operating components. Where marketing is dominant, the technical component may not even be the best route in. Engineering may well need reinforcement in building commitment to the development, and it may even resent the intrusion of R&D into its territory.

One perceptive lab manager, who took over a laboratory in operations that was locked in competition with the product engineering groups, quickly recognized that his most important potential allies were in product planning in marketing. The business was marketing-dominated, and therefore the support of marketing could more than counterbalance the resistance of engineering. Product planning's time horizon and interest in new product opportunities matched very closely with those in R&D—much better than did engineering's. This is a different illustration of impedance match, in this case involving time scale and business focus.

At the outset, an obvious first option is to look for a potential champion in operations—some figure of sufficient stature to command respect, who can be motivated to push a project. The temptation is to seek support among senior executives—somebody who would not be threatened by the development and who, from his lofty vantage point, would regard it as a minor disturbance at worst. Although support at high levels is clearly desirable, it is not enough. It may well be resented at lower levels by people who feel they are being pressured into doing something they do not want to do. These people in the trenches will have many opportunities to sabotage the development and see that it fails, if they wish to do so. I know of two important and promising developments that foundered when support was stimulated at high levels in the receptor organization, but resented by people at lower levels, who killed the projects with relative ease.

Should R&D Undertake Business Development?

One of the most difficult operating decisions an R&D manager must make is what to do with a development for which no sponsor can be found in operations. His inclination often may be to pursue it anyway, but in my experience, the chances of success are very slim. The development is, more or less by definition, a long shot. It will have relatively high visibility, and it will have many doubters waiting to say, "I told you so!" Painful though it may be, in most cases the wise course is to shelve the development or provide only minimal resources until a more promising relationship with operations can be developed—often a change in just one key management position is all that is needed. One alternative that must not be overlooked, however, is the possibility of licensing or selling the technology. Sometimes the threat of such action can galvanize new interest.

The frustration of having to live with these situations sometimes leads R&D management to seek a change in scope so that new business development becomes an adjunct to R&D. This step can be successful, especially if a seasoned operating executive is in charge of R&D and asks to take on the work. But it is no panacea. Success is still critically dependent on the entrepreneurial capability of the people put in the program. Although the people in R&D will have the sophisticated knowledge and the enthusiasm and vision so vital for success, they will rarely have the skills needed in other areas. Consequently, a quite different type of team from that typically found in R&D must be assembled. There may well be considerable tension between the business development team and the people in R&D. Furthermore, the competition for resources that exists in operations between the ongoing business and the new venture is now transferred close to home—between R&D and the venture.

A new business development component does give R&D the charter and resources to start businesses, but sooner or later a fledgling business must be incorporated within the operating structure. Even though delayed, that transition may still be fraught with difficulty. Uncertainty over the potential of the business may well be diminished, but the resources required to support its growth are now greater. Consequently, it may still be "the enemy" as far as present businesses are concerned. It requires continuing investment, which some may feel is taken from more deserving projects that support present businesses. Perhaps even more, it diverts valuable managerial time and attention.

Organizational Change Is Not Sufficient

What this suggests is that organizational steps by themselves are not sufficient to resolve the inherent dilemmas in technological innovation. The "foreign body" problem is simply moved about within the enterprise. Additional management mechanisms (rejection-suppression mechanisms, if you will) must be fostered if systematic, continuing support for innovation is to be available. These mechanisms must address the motivational aspects of technology transfer. The challenge is to transfer the commitment and enthusiasm of those who are creating a new technology to those who will develop and apply it.

One must distinguish between situations in which the recipient of a new technology is readily apparent and his receptiveness is high and those where the recipient is unclear or unenthusiastic-to-hostile. The actions that need to be taken are the same in both cases, but the care and attention devoted to them will vary.

Soliciting help from operations

Honestly soliciting the help of people in operations can be the first mechanism for building commitment. They can make important contributions even from the early days of a project. One must not simply attempt to seduce them into becoming involved. Developers and appliers differ from inventors in their knowledge and experience. Their perception of attractive features in a new technology can help guide its development, and their recognition of potential downstream problems can ensure that they are addressed early. The success of Post-it (3M's Scotch-brand note pad) occurred only after the persistent inventor of an accidentally discovered, very poor adhesive encountered a marketing man, who asked, "Can you coat it on paper?" As a development progresses, the skills and equipment that are available in operations become valuable in performing some tests and measurements. Consequently, the program can become a joint development well before it officially transfers to operations. Meaningful, substantive participation in the development early on is indispensable for nurturing commitment.

Money as a motivator

A second mechanism, which warrants more extensive use, is to use financial resources—money—as a motivating force. One reason people in operations are sometimes not enthusiastic recipients of new technology is that they operate on a very short financial leash. They literally do not have the funds, because they are not permitted the luxury of uncommitted resources. Furthermore, even if they are willing, obtaining resources takes time. After being involved in dozens of studies that recommended reallocation of resources in businesses facing only ordinary problems, I concluded that the expected time constant for action is three years. The current year is already locked up in terms of both objectives and resource commitments. Another year is usually required to sell management on the idea and nurture plans for reallocating resources. Action can begin in the third year. Obviously, this sequence can be compressed—given sufficient urging or organizational power. However, it helps to recognize the normal time constraints for action when one sets out to push a development program. Asking for quite unrealistic responses is not productive!

If the R&D organization has available a source of funds to sponsor work in operations, this time sequence can be short-circuited. The money serves two functions. Not only does it provide relief from the budgeting bind of fully committed resources, but also it presents tangible evidence of R&D's belief in the opportunity. The old adage "put your money where your mouth is" applies. If the potential local champion is able to go to his management and say R&D feels strongly enough about this to contribute some of the funds, he is in a much more powerful position.

Mechanisms for providing such funds are diverse. The President's Projects in Hitachi are one example, and the practice is used elsewhere in Japan as well. GE used such an approach for over twenty years. In CRD a separate budget item, labeled "Support Programs," represented funds that had to be spent in operations; they could not be spent in-house. Although these funds supported a variety of activities, including work originating in operations, they were used to help transfer technology. The purposes and activities varied; for example, purchase of a needed piece of equipment, special evaluations requiring operating inputs, paper designs of potential applications, or cost projections. The common theme was that CRD funds provide the flexibility to pursue unanticipated opportunities for which operating components rarely had available reserves. These funds frequently ended up being seed money for stimulating much larger programs in operations.

Using money as an enabling mechanism has a number of advantages. It is transferred more easily than people and is easily divisible into small increments. It can be given as much or as little publicity as one wishes. Termination is less traumatic than terminating an organizational component. The funds represent a tangible expression of confidence on the part of CRD management that can induce a doubting operating manager to invest.

There are, however, dangers that must be guarded against. Probably the most pernicious is the temptation to use subterfuge at a time of budget stringency. A manager strapped for funds may offer to make support program funds available to an operating component if it in turn will pay for certain work being done in R&D. Despite the potential for abuse, these funds can be very high leverage vehicles for accomplishing technology transfer.

Commitment to make it work

Perhaps the most important factor in technology transfer is an attitude, reinforced by action, that says: our goal is not just to transfer the technology, but to see that it succeeds in your environment. The Major Appliance Laboratory in GE has been one of the most effective in the company, and its reputation, carefully nurtured over many years, is one of total commitment to partnership with engineering and with manufacturing in introducing and debugging new materials and processes. It understands operating needs and constraints, the information it provides is couched in language that communicates, and perhaps most important, engineering and manufacturing know lab people will come out on the factory floor at any hour necessary to resolve problems.

The psychology of change

One very important ingredient, which is frequently violated in all innocence, is the necessity of understanding the psychology of change.[3] The inventor or discoverer of something new quite naturally becomes excited about it. He believes the way to recruit supporters is to encourage them to share his vision. He emphasizes the newness and differentness of the discovery. What this approach ignores is the disruptive and unsettling characteristic of change. In order to become involved, one must have faith that the discovery will in fact succeed and be worth the effort. Change may also alter or destroy conventional modes of operation—a consequence that is obviously and legitimately threatening to many. Thus the inventor's vision can be the user's or the developer's nightmare. Effective transfer of technology requires dealing with two powerful negative forces: it may not work or it may hurt the user rather than help him. Therefore, effective nurturing of change must address anxiety as well as seek to stimulate vision. It must systematically emphasize regions of continuity, aspects of the environment or of the process that will still be needed, skills and experience that will still be valuable, and capital investment or marketing assets that will help ensure success. Brave new world must be embedded in comforting old world.

Coping with anxiety also means responding to the dynamics of technological change. One must never ignore the possibility that a vision *can* turn into a nightmare. Both anxiety and enthusiasm are always present, and people can be motivated to commitment when the latter is dominant. But as we have seen in our discussion of risk, the resource requirements for a development tend to escalate rapidly. Consequently, even though the development is progressing satisfactorily, the perceived risk can also expand rapidly. The balance between vision and anxiety can shift adversely for the original sponsor, if only because it outruns his tolerance for acceptable losses. Thus the need to stimulate enthusiasm and allay anxiety is a continuing one. As noted in the discussion of risk, the possibility of needing to shift sponsorship to a higher level, where the resource base is larger, must be recognized early on and missionary actions initiated that repeat, in similar manner, the work leading to the initial transfer. A successful technology development is often analogous to a relay race in which the baton must be passed as the race progresses; and just as in a relay race, the runners need to have been running in parallel long enough to synchronize their motion before the baton is passed.

Leonard-Barton & Kraus call attention to additional affective factors that are important to successful technology transfer.[4] The typical enthusiasm of the developer, noted above, can easily lead to overselling either the virtues or the timeliness of the project. Overenthusiasm can easily turn to disillusionment when unrealistic expectations are not fulfilled.

Practical Problems in Technology Transfer

A common problem arises from initiating several changes in process or procedure simultaneously; then when difficulties ensue, the source of the trouble is obscured. The first superconducting magnets to be used in GE's magnetic resonance (MR) machines were built in CRD. They were sent to the new plant built to supply commercial quantities to act as models for quantity production. The plant changed a number of features, and its first output would not work. Since several changes had been made simultaneously, it took much effort to identify the problem and re-establish a workable process. The success of a new technology is critically dependent on the validity of its first test in application. If the application is regarded as unrepresentative or not sufficiently demanding, doubters can too easily discount the results. Conversely, if a fledgling technology is first tried in an exceedingly demanding application and fails, it may unfairly be rejected even though its promised new capability does in fact warrant application.

Frequently overlooked is that the criteria for judging new technology must be valid. If a new process is intended to improve quality but is measured for effect on quantity of output—because that is what has traditionally been measured—the results are not valid. Also sufficient time must be given for a shakedown and a return to stability. A new product or process almost invariably intrudes on habituated behavior, which complicates the task of evaluating performance.

Because possible benefits are often difficult to identify in advance, important outcomes may not be foreseen, or worse, may never be perceived. One laboratory took credit for a small, but attractive cost reduction that resulted from substituting silicone for gold in a seal. The laboratory did not think to ask and never knew—until years later and quite by accident—that a much larger cost reduction had resulted from the greater reliability of the new sealant, which significantly reduced field failures and customer dissatisfaction.

Finally, technology transfer requires the existence of an adequate infrastructure to support a new technology. The successful major appliance laboratory, mentioned earlier, must supply relevant criteria and methods of measuring conformance with specifications for evaluating incoming material before it can expect manufacturing to consider adopting a new material.

The Challenge of Manufacturing

A major barrier to the current need for improving technology transfer is the absence of these culture-building activities between manufacturing people and R&D people, because often they literally do not know how to talk to each other. They lack a common vocabulary, and neither appreciates the constraints and pressures the other faces. Furthermore, manufacturing people frequently lack the tradition of interchange among themselves that is common in product engineering. In technical meetings involving representatives of both manufacturing and product engineering, I have been aware of a greater sense of isolation in those from manufacturing. The common lore is that in manufacturing, technology is exceedingly "site-specific" and that the applicability of external technology without substantial adaptation is highly questionable.

After decades of neglect, improving the incorporation of new technology in manufacturing is beginning to receive the attention it warrants. In view of the importance I attach to overcoming fragmentation within technology, let us now turn to technology transfer into manufacturing.

The first step is to recognize that two frequent barriers which are invoked as reasons for ignoring manufacturing are really excuses: (1) you get no credit for working on process problems; and (2) the only meaningful experiments that can be done are on the shop floor—there is no place for R&D to be useful.

I have seen "you get no credit" used as an explicit reason for R&D managers' refusal to work on manufacturing problems. In my view, no manager of R&D should be permitted to stay in his job after such a statement. First, it isn't true, because I know of several highly regarded laboratories that focus heavily on manufacturing technology. Second, "no credit" is inexcusable as a reason for not doing something.

The second excuse is more forgivable, but no less wrong. It is true that one can rarely use an entire factory for an experiment (although the Pilkington float plate glass development did just that), but manufacturing comprises dozens of different technologies, and some do not require factory experimentation to validate them. Furthermore, arranging to work only with certain operations or certain shifts can often reduce any perturbations created. For example, a behavioral research study conducting experiments on the "effects of perturbations in habituated behavior on worker output" attempted to limit the perturbation it created by working only with the night shift.

Although these are simply excuses, improved contribution of R&D to manufacturing does require recognizing some of the valid constraints that influence both the choice of programs and the modes of interaction with manufacturing.

Understanding the manufacturing environment

First, R&D must recognize the enormous and overriding commitment that manufacturing people make to achieving uninterrupted output. Success in that endeavor is the first criterion by which manufacturing judges itself and is judged by the rest of management: Does it meet its production quota and schedule? Meeting that goal always has highest priority on the attention and energy of people in manufacturing. Given enough promise in the benefits accruing from any interruption and enough confidence in the person proposing it, manufacturing people may agree to disrupt production. But one had best start by accepting the immutability of that priority rather than attack it as a barrier to any collaboration.

Second, manufacturing *must* be conservative in incorporating new technology. I know of three factories that were built, but did not work as planned—two were never even started up. They instantly became monstrous embarrassments to management. Sick products can be embarrassing and painful; sick factories can be life-threatening. This great difference in scale of perceived risk—it is not just the investment, it is the future of the business—is too often not appreciated by R&D people.

Third, the shop environment poses much more severe constraints on adopting new manufacturing technology than on new product technology. The most common complaint I have heard from manufacturing managers is "engineers don't recognize the people factor." New manufacturing technology not only requires learning new skills, but also creates anxiety and resentment because it changes status relationships and alters communication patterns. These factors may appear secondary to the developer of new technology, but they are central to success on the shop floor.

Fourth, manufacturing, at least in the United States, has a very limited experimental tradition. The idea of doing something just to learn from it is not natural to most manufacturing people. Since they have had virtually no experience in trying something and then changing it in hopes of "getting it right," they tend to be quick to give up. Their reaction is, that didn't work; let's drop it and go back to something we *know* will work. There are, of course, exceptions—and increasingly so—in more technology-intensive industries. It is also true that manufacturing people seem somewhat more comfortable with the prospect of experimenting with purchased equipment—possibly because they can hold the vendor's feet to the fire if it performs inadequately.

Fifth, R&D people and manufacturing people lack a common language, especially manufacturing system technology. This makes it difficult for R&D people to identify and characterize the problems whose improvement would have the greatest leverage. The processing and fabrication technologies appear to be ones that R&D people feel more comfortable with, and that may be why these technologies are more likely to be addressed by R&D programs, although they often are not the highest priority in manufacturing.

I was astonished to learn that manufacturing electric ranges was regarded as the toughest manufacturing assignment in GE's Appliance Park. Ranges have no moving parts and involve less sophisticated and diverse fabrication than refrigerators, washing machines, or dishwashers; *but* they come in a much larger variety of configurations and sizes. Monitoring control of this diversity was regarded as more difficult than manufacturing refrigerator compressors or washing machine transmissions.

The lack of a common language and techniques for identifying high-leverage costs requires R&D to recognize that it truly is beginning at square one when it seeks to help manufacturing. I once waxed enthusiastic about the use of a sophisticated, remote-controlled robot to move a heavy product from one conveyer line to another. An experienced manufacturing man just snorted, "That's a Rube Goldberg solution. If they had set up the material flow properly and balanced the line for output, they wouldn't have had the problem in the first place."

The final constraint is that manufacturing people lack the tradition of broad professional interaction that is common in R&D and in product engineering. In part, this results from a widespread belief that effective manufacturing technology is very site specific, as noted earlier. Manufacturing people believe that technology must be carefully tailored to the requirements of a local environment, and understanding that local situation in detail is a prerequisite to applying new technology.

Overcoming barriers

Having listed all of the barriers to effective transfer of technology to manufacturing, what can be done to begin making contributions? One action, which should be obvious but is sometimes not taken, is to involve people with a manufacturing background in technology development and thus avoid the impression that R&D is simply extending its reach into new territory. Given manufacturing's long history of lower status and isolation from product technology (excluding the process industries), there is an implicit, and much resented, arrogance in proposing to solve its technical problems from the position of no experience that has existed in many laboratories. It is noteworthy that when four world-class manufacturing companies—General Motors, Philips, Western Electric (AT&T), and Siemens—decided to create a corporate-level manufacturing technology component, each organized it separately from the traditional product-focused corporate R&D laboratory.

Two additional people-focused steps must be taken. First, insist on cross-functional experience as a requirement for claiming technical professional qualifications. The Japanese make it clear that technical graduates must have such experience in order to be regarded as fully qualified professionals. Furthermore, they believe that good manufacturing requires more engineers present on the factory floor, and they put them there. Second, design some in-company training programs explicitly for joint attendance of people in both manufacturing and engineering.

In all cases it is better to start by working on carefully limited problems. Manufacturing people are, for good reason, very wary of total-solution, factory-of-the-future approaches. In the first place, most upgrading will have to occur in existing plants. Technical advances that require starting with a new plant—so-called green-field solutions—are not realistic for most situations. Manufacturing people are much more comfortable with "islands" of progress that may not be optimal in terms of the factory of the future, but that work and pay their way, than beautifully optimized systems that do not work. Unfortunately, the latter are all too easy to construct—witness the difficulties of General Motors with new high-technology factories, as far back as Lordstown.

The final step, and probably the most important of all, is to recognize the need for establishing credibility with manufacturing. This means "going the last mile" and beyond, to ensure that new technology will work as promised. One lab manager, with long experience in dealing with manufacturing, went so far as to guarantee the improvement in mold cycle times that his modifications in the molds would make *or* to restore the molds to their original state. He was trying to enlist a new client, and he knew the conservatism and doubt he would have to overcome.

Manufacturing people lead harried lives. An Industrial Research Institute workshop for joint participation by senior R&D and manufacturing managers had trouble encouraging attendance because the manufacturing people felt they could not be away from the job that long. What we have been saying, in effect, is that improving the contribution of technology to manufacturing is in part a marketing challenge for R&D. Like all effective marketing, success requires understanding the customer, demonstrating that you truly offer him value, and finding ways to predispose him favorably toward your offering.

Conclusion

The picture that emerges from this discussion of technology transfer is one of a complex, multidimensional process. The most visible part is the series of discrete, focused programs that are intended to transfer specific developments to a given operation. However, this process occurs as an overlay on a mosaic of continuing activities which transfer technology, probably in smaller increments, but which also foster networks and encourage informal communication. Without such "culture building," the success of the larger, more discrete programs is problematical.

Notes and References

1. Dr. Eduard Pannenborg, retired Vice Chairman, Board of Management, NV Philips Gloeilampenfabrieken, Eindhoven, The Netherlands.
2. J. Morton, *Organizing for Innovation: A Systems Approach to Technical Management,* McGraw-Hill, New York, 1971.
3. William A. Kraus, a personal communication.
4. Dorothy Leonard-Barton and William A. Kraus, "Implementing New Technology," *Harvard Business Review,* November–December 1985, pp. 102–110.

Part

3

Strategic Management

This next section will address the strategic management of technology. It develops the concepts relevant to the forces in technology management which oppose operational management: the irreplaceable need for change, for incremental improvement in present practice, and for the major steps that introduce new levels of capability. Thus the other anchor of the technology dyad of tension lies in strategic management, which in all areas (not just in technology) must address the issue of change mandated for sustained competitive advantage and for survival.

In Chap. 1, I pointed to the looseness with which people used the term technology. Strategy, strategic planning, *and* strategic management *warrant similar warnings. They are applied to a variety of different activities, each of which has a legitimate claim to be labeled as "strategic planning" or "strategic management." Since I have no desire to try to invent a new set of more carefully defined labels, I will attempt to clarify the ways in which I will use these terms.*

Two important differences in usage need to be delineated. First, we need to distinguish between viability strategy—the complex of actions and priorities designed to ensure that an enterprise will survive—and competitive strategy—the complex of thrusts and priorities that constitute one's mode of competing. The two concepts are not a dichotomy, but rather ends of a spectrum. The term survive *implies two important implicit subthemes. The first is a requirement to "prosper," because failure to do so imperils survival. The second is to "change," because a combination of internal dynamics resulting from growth and maturation, and external dynamics resulting from changes in competition and the environment, mandates change as a condition of survival. My choice of the label "survival" may seem unduly negative or stark, but the choice is deliberate, because ultimately that is just what is at stake—the survival of the enterprise. Note that I have specified that in order to survive an enterprise must prosper, i.e., it must achieve sufficient profitability to continue to attract capital and to earn enough return on its assets to protect it from dismemberment or attack. Survival demands more than a performance not quite bad enough to terminate the enterprise. The process involved may warrant more attractive labels, such as "revitalization" or "renewal," but the goal is survival. It cannot be taken for granted.*

In the competitive dimension of strategy, one must distinguish between single-business (or closely related business) companies and multibusiness companies. Strategic planning at the corporate level for a multibusiness enterprise cannot be the same as for a company with a single line or closely related product lines. Competition for a single-business entity is in the marketplace, offering superior value to customers. If it does that effectively, its performance will be satisfactory—provided that its markets permit an acceptable rate of return. Competition for the multibusiness enterprise is in the capital markets: Do its present portfolio of businesses and mode of management produce a competitive rate of return? Since multibusiness companies confront market competitors only through their various separate businesses, it rarely makes sense to talk about competitive strategy at the corporate level for such an enterprise. As we will see in Chap. 8, there are other components of strategic management that must be tackled at the corporate level, but competitive strategy per se must be developed and implemented by those who face competitors in the market.

Thus in discussing strategic management, we must keep in mind differing points of view. A single-business entity must be concerned with both competitive strategy and viability; a multibusiness enterprise must be concerned primarily with viability.

Strategy is concerned overwhelmingly with questions of change. How much must the enterprise change in order to survive and to continue to prosper? How much change can it finance and manage? How fast can it change? These are profoundly difficult questions. We know very little about the management of change—how to perceive the need for it, how to plot a new trajectory, and how to implement it—and much of what we know is not very reassuring. The topical media are full of evidence of mistakes, trauma, and failure.

Competitive strategy comprises the options chosen to offer attractive value to the customer—cost, product attributes, segmented customer focus. It reflects the strengths and weaknesses of a firm, perceptions of competitors' strengths and weaknesses and modes of competition, and projections of market characteristics and customer preferences.

Multibusiness strategy focuses first and foremost on portfolio optimization—what mix of sources of revenue is desired and what allocation of resources will best bring about this preferred mix. Multibusiness strategy must include other components, such as corporate organization and culture, management style, the conventions that guide behavior, and the acquisition or development of the new resources that will be required to support a different business portfolio.

Technology plays an important role in both kinds of strategy. Technological dislocations associated with major innovations are one of the major drivers of change. They can be either threats to survival or opportunities for growth and often are both. Technology is an intrinsic element of competitive strategy. It is the primary enabling resource by which one provides value to the customer. The way in which this resource is used to provide value—performance, cost, convenience, assurance of certainty in delivery and service—constitutes a major ingredient of competitive strategy. Technology is also a major consideration in multibusiness strategy because advances in technology play a primary role in determining the future prospects of a business. Consequently, they strongly influence the relative attractiveness of various businesses in a portfolio.

Chapter 8 sets the stage by examining strategy, strategic planning, and strategic management. It summarizes items on the menu from which competitive strategy is devised. It focuses, however, on strategic planning and the dynamic changes that occur as strategic planning is implemented. My involvement over a dozen years with strategic planning in GE made evident to me how much the process and the issues change. Strategic planning creates its own momentum, which influences management, but which also necessitates midcourse intervention by management. The process considerations in developing and implementing strategy, which are so important to practitioners, receive careful attention. Strategic planning must evolve into strategic management or it becomes a sterile exercise.

Chapter 9 then takes up the role of technology in strategic planning. It explores the utility of maturity curves as a tool for guiding both competitive and viability strategy and describes the process considerations involved in creating the technological vision and perspective that are indispensable to viability strategy. My focus is on the interplay between the continued application and extension of conventional technology, and its replacement by new technology—the centerpiece of strategic technology management.

Chapter 10 examines more specifically the work involved in integrating technology into strategic planning—one of the most oft-expressed needs of technology managers. It describes what a technology strategic plan should include, some of the tools used in strategic planning and in developing technology strategy, and proposes the basic technology issues on which the technology manager and the general manager must agree.

Chapter 11 is devoted to innovation—the core ingredient in technology's contribution to viability strategy. I examine the inherent difficulties in accomplishing innovation, the reasons why resistance is a rational response of those affected by it, the benefits it can provide, and the actions and environment needed to nurture it.

Chapter 12 examines the challenge of managing change in a broader context than that conventionally associated with innovation. Management is now held explicitly responsible for the survival of a corporation in perpetuity. No management has ever felt it was not *responsible for the survival of the enterprise, but the pressure to look into the future and initiate whatever changes in portfolio, location, culture, organization, and external relationships that are necessary to ensure survival is more explicit than ever before. I call this managing proactive change. It is one of the most difficult tasks in management. Examination of experience is not reassuring—the trail is littered with wreckage. And yet there are successful examples and options available which suggest that skilled, persistent managers can succeed.*

Chapter

8

Strategy—Strategic Planning—Strategic Management

It is tempting—tempting but wrong—to begin a discussion of the strategic management of technology by plunging directly into an examination of the role of technology in strategic planning. Such an approach assumes that there is a common understanding of what strategic planning is and how it is implemented through strategic management. My experience does not support this assumption. One of my major objectives is to improve the overall understanding of technology and its place in an enterprise. In order to do that, we must start by distinguishing among strategy, strategic planning, and strategic management.

> *Strategy* is the array of options and priorities with which one elects to compete (offer superior value to the customer) and to survive (sustain a level of financial performance that will continue to attract capital and to retain the autonomy of a business—in other words, protect it from financial attack).
>
> *Strategic planning* addresses the continued viability of strategy; it probes the need for change. Is change necessary? In what direction? At what rate?
>
> *Strategic management* is the implementation of modifications in the fundamentals of how one competes and survives. It controls the actions and behavior required to implement change.

This chapter explores the subject of strategy, covering the spectrum from market options to the survival of an enterprise. It then examines what is required for strategic planning to address effectively the entire spectrum of change, from change in market options to change in portfolio to change in the fundamental character of the enterprise. The term survival may seem a bit grim. A more upbeat term, such as renewal or revitalization, might seem more appealing, but starkly put, survival is exactly what is at stake—even for large, and supposedly impregnable, companies. Notice that I have defined survival as performance that will continue to attract capital and protect from financial attack, which is intended to connote more than just barely getting by. Many businesses whose managers *never* dreamed that it might disappear have done just that! Contemplating such a possibility does concentrate the mind.

Historical Perspective

The decade of the seventies will no doubt be regarded as the era of strategic planning. The concept appeared, flowered, and by the end of the decade was generating a backlash in the business press. It was accused of overpromising, of being simplistic, and of failing to achieve implementation.

I have participated in strategic planning from its early days in industry, and much of the discussion sounds distressingly incomplete, simplistic, and distorted. Strategic planning, like technology forecasting, was being conducted in some fashion before it was so labeled. It continues to be performed in some guise, because it has to be. Discussions with dozens of executives, in companies large and small, indicate that many of the issues they wrestle with fall under the rubric of strategic planning, irrespective of how the topics are labeled. Although the attention during the seventies helped greatly to systematize the activities involved and to provide tools of increased rigor, the literature somehow failed to make evident the dynamism inherent in the process—the way it evolves over time, the way it must accommodate to different managerial styles, and the way both internal and external forces change the focus and the priorities.

The strategic planning movement began with emphasis on three powerful strategic concepts that are useful, but not sufficient: the need to define and characterize the business segment for which it made sense to talk about strategy (i.e., the strategic business unit or some such entity); the leverage of experience curves in achieving market dominance and thus the value of high market share; and the matching of business strengths with market opportunity in some sort of matrix.[1] In retrospect, these concepts would better have been regarded as helpful diagnostic tools to initiate strategic planning, because they did help bring order out of a chaos of information. They did not provide much guidance for what to do.

The nitty-gritty of strategy has come along much more slowly. Knowing about the leverage of experience curves is not an adequate foundation for establishing market dominance. What does one do when a business does not have dominance, and probably cannot get it? How does one deal sensibly with the enormous variety of actual business situations instead of simplified abstractions? That has required much slow and dearly bought learning.

The Planning Process

In contrast to this delineation of strategic concepts, which provided a basis for greater rigor in thinking about strategy, the *process* by which strategy is devised and implemented has been a virtual black hole. And yet shop talk among practitioners almost invariably concentrates on strategic *planning,* i.e., on the process. Much of the discussion in the literature has had a curious static quality, as though developing corporate strategy were something done once, by some mysterious process, before moving on to other tasks. The discussion in this chapter will draw heavily on my 12 years of continuous involvement in strategic planning in GE, beginning with the McKinsey study that antedated the start of strategic planning and including dozens of discussions with executives in other companies.

Strategic planning takes on a life of its own. With the passage of time, strategic planning addresses issues that were never even conceived of at the start. In part, this is because strategic planning is a powerful tool for management learning. As initial issues are resolved, new ones appear or attention can be turned to problems less easily addressed. In part, this results from the fact that strategic planning generates a new kind of data base that reveals features of the business not known before. These new insights call attention to new issues. In part, it results from internal dynamics within the planning community that can be counterproductive if not policed. Finally, of course, strategic planning also accommodates to changes in the external environment.

In order for a strategic planning system to function effectively, it is necessary to devote careful attention to the process, as well as to the substance, of corporate strategy. Certainly, in the case of GE much thought was devoted to the process throughout the first several years of use. The large numbers of people from other companies who visited to learn of our approach were always vitally concerned with process. All of us were interested in such questions as: How can we establish a strategic plan? Even, what *is* a strategic plan? What are we trying to accomplish? What has motivated us to action? Who should be involved? What should they be asked to do? How can they communicate effectively? In what sequence should things happen? How can we build consensus for the output? How can we convert the plan into action? I do not mean to imply these questions were written down and addressed specifically. But if one listened to our discussions and observed what we did, it would have become apparent that these process questions were always present subliminally, along with the substantive ones of strategy.

To begin with, what is strategic planning? How does it differ from other kinds of planning that have been carried on for centuries? Once strategic planning gained prominence, it usurped the stage and seemingly swept all planning under its umbrella.

Operational Planning vs. Strategic Planning

Perhaps one can best understand strategic planning by distinguishing between it and operational planning—the conventional planning that is a part of every human endeavor. Operational planning seeks optimal efficiency in attaining a given objective or set of objectives; its goal is to do better what is now being done.

Three features of operational planning are important. First, the objective becomes the start of the process; the goal of the plan is optimal attainment of that objective. Operational planning emphasizes the identification of work to be done, the time required, the resources needed, and sequencing. It focuses on flow so as to ensure that no unnecessary delay impedes progress at any point. Second, the resources available are fixed in character. They may be augmented in magnitude, but the assemblage of skills and experience, the facilities, and the management techniques are already largely in place. Third, and most significant, the character of the enterprise itself is taken as a given. The interaction of its salient features with what it is setting out to accomplish is not incorporated explicitly into the planning process. Although successful fulfillment of a series of operational plans may change the character of the enterprise, any such change is an incidental by-product rather than central to the plan.

Strategic planning is broader in scope and adopts a longer time frame. It introduces three additional elements. First, and most important, it addresses explicitly the issues associated with raising the performance (capability) of the enterprise to a higher level or with ensuring its continuing, profitable survival. These issues go well beyond improving its effectiveness in achieving operational objectives, and focus on improving its capability to perform either by changing its portfolio or changing its modes of operation. Second, strategic planning incorporates explicit consideration of the character of the enterprise into the planning process. I choose the broad term "character" deliberately because strategic planning should not be constrained by arbitrary limits of what is permissible to question: business portfolio, modes of competing, organization, management style and competence, corporate culture, and so on. It is better to think the unthinkable for oneself than to be forced to do so from the outside. That is not to say that the character of the enterprise will be deliberately altered, although frequently it will; rather, that the planning process directly raises the issue of whether or not the enterprise *must* be modified in order to ensure its continuing survival or improve its performance. Obviously, identifying and evaluating the significant features of the enterprise become crucial. Finally, the two issues above force examination of whether the configuration of resources presently available—tangible assets, human skills and experience, and relationships with external organizations—is adequate to achieve strategic objectives. Operational planning looks inward and focuses on improvement of the present business. Strategic planning looks both inward and outward and asks, in the light of both internal and external trends and events, where is this business going? Do we want to keep on doing what we have been doing, or should we be seeking change? If so, in what direction, how much, at what rate?

A separate consideration that cannot be ignored is the attitude and aspirations, even the style, of management. Effective strategic planning always occurs in a specific management context: its attitude toward risk; its aspirations for achievement; its knowledge of markets and technology; and the constraints it perceives from investors, employees, unions, and customers.

Competitive Strategy

Michael Porter has developed a detailed and highly regarded paradigm of competitive strategy.[2] Its entire structure is based on the proposition that business success rests on satisfying customer needs. Therefore, the objective of competitive strategy is to devise a route to a sustainable advantage in offering superior value to customers. Three options are available: cost, differentiation, and focus. A cost advantage may be attained from some combination of economies of scale, proprietary technology, or access to low-cost inputs. Differentiation seeks uniqueness in the eyes of the customer that will result in a premium price. Although uniqueness is perhaps most often associated with perceived product attributes, it may also come from the delivery system, marketing approach, service, or other similar factors. Focus is achieved by segmentation, i.e., identifying narrowly defined markets with specialized characteristics that can be specifically targeted either through cost or differentiation.

Although competitive strategy requires choosing to emphasize one of these options, the alternative routes cannot be ignored. Product attributes must be regarded as approximately equivalent with competing offerings in order for cost to influence the customer. Similarly, costs must be near parity in order for a price premium for unique features to yield advantage. A broad approach to the market cannot overlook the potential of specialized markets and the opportunity they may provide for a competitor to attack. Xerox encountered this situation in ignoring the low-cost, small-volume copier market, which had large potential for growth. So, concludes Michael Porter, in devising a competitive strategy, an enterprise must analyze two questions: How will it seek to create value, and how will the value be distributed in its industry among suppliers, customers, competitors, and itself? The latter, of course, influences the choice of strategy, because some may offer a more attractive distribution than others.

The impact of technology is pervasive in competitive strategy. It is a major factor in determining costs and establishing uniqueness, whether in product or other aspects of a business. It is often decisive in seeking advantage through focus without incurring undue cost penalties for smaller-scale operations. Even more important, technology can change the rules of competition; therefore, it is a major consideration in allocating the value produced among the participants in an industry. Technology can be not only a powerful competitive weapon, but also the most latent threat. Although other changes in strategy—seeking lower-cost inputs, changing distribution or service, improving internal operation, and so on—are difficult to disguise, advances in technology, whether in the product or in process, can be implemented with little warning.

Is More Fundamental Change Needed?

Michael Porter and others have illuminated admirably the basic options of one end of competitive market strategy. The territory further along on the spectrum has received less attention. A focus on what combination of cost, performance, and market segmentation one should adopt as a competitive market strategy carries with it a subliminal assumption that a well-chosen and well-executed strategy from among these options will provide satisfactory performance. Does one dare make that assumption? There are situations which may call for more drastic action. A major change in technology, such as the appearance of CAT scanners, is one example. Shortly after CAT scanners appeared, the market prospects for conventional medical X ray underwent a dramatic decline, and I can think of no conventional strategic response that could have counteracted the change. A much more fundamental change than simply retuning its competitive strategy was necessary if GE's medical X-ray business was to survive. Semiconductor memories created a similar problem for magnetic-core memories.

A gradual deterioration in the long-term prospects of an industry also creates a situation requiring a more fundamental change. Persistent, large-scale global overcapacity in steel can be met only by major changes in individual companies. The attempts of steel companies to diversify, whether or not they are well-advised, reflect their recognition that an apparently irreversible shift in prospects has occurred.[3] The large changes Alcoa initiated in its portfolio showed that its management believed that the long-term prospects for aluminum no longer held the promise of opportunities for adequate return on investment, much less for growth.[4]

No enterprise can ignore the questions: Are we truly viable in our present form for as long as we can credibly project into the future? Does an "inertial forecast" that extends present trends and relationships into the future (fine-tuned for competitive advantage) lead to a destiny with which we are comfortable? It is noteworthy that the initial foray of GE into strategic planning pursued two paths simultaneously. First, if performance were improved to the maximum reasonably attainable, what could it expect to achieve in profitability and growth? Second, as it evaluated its competitive and socioeconomic environment and the criteria of the investment community, what performance was necessary to meet management's criteria of success? If there were a gap (and there was), how should it go about closing it? Without any particular guide to follow, GE was more or less intuitively addressing the full spectrum of change that a company will sooner or later face if it is to survive and produce consistently acceptable or superior performance.

Strategic Planning

Strategic planning is typically initiated in response to one of several searching questions:

1. What's wrong with us? Why don't we do better? How can we do better? Have we reached a plateau in growth?
2. How can we surmount this critical vulnerability—limited market, single product, dominant customer, high cost, etc.?
3. How can we maintain the very rapid growth we have been enjoying?

In beginning to develop answers, management in turn addresses four deceptively simple questions:

1. What kind of enterprise are we?
2. What kind of world do we expect to be living in?
3. What do we as a management wish to accomplish with this enterprise?
4. What has to be done to convert those three questions into a coherent course of action?

To answer those questions, management must face the fundamental issue underlying all strategic planning: how to balance the continuity and stability that foster efficiency and effectiveness against the introduction of disruptive change that may lead to a higher level of performance or ensure continuing viability. How much change can be managed? How much can be financed? Thus once again we see that a critical ingredient of management is the balancing and rebalancing of incompatible requirements, as suggested in Fig. 8.1.

The character of the enterprise

Obviously, any process that contemplates changing the character of the enterprise must start by understanding its present character well. What do we mean by the character of the enterprise, and in what ways could strategic planning seek to alter it? The "concept of the enterprise" I have in mind is much broader than management style and corporate culture, and was examined in detail in Chap. 3.

The fundamental first step in strategic planning—and a touchstone that must be revisited continually—is to understand as rigorously as possible just what business you are in. It is all too easy to delude yourself both with respect to what business you are really in and what skills and capabilities are in fact the core strength of the company. The managers throughout the Insulator Department in GE said consistently that their business was providing electrical isolation to the power transmission industry. But looking at their skills and facilities,

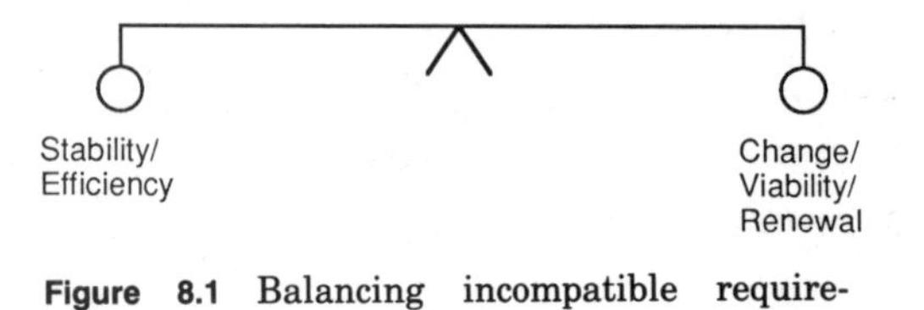

Figure 8.1 Balancing incompatible requirements.

and even more, examining the moves they contemplated to expand the business, made clear that they were really in the ceramics business.

The mission statement

The concept of the enterprise is a mixture of obvious features and subtler characteristics, which could be almost impossible for an insider even to perceive in their entirety. In many ways they are analogous to the entities that an anthropologist would use to describe community or other sociocultural groupings. For purposes of strategic planning, a business should prepare a mission that states what business it is in, what markets it serves, and how it proposes to serve those markets. For a multidivisional business, a mission statement, of necessity, must become more general. Nevertheless, an attempt to prepare one must be made; otherwise, a business runs the risk of becoming so unfocused that no unifying theme to guide behavior can be identified.

Some examples may help illustrate this point. For many years GE's lamp division carefully avoided saying it was in the lighting business. To have done so would have included not only lamp bulbs, but fixtures, controls, and ballasts. They all employ different technologies, are sold through different channels, face different competitors, and respond to different competitive forces. The division did make clear, however, that it served both consumer and industrial and commercial markets for bulbs, even though the latter was different in distribution and customer preferences. The division also made clear that it intended to be a full-line supplier and to be a leader in developing new technology.

As Xerox grew in size, its management began to ask whether just the copier business would sustain its remarkable growth trajectory. Obviously, the second generation of management did not want to go down in history as the team that managed the transition to lower growth. Consequently, the company's mission was broadened to include office systems, and Scientific Data Systems (SDS) was purchased to provide entree to this broader arena.

Reasons to change the character of the enterprise

A company may seek to change its character because of an inherent vulnerability it perceives now or for the future. I noted in Chap. 3 how the long-term specter of the depletion of petroleum reserves began to motivate the oil companies to diversify and the difficulties they encountered in doing so. The inherent vulnerability of heavy dependence on government orders has impelled both United Technologies and

Emerson Electric, at an earlier time, to change their business portfolios.

Alternatively, management may seek to alter the character of an enterprise in order to achieve a higher level of performance, again by changing its business portfolio. With time, this can be accomplished by sharply altered allocation of investment so that the internal mix of sources of earnings is changed significantly. However, management may also alter the mix by judicious disposition and acquisition of businesses. Harris Corporation is renowned for pursuing such an objective. In fact, it is likely that this goal is characteristic of most companies in mature industries which embark on strategic planning. The business press provides weekly evidence of such endeavors.

The most obvious aspect of a mission statement is the portfolio of businesses an enterprise proposes to pursue, but the internal modes of behavior and management conventions should not be overlooked. Companies encountering a drastic deterioration in performance—the auto industry, steel, and construction equipment, for example—are forced to examine their fundamental viability as an enterprise, to find bottom as it were. Although this examination must cover many fronts, among the most complex are the conventions and practices by which management carries on its work. Strategic planning, even in less traumatic circumstances, should always examine those same behavior patterns.

Perhaps the happiest circumstance is the need to alter the character of the enterprise in order to maintain a dramatic rate of growth. Apple Computer faced the challenge of making the transition from being an entrepreneurial free spirit in the industrial ranks to being a disciplined supplier and supporter of a family of products—a natural and inevitable evolution for a successful business. This problem is endemic among the high-technology companies along Route 128—what do they do for an encore? Small companies that have enjoyed dramatic growth in niche markets—such as Floating Point Systems, Sensormatic, or Chyron—face the challenge of finding additional products to maintain the growth rates of earlier years.

A study of 100 high-growth, high-technology companies by Booz-Allen[5] showed that periods of crisis are surprisingly common. They result from a combination of three forces: expansion, diversification, and changes in the basis of competition. The situations encountered require changes in strategy, structure, operations, and management—in short, in the character of the enterprise.

Virtually every successful company, be it Hewlett-Packard, Intel, Texas Instruments, Digital Equipment, or DuPont, American Express, and Sears, periodically finds itself forced to reexamine the sources of its growth and to consider the necessity of entering new fields. Such a strategic step often also changes the character of the enterprise in ways that are not anticipated. Proctor and Gamble, 3M, and American Home Products are justly renowned for their success in gradually expanding their portfolios without placing undue stress on their ability to manage their enterprises. In other words, they have been consistently skillful in balancing continuity and change, even though there are periods when they seem to be undergoing one more passage to another stage in their existence.

The Process of Strategic Planning

Given the importance of understanding the character of the enterprise and the likelihood that strategic planning will lead to a decision to change its character, it is not surprising that practitioners of strategic planning place their emphasis on process, because process determines the nature of the relationship between strategic planning and the people in the enterprise. Strategic planning starts a chain of events intended to place the business on a new course; yet many of the events that this change will bring about cannot even be imagined in advance. This process has been described metaphorically by saying, strategic planning is a journey and not a destination. Unless strategic planning in turn energizes the strategic management that sets in motion all the actions to begin and sustain that journey, it is an empty exercise which will atrophy.

If strategic planning becomes a journey, the landscape should change, and it does. My experience with strategic planning, augmented by discussions with colleagues in many other companies, has led to my identifying four different components of strategic planning: portfolio optimization; resource strategy; sources of long-term growth; and management staffing, organization, and culture. They could be labeled "phases" because they occurred in sequence in GE, but it is not clear that the sequence is fixed. While the term phase connotes movement from one to the next, these components appear more in the form of overlays, which are superimposed on those already in place. Their appearance on the scene tends to change managerial priorities regarding various aspects of strategic planning, but does not supplant old concerns with new ones. These various components are probably more typical of companies that have grown largely from within; there may well be others for businesses in which acquisition has been a major factor in growth. For example, I will not address the integration required following a period of growth through acquisition. Nor will I discuss restructuring—typically a euphemism for downsizing an enterprise so that assets and resources are commensurate with business potential.

Portfolio optimization

The component of planning that a company chooses to start with will reflect the needs its management perceives. Nevertheless, portfolio analysis is undoubtedly the most common starting point. The questions noted earlier that typically initiate strategic planning for survival generally lead first to a critical examination of the present business portfolio. For the multibusiness enterprise, the entity to be evaluated is the individual business (or strategic business unit, as it is sometimes called). For the undiversified business, the entity is the product or product line. The objective of the analysis is to achieve a new level of discrimination in determining the prospects of the various businesses (or products) in the portfolio. Although the focus will generally be on the prospects for continuing profitability and growth, it will or should include other elements, such as cyclicality, strength and nature of competition, maturity of technology, or vulnerability to changing sociopolitical forces.

The objective of portfolio optimization is to determine the appropriate strategic role of each component of the portfolio. This step has to be completed before settling on a competitive strategy. However, the process involves an iteration between possible competitive strategies and the future prospects for each business, both of which may vary because of external considerations or internal resources available to implement strategy. In other words, one may decide to deemphasize or exit a business not because it is unattractive, but because one lacks the resources to exploit it as well as other businesses that are even more attractive. A secondary objective of portfolio optimization is to consider the need for, and manageability of, deletions and additions to the portfolio. The intellectual challenge in portfolio optimization is to devise a variety of perspectives for examining a business (or a group of businesses) in order to adequately characterize its present position and future prospects for growth, financial performance, and survival, as well as its fit with the overall portfolio being sought. The objective is to make certain one truly sees the elephant rather than tusks, or trunk, or ears, or body.

Matrices are frequently used as convenient methods for displaying strategic relationships. One of the most useful is to compare the attractiveness of an industry with the capability of the enterprise to serve it. Factors can be combined in a quasi scoring system for each dimension. For evaluation of industry attractiveness, factors that lend themselves to quantification include market size, rate of growth, profitability of those serving the industry as well as profitability of those participating in the industry, extent of fragmentation among customers, cyclicality, inflation, importance of non-U.S. markets, and importance of international competitors. Additional factors, less easy to quantify, should not be excluded, e.g., vulnerability to regulatory constraints or sensitivity to sociopolitical changes. Factors determining current business strength include market position, which can be evaluated by market share, share relative to major competitors, and trends in share; relative profitability versus major competitors; and competitive status, such as breadth of product offering, product quality, technological status, cost and productivity, and distribution and marketing.

In constructing a matrix, an individual business would examine products or product lines and a multibusiness entity would compare businesses in order to visualize an entire portfolio, its present prospects, and possible changes. Since the initial emphasis is on portfolio optimization, management concentrates on the change in financial performance that is projected to result from a proposed change in portfolio and the financial resources needed to implement it. By a process of iteration between possible strategic moves and resources required for implementation, a manager can converge on the preferred strategy.

There are relatively obvious strategies associated with the numbered cells in Fig. 8.2. Businesses in cells 1, 2, and 4 are usually slated for growth and preferred investment. Those in cell 3 can be treated in several different ways: increase investment to strengthen position in an attractive industry; since this situation often exists in an emerging industry, choose to wait until likely courses of action become more evident; conclude that internal resources are inadequate to exploit the

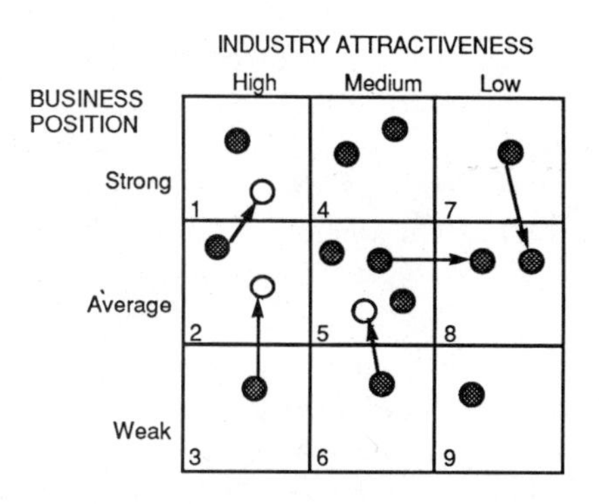

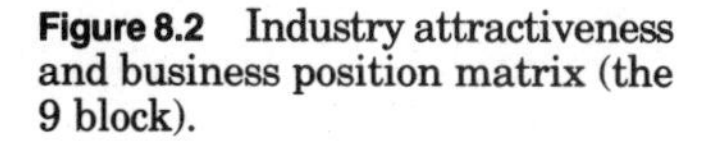
Figure 8.2 Industry attractiveness and business position matrix (the 9 block).

opportunity and seek a partner; or decide that a powerful position will be difficult to achieve and sell the business as an attractive opportunity for somebody else. Businesses in cells 5 and 7 are often regarded as important cash generators whose present position should be protected. Those in cell 8 are usually put on a harvest trajectory, and those in cell 9 are candidates for disposition.

Although the construction of a matrix appears to focus on quantified factors, other considerations are important, often dominant. Management's attitude toward risk is undoubtedly the most significant. Another fundamental and related choice is between one or a few large thrusts and many small ones. For example, at the outset, GE's managers were highly sensitized to the portents of major thrusts because of the ongoing experience of simultaneous commitments to computers, nuclear power, and commercial jet engines. As a result, there was a reluctance to embark on additional moves of similar scale.

In the case of GE, the first three or four yearly cycles of strategic planning were focused primarily on portfolio optimization. The initial effort emphasized a quantification of market prospects and financial potential, together with an evaluation of the opportunity for technological advance. With the passage of time, the company's top executives became increasingly concerned about the coherence and validity of the data, and, in particular, the assumptions and thinking that underlay the numbers which were so prominent in strategic plans presented by operations. Consequently, the two aspects were separated, with consideration of numbers being preceded by consideration of concepts, assumptions, adequacy of data, and possible alternatives. For each business, the objective was to reach a common understanding of its basic nature, the salient features of its environment, the key strategic issues it faced, and the options available to it. Only after this, was it sensible to consider a quantified strategic plan in terms of financial costs and benefits. Thus one significant and often overlooked benefit of strategic planning is that it can be an effective learning exercise for the entire management team.

The effect of this portfolio optimization process on GE's sources of earnings was dramatic. The change during the period 1968 to 1986 is shown in Table 8.1.

Resource strategy. One effect of portfolio optimization is to change the mix of businesses. Therefore, a natural outcome of progress in this dimension of strategic planning is to change the character of the enterprise in such a way as to modify the resources needed by the emerging

TABLE 8.1 Sources of GE Earnings

	1968	1979	1986
Electrical equipment	80%	47%	52%
Materials & natural resources	6%	27%	10%
Services	10%	16%	18%
Transportation	4%	10%	20%

SOURCE: Data for 1968 and 1979 from "Integrating Social, Economic, Political and Technical Forecasts into Business Strategy," Standley Hoch, *Research Management,* November 1981, p. 11. Data for 1986 from "The Welch Years: GE Gambles on Growth," *Industry Week,* April 20, 1987, pp. 30–32. Classification varies slightly from 1968 to 1979 data, but the effect is trivial.

company. This in turn will modify the issues that strategic planning should address. Strategic planning creates its own agenda!

Human resources. What is meant by resources in the strategic sense? Most obvious is human talents and experience. A domestic company attempting to internationalize will require skills and experience different in character from those presently on board. A Texas Instruments or Hewlett-Packard trying to enter consumer, as opposed to industrial, markets faces an analogous need. Perhaps more subtle, a company supplying components that wishes to integrate forward into systems or subsystems will also need different skills in design, application engineering, and marketing. Hewlett-Packard and Intel have both been facing this challenge.

Facilities, equipment, plant location, infrastructure. The nature and location of physical facilities is another strategic resource that may well be affected by changing the character of the enterprise. This applies not only to manufacturing, but to distribution and service as well. When IBM decided to enter the personal computer market, which has great volatility in products and competitors, it departed from its traditional approach of depending largely on internal fabrication and assembly. Instead, it improved its response time, reduced its capital requirements, and spread its risks by relying heavily on vendors for components and subsystems. As GE pursued a global cost advantage, many businesses located more and more manufacturing facilities in the Far East, until control over the location of offshore facilities itself became a strategic issue. The need to develop a global service and logistical infrastructure was a major barrier to Airbus in setting out to become an international supplier of commercial air transports. Conversely, the logistical and maintenance capabilities of Caterpillar are an important competitive resource.

Technology. If technology is defined as the capability to provide the goods and services an enterprise seeks to offer, then changing the character of the enterprise by changing its operating conventions—set points—or by changing its portfolio has a high probability of changing its requirements for technology. Thus technology becomes an important resource strategy consideration. The relationship between a different product mix and a need for different technology in product engineering and manufacturing is obvious. The relationship with a change in operating conventions may be less so. Conventions are powerfully influenced by the quantity and timeliness of information, and the ability to process it and communicate with others. Consequently, as we have noted repeatedly, information processing is rapidly becoming an equal, if not a dominating, dimension of technology, along with product engineering and manufacturing, and it is central to any need to change conventions.

When Richard Shea, CEO of Pepperidge Farm, set out to revive the company, he determined that changing the conventions on acceptable freshness could provide an important advantage.[6] The critical barrier in reducing delivery time for bakery products, from oven to store shelf, was information. The more rapidly Pepperidge could know what was taken from supermarket shelves, the more rapidly it could adjust production schedules. Installing information and communication systems that provided much more rapid and detailed information on the exact state of inventory was a key ingredient in success.

Of course, advances in technology itself can create both strategic threats and opportunities, which must be evaluated and incorporated into the plan. We will examine the role of technology in strategic planning in Chap. 9.

Availability of capital. The adequacy and cost of capital, together with the timeliness of its availability, are obviously indispensable elements of resource strategy. Inadequate funds frequently constitute a major barrier to implementation of strategic alternatives. But occasionally the challenge is to discover sufficiently attractive opportunities to invest a veritable plethora of funds. Most would call the latter "good trouble."

Corporate culture. Most subtle and most intractable of all, the culture of an enterprise represents a critical resource. The body of shared knowledge, mutual expectations, modes of interaction, common values, systems of reward and retribution, and modes of organizing work that enable an enterprise to be effective is both a priceless asset and a daunting barrier to change. The task of even perceiving its relevant features, much less changing them, is perhaps the most difficult in all of management. It is not surprising that one of the most widespread and persistent topics at lunch in most companies revolves around these themes: What are we like? Why are we successful? What are we good at? What are our weaknesses? Why should we not attempt to do certain things? For example, a company such as DEC, which is used to dealing with relatively few, but technically sophisticated customers who buy in large quantities, does not have very effective instincts or very useful systems for selling personal computers one at a time to virtual amateurs.

An example of resource strategy

As strategic planning proceeded in GE, its management became increasingly troubled by the question of whether sufficient attention, at both corporate and operating levels, was being given to the identification, acquisition, and deployment of the resources needed to support the emerging company. Attention to these subjects was mandatory for effective strategic management. Consequently, a new dimension was added to strategic planning—resource strategy. Four resources were singled out for focused attention: finance, human resources, technology, and manufacturing capability and capacity. Corporate culture was not included, perhaps partly because there was no organizational component to assign responsibility to. This new activity was initiated at the corporate level, but the basic responsibility lay with operations. The corporate role was twofold. First, it called the attention of operations to potential resource issues that warranted careful scrutiny. Second, it then determined whether any additional corporate activity, e.g., change in recruiting or university contacts or new training programs, was indicated by the aggregate of the responses from operations.

In undertaking to develop resource strategy, GE was breaking new ground, because no prior experience or external guidance was available. Although differing in detail, strategic planning for each resource involved the same basic set of activities. The first was to characterize the present state of the resource in strategic terms. Second was to evaluate the differing demands that would be placed on the resource by the "new" company as it emerged. Third was to examine external trends and events likely to have an impact on the resource. Out of this, planning attempted to identify potential mismatches between the resources likely to be available and the need for them by the evolving GE.

Financial resources. In finance, attention was devoted to long-term projections: growth in earnings and return on investment, requirements for capital to support the growth, expected payments of dividends, the borrowing capacity of the company, and the likely cost of capital. The basic objective was to evaluate whether funding would be a strategic limitation on growth or, conversely, whether funds generated might in fact exceed the requirements for growth and thus constitute a key strategic issue: what to do with the rapid increase in available funds. Indeed, the rapid acceleration of oil prices during the 1970s posed just this problem for the integrated oil companies. Standard Oil (Ohio) was confronted with this issue in dramatic fashion after the discovery of North Slope oil. A defense contractor making deliveries on a large contract faces the same problem. Although some might label this as "good" trouble, failure to devise effective strategies for deploying this cash flow has led to serious criticism of some management.

Human resources. Thinking strategically about human resources in GE involved a sequence of steps that started with characterizing the configuration of skills present in the work force, especially the professional and managerial work force. Then by extrapolating the impact that changes in resource allocation would have on the future mix of businesses, it was possible to analyze the modifications in the configuration of skills that would be required to meet the "new" business needs. Finally, came an examination of the new problems, such as recruiting and training, that might arise in altering the skill mix.

For example, as the company began to examine the possible changes in its business portfolio, it became apparent that the existing mix of technical skills, which primarily reflected the decades-long growth in businesses associated with the generation, transmission, control, and use of electricity, would not correspond with emerging needs. The company would need increasing numbers of people skilled in integrated circuits, software for both operating systems and applications, architecture of information systems, and factory automation.

This mismatch in turn illuminated possible alternatives needed in the company's recruiting practices, relationships with science and engineering departments in universities, and company training programs. In response, the company commissioned a study of college students' perceptions of career opportunities. How attractive was the company compared with its new competitors for technical graduates? What factors did students weigh in arriving at their perceptions of companies as potential employers? Using this information, a program was established to begin strengthening GE's contacts with university departments that would be important future sources of graduates.

A parallel program was undertaken in in-house training. Two corporate-level training programs were established at entry level—one for engineers to receive training in the new skills associated with hardware and another in software. A management training program in managing electronic businesses was started, and cases of internal businesses that had made a successful transition from traditional technical skills were developed and publicized.

Technology resources. Formulating technology resource strategy meant concentrating on the internal dynamics of technology itself as well as the impact of changes in the company's requirements for technology in the future. With respect to the former, it was necessary to identify the classic conventional technologies that were the current backbone of corporate effectiveness and to assess the company's capacity for maintaining competitive strength in those technologies. In parallel, it was necessary to identify the emerging technologies that might supplant the present conventional technologies and assess the company's ability to respond appropriately. In this example, the link between human resource strategy and technology strategy is apparent. Effective strategic planning for resources requires a careful attention to the cross-functional implications of other elements of strategy and the formulation of integrated efforts where appropriate.

Manufacturing resources. In manufacturing, it was necessary to consider present plant locations and attractive new sites for the changing company; to determine the competitive state of equipment and methods; to evaluate trends in productivity improvement both internally and externally; and to assess the strategic viability of present procedures and guidelines for make-buy decisions in the light of the changing character of the company.

Of course, an essential element of resource strategy is to examine the strategic plans of operations from the standpoint of resource planning. Do the plans demonstrate an awareness of the resource implications of the strategy? Where significant modifications in the resources are required, are plans being developed to put them in place? Are strategic plans in operations converging on some common resource issues? Is meeting these needs something that can be left to operations, or should corporate initiatives be considered? Is management alert to the magnitude and dynamics of the problem, or are special programs to heighten management awareness needed?

Sources of future growth

Generating long-term sources of future growth is an indispensable component of strategy. One hidden benefit is that strategic planning can increase management's sense of ease with the state of the business. If managers believe that businesses are on the right strategic trajectory and are implementing plans effectively, then they can turn to other long-term problems. Thus in GE, as evidence accumulated that questions regarding resource strategy were at least receiving increased attention and that responses were being devised and implemented, a new set of concerns began to appear.

Examination of many strategic plans in operations disclosed a common pattern. Typically, the plan indicated that actions under way would, if successful, provide acceptable growth and performance for the next 5 to 10 years; the period beyond that appeared to be very uncertain. The challenge for corporate management lay in inducing operating management to invest more effort and resources in planting the seeds for long-term growth, without attempting to do the job for it.

Developing mechanisms for focusing management attention and resources on sources of long-term growth is a universal problem. Although long-term growth is a principal responsibility of top management, the on-site perceptions and ideas of operating management can constitute an important input. Attempts by headquarters to point the way to opportunities in operations are unlikely to be effective unless other supportive actions are taken. For one thing, the time scale for implementation far exceeds the typical incumbency of management positions in operations. For another, evaluating prospects and progress is a minefield of opinions and prejudices. Consequently, a corporate initiative to stimulate greater attention to growth is unlikely to be effective unless it elicits the participation of operating management.

In GE, a deceptively simple and ingenious solution was devised, a solution that invoked the wisdom of an old truism: form influences substance. Long-term business development was officially declared to be a component of strategic planning, and a virtual blank sheet was included in the planning documentation, with little in the way of guidance or constraints provided. The effect was immediate. Since a plan obviously could not be submitted with blank pages, operating management began to scramble to develop responses. Corporate R&D began receiving requests to participate in brainstorming sessions to identify possible new opportunities. Responses from operations ran the gamut: greater international activity, direct foreign investment, acquisition or joint venture, increased vertical integration, penetration of new markets, and internally based technology development—all appeared in one form or another.

The extent to which this planning for long-term growth is assigned to an individual strategic business unit will depend in part on the way a company is organized. Given the strong pressure in operations to achieve short-term performance targets, it is not uncommon to assign primary responsibility for this investment in long-term growth to a higher level of management, e.g., group or sector level, if they exist. In fact, sometimes such a level of management is created specifically to ensure effective attention to growth.

No one would argue that all the proposals were sound or that all operating managers had suddenly been converted to entrepreneurs. However, discussions with planners and line managers, as well as reading the plans before and after, showed that management had been induced to devote much more attention to the long-term future of the business. And in fact many new business development efforts were initiated.

Organization, executive development, and staffing

The final component of strategic resource planning that must be addressed involves its impact on organizational structure, management development and staffing, and corporate culture. Just as success in other dimensions of strategic planning creates both a need and a capability to address new issues and concerns, an enterprise that is changing its character in planned, systematic ways is creating a need for new management skills and styles and possibly for new forms of organization.

With a growing cadre of managers who are skilled and experienced in strategic planning, the attention of top management can turn increasingly to consideration of the managerial and organizational implications of strategy as it evolves. The type of management needed for businesses targeted for rapid growth obviously differs markedly from that for mature businesses targeted for generating funds. More important, if a business is being asked to change its character, it may well need new management. In GE consideration of these requirements became more systematic and explicit as strategic planning matured and became more deeply embedded in the company. Eventually, explicit consideration of the management development and staffing requirements implicit in a chosen strategy became the final step in strategic review.

One of the advantages of a multidivisional company is that corporate management can initiate changes in management to better match strategic trajectory. If a manager with a strong entrepreneurial bent is running a business placed in a harvest mode, he is very likely to produce turnaround strategies that are not accepted. Even worse, his own behavior will be at odds with the signals from headquarters, and the entire organization will be confused.

Implementation: Strategic Management

What attributes contribute to the successful implementation of strategic planning? Most important is the immediate introduction of steps for implementation. In other words, line management must assume ownership of the plan and begin to act accordingly. If the plan is put on the shelf until "things are better," either it is an inoperable plan that must be reworked or it is doomed. In GE, line managers were promptly assigned yearly performance targets that were consistent with their strategic trajectory. Allocation of resources was adjusted accordingly, and as noted in Table 8.1, the sources of earnings changed dramatically. Just as important, managers were held accountable for initiating nonfinancial strategic programs—for example, changing the product mix, the served market, the resource capability, or the operating conventions.

Another important consideration in implementation is choosing the mode of interaction between top management and management at the operating level. If a management is highly centralized and directive, evoking the participation of a management that is accustomed to taking orders may be difficult. Of course, top management may conclude that its vision and response are so solid that no inputs are needed from below; however, it still must effectively communicate complex changes in direction and in behavior that are hard to transmit accurately.

Guiding without usurping responsibility

GE faced a different problem. It operated in a decentralized mode in which there was a tradition of constructive tension between operations and headquarters regarding targets for growth and performance, allocation of resources, and priorities for the future. In this environment the challenge was to develop a mode of interaction that captured the attention of operating management and influenced its priorities without usurping its responsibility for strategic planning and implementation at the level of the strategic business unit.

The eventual ingenious response, after some experimentation, was the use of "strategic guidelines." These guidelines said, in effect: next year, as you do your strategic planning, in addition to the normal activities and routines that you engage in, we would like you to pay particular heed to the following issues. The guidelines were few in number, issued under the imprimatur of the CEO, and carefully selected to have broad company import. Acceptable responses could vary. For example: "We've looked at it and see no significant implication for our business," "We see a need to develop a better strategy and here is our preliminary response and plan for further action," or "We've foreseen this issue and here is the plan we are already implementing." The guidelines were intended to cover situations where greater management sensitivity and concern seemed to be warranted, i.e., management priorities needed rethinking, or where responses to date seemed to be inadequate.

Establishing legitimacy

Perhaps the most subtle requirement for implementation is simply enough time for the activity to achieve legitimacy. Successful strategic planning inevitably leads to differential treatment for businesses. Some GE managers had easy access to the investment window; others received little or no investment support. Some businesses had high visibility as targets for rapid growth; others faced the more prosaic challenge of running a very tight ship.

In order to accede to such differential treatment, a management cadre must come to accept the validity of the criteria and the fairness of the process by which investment allocation decisions are made. Legitimacy cannot be mandated—it must be earned, and that takes time. Several planning cycles need to transpire before managers can see the way resources are allocated among various businesses. Even more important, they themselves must experience the process by which those decisions are made and see the criteria that are used for doing so. Only then can they determine whether the results are in the best interests of the company, whether the process gives operating management its fair day in court, and whether strategic achievement will be rewarded even in differing circumstances.

Remaining tentative

An important corollary is to remain tentative—even though the strategy will need to be presented with specific objectives and directions established. The dictum would seem to be: act positively and forcefully in order to inspire confidence and elicit participation from the organization, *but* think tentatively and even defensively in case you are wrong!

The eventual determinant of the success of strategic planning is the soundness of the strategic goals and trajectories established. The goals inevitably are reduced principally to financial terms: growth in earnings, return on investment, and related measures. Clearly, the enterprise wishes to achieve all that its resources and resourcefulness will permit within the options available. However, laying on unattainable performance goals eventually proves destructive. GE's management retained a continued questioning attitude toward the realism of its growth targets. In fact, it did not attempt a detailed and rigorous definition of its targets. They were regarded as goals to inspire and motivate, more than as weapons to punish.

No management has the wisdom and vision to establish immutable targets that are appropriate from the very beginning. The designation of investment attractiveness for individual businesses and the formulation of earnings targets must include, in the minds of those setting directions, a quality of tentativeness. This reduces the likelihood of large error and retains the flexibility to introduce change. Such an approach is consistent with the "logical incrementalism" that James Brian Quinn argues is characteristic of successful strategizing.[7] This tentativeness also permits time for a consensus to emerge concerning the validity of the strategic role assigned various businesses.

In the case of GE, some three or four years elapsed before strategic determinations began to solidify. The recession of 1974 to 1975 tested management's resolve in maintaining differential treatment of businesses. Although no business was exempt from pressure to reduce expenses, strenuous efforts were made to maintain momentum in businesses deemed critical to the future. Thus even in times of stress, management was clearly committed to the actions called for by the strategic plan.

Attracting competent people

One of the most important requirements for success is for bright, ambitious people to perceive that a stint in strategic planning is valuable to their careers. GE's future CEO and vice chairman helped initiate strategic planning. Some strategic planners were promoted to senior executive positions, and throughout the company it became apparent that manager of strategic planning was a fast-track assignment.

Nurturing constructive tension

Further, successful implementation must include the nurturing of creative tension among elements of the business. Successful strategy does not emerge from a lovefest, nor does it emerge from a *diktat.* One source of tension is between line and staff. In my experience, the staff tended to take on the role of urging consideration of sweeping alternatives and arguing the validity of analytical techniques for rendering judgment. Line managers, in contrast, invoked realism and pragmatism—the need to maintain balance, and a fallback position if plans proved overly ambitious.

Another tension is between corporate management and operations. Managers advance in part by becoming skilled advocates. Valuable though this talent may be, it does diminish and filter the richness of the information made available to top management for rendering decisions. Naturally, managers make the best possible case for their recommended course of action. Other options they might mention are usually straw men for their preferred course. Review sessions can become sparring matches in which top management probes for the weaknesses carefully obscured in a proposal. The nurturing of tension in GE was explicitly provided for by initiating both top-down and bottom-up planning and by requiring formulation of alternatives by operations. The stipulation to offer genuine alternative strategies, each of which operating managers must be prepared to implement, encouraged dialogue, enriched the information available to decision makers, and at the same time, ensured effective participation by operating managers.

Avoiding pitfalls

Planning systems and the planners who devise and operate them seem to be prone to certain pitfalls. A wise line executive will be on the lookout for them. Most important: don't let the planners do all the planning! Planning is a fundamental line management function. If planners produce the plan, the result may be an impressive-looking book that is superficially more coherent and imposing than one that line managers could produce. Unfortunately, these managers probably will not feel any commitment to the planner's output. They may like it; they may use it to impress the boss or customers or investors. But they will not feel that it is theirs and that they must fulfill it. A planner's plan is likely to become no more than a tribute to his own ego.

Then what do planners do? They are architects of the planning system, and just as in construction, their task is to employ their specialized skills to develop a structure that meets the client's (in this case, line management's) needs. And just as in architecture, the system works best when there is constructive interaction between the planner and his client (management). Planners can devise techniques for analyzing and evaluating data. They can assemble information. They can critique results. They can help package the output.

The planner's target should be next year's plan, not this year's. A planning process typically encompasses three classes of work: (1) a more or less finished product that is presented as a proposed analysis and course of action for approval; (2) a class of problems or issues with which the line management feels sufficiently comfortable (or sufficiently in agreement) to surface for discussion with higher management, but without proposing a resolution; and (3) a hidden agenda of problems or issues that are under study or are being debated, but which management is not yet ready to bring up with higher managers. A planner can do little, except cosmetically, to the first category. He should concentrate his attention on the second and third, which will emerge later as plans. His goal is to broaden his management's awareness of issues, to provide it with additional data, to lead it to question its priorities—in short, to raise its discomfort with the current year's plan enough to do the necessary homework and soul-searching to produce a different output for the next cycle.

A second pitfall to guard against is the endless expansion of requirements to generate additional information. Preparation of a plan always makes one aware of deficiencies and limitations because adequate data were not available. Unless it is watched carefully, data gathering—which typically falls on operations—can become a monster, generating information that makes the planners feel better, but that has limited usefulness in clarifying issues or aiding decision making.

Another pitfall: allowing too many layers in the process. Each organizational level tends to add its own interpretation of instructions from above. Even worse, it introduces its own special requirements. Conversely, each level of review introduces further remoteness and carefulness in stating issues and plans in the name of consistency and coherence. Worse yet, intervening reviews can become so extensive that the time allowed for preparing a plan can be less than the time required for the multiple reviews. One disgusted manager described it as "rousting out the troops at 4 o'clock in the morning for the 5 p.m. review by the general."

Many of these pitfalls are related to the makeup of the planning group and the length of stay of its members. Planning should provide for some movement of people. The group should include some long-term residents who can preserve institutional memory. But it must also include some new people who will question, who will need to be convinced, and who will suggest new approaches.

Conclusion

Perhaps the best evidence that planning is succeeding is that the process itself changes. Strategic planning should introduce an additional dynamic into a business, and that dynamic will inevitably reflect back on planning itself. Recent experience indicates that as strategic planning becomes embedded in the management culture, the formalisms can be reduced. There is no need to continue acting as though strategic planning is being done for the first time. Not every aspect of every plan has to be reviewed every year. New management teams will introduce new terminology and new priorities. A former colleague recently commented, "Oh, we still do strategic planning. The terminology has changed and the rhetoric is different, but it is still strategic planning." Left to its own momentum, strategic planning, like any management tool, will tend to become bureaucratized—just another exercise the system mandates. Yet without a sustained, systematic program in the early stages, the necessary learning and inculcation into the culture will not occur. The need for strategic planning does not diminish. However, the manner in which it is carried out changes, partly as a result of years of learning in a more explicitly formal and systematized structure. If strategic planning does not result in raising the competence of the entire management cadre, it is not fulfilling its promise. In fact, the whole process is likely to be more effective if its potential as a tool for management development is recognized in the early days of formulating the planning process itself.

Some have argued that as an enterprise moves beyond portfolio optimization, the term *strategic management* rather than *strategic planning* would be more appropriate. Irrespective of the labels, there is no doubt that unless strategic planning becomes embedded in the organization and culture of the company, unless it becomes a way of life for management, it is a sterile activity that does not warrant support. In order to achieve that kind of penetration, as much attention must be devoted to the process as to the substance of planning.

Strategic planning is a dynamic process. The issues, the priorities, and even the data base evolve with passage of time, changing circumstances, and changing management styles. Managers of technology cannot participate effectively in this process unless they perceive the present state of planning and develop the responses that are appropriate to emerging and felt needs. A focus on innovation and venture when management is deeply concerned about productivity and cost competitiveness may not be wrong, but it is unlikely to be effective without a lot of preparation and education first. In other words, technology strategy need not slavishly mirror more general business strategy, but it must be sensitive to the need to justify any contrarian posture. Even in a contrarian mode, much of technology's resources must be devoted to the support of strategy, and that support must be consistent with overall management concerns and priorities.

Notes and References

1. "Corporate Strategies Under Fire," *Fortune,* December 27, 1982, pp. 34–39.
2. Michael E. Porter, *Competitive Advantage: Creating and Sustaining Superior Performance,* The Free Press, New York, 1985.
3. *The Competitive Status of the U.S. Steel Industry,* National Academy Press, Washington, D.C., 1985.
4. "Alcoa Hedges Its Bets, Slowly," *The New York Times,* Business Day Section, October 24, 1985, p. D-1; "Alcoa: Recycling Itself to Become a Pioneer in New Materials," *Business Week,* February 9, 1987, pp. 56–58.
5. "Critical Transitions—Managing Change in High-Growth Firms," Booz-Allen & Hamilton, Inc., 1985.
6. "A Smart Cookie at Pepperidge," *Fortune,* December 22, 1986, pp. 67–73.
7. James Brian Quinn, *Strategies for Change: Logical Incrementalism,* Richard D. Irwin, Inc., Homewood, Ill., 1980.

Chapter

9

Role of Technology in Strategic Planning

One of the central themes of this book is the notion that every enterprise develops a distinctive set of features that I term *the concept of the enterprise*. These features start with the idea of a product or service that will be offered and the function it will provide for users. They also include the mode of competition adopted, internal behavior, criteria for measuring success, and characteristic response to the outside world. In abbreviated form they constitute a mission statement, the starting point of a strategic plan.

Technology, inescapably, is a central and pervasive ingredient in the concept of the enterprise. As I have defined it, it is the system of capability that enables the enterprise to serve its customers. Thus whether its manufacturing involves discrete or continuous processes, whether it offers components or systems, and whether it offers custom-designed or mass-produced products powerfully affects the character of the enterprise, a concept discussed in detail in Chap. 3. Any attempt to change these market entities will prove complex and traumatic and has profound implications for technology.

The mode of competitive response is also deeply intertwined with technology. Does a business attempt to lead through technological advance or achieve competitive success by being an effective follower? Will it focus on superior product attributes for the most demanding application or instead emphasize lower cost and undistinguished product performance? Perhaps most subtle of all, will it choose a unidimensional approach to success through one or a very few distinctive technical features, such as product performance, or a multidimensional one of many different elements that provide a menu of potential values to the customer. Virtually all businesses start with the former, but most must segue to the latter at some point.

The centrality of technology to manufacturing enterprises has been apparent for centuries. Its significance for services is much more recent—with the exception of telecommunications. The advent of information processing has elevated the importance of technology to service businesses as well. For instance, the sophisticated passenger reservation systems developed by American Airlines, United Airlines, and others are totally dependent on complex and powerful information processing systems. These systems have proved to be profitable in their own right; but even more, they have been strategic weapons of enormous power, and they have broadened dramatically the range of additional services that can be offered, not only to increase revenue but also to foster customer loyalty. The entire business of Federal Express is predicated on being able to know in great detail, and in close to real time, the status of packages. In American Hospital Supply a sophisticated order-entry system facilitates very tight inventory control by customers.

Technology Roles

It follows, then, that technology must play a central role in strategic planning, which addresses the fundamental questions of how to establish a sustainable competitive advantage and how to ensure the survival of an enterprise. In many cases, technology is a forcing function, creating the need to change; in virtually all cases it is an enabling resource without which change is impossible.

In this chapter we will explore in more detail what roles technology should play in strategic planning and then examine how management should go about ensuring that technology fulfills them. There are three roles: keeping present products healthy, developing new products, and providing significant advances in operating effectiveness, advances which raise the competitive capability of the enterprise to a new level.

Keeping products competitive

The most obvious role of technology is in keeping products competitive. The conventional elements of competition in products are features, performance, and product cost. Success demands attention to technology in both product design and manufacturing processes.

The technical work involved in keeping products competitive requires addressing two kinds of issues: Are we ensuring that conventional technologies currently incorporated in our products, processes, distribution, and service are being used and extended appropriately and effectively—that is, exploiting technology maturation? Are we aware of, appraising dispassionately, and developing when appropriate emerging technologies that could supplant those now being used—undertaking technology substitution?

Technology maturation. In order to explore the subject of keeping products competitive, we need to consider more critically the problems of applying the concepts of maturation introduced in Chap. 2. Studies of both industries and technologies indicated that they go through a process of maturation that follows the familiar S curve (Fig. 9.1), but that their dominant features change in different parts of their life cycle—from early emphasis on product attributes to later emphasis on costs and process technology, to final emphasis on financial management of capital-intensive industry. Applying this understanding, gained from retrospective studies of individual technologies, to present circumstances, and especially applying it prospectively, is difficult.

First, evaluation of where a technology stands in comparison with theoretical limits is not always easy. Many technologies—even quite old ones—are still below theoretical limits, but the practical barriers to approaching those limits are formidable. For example, the magnetic properties of a so-called Epstein strip used in laboratory evaluations are far above those attainable from a steel mill producing strip for magnetic sheet, but closing the gap is still a tortuous, formidably difficult process. (Metglas, a new class of metals that are amorphous rather than crystalline, may solve the problem by a radically new approach, but one quite remote from a direct attack on the problem.)

Nickel-based alloys have been regarded as a barrier to increased operating temperatures in jet engines for thirty years, but ways continue to be found to increase their temperature capability. Furthermore, according to the conventional wisdom, the way around the temperature barrier was to go to metals, such as chromium or tantalum, with a higher melting temperature. Instead, a more cost-effective

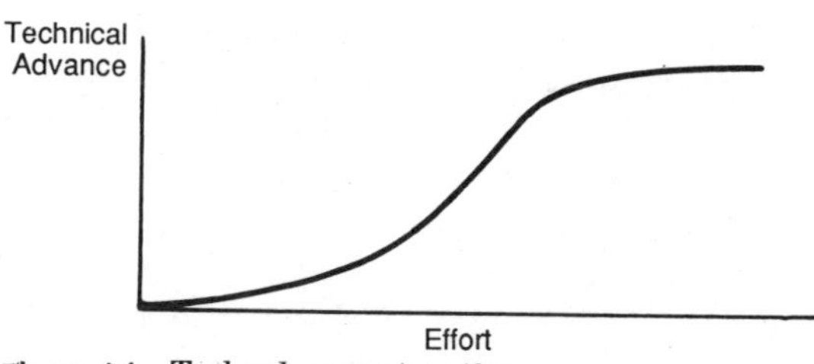

Figure 9.1 Technology maturation.

solution proved to be blade cooling, a design concept that had to come from a different technology. In addition, it also required the development of an entirely new technology—laser drilling—which could not have been anticipated in the mid-fifties.

In the early seventies, as people began extrapolating ahead to the appearance of micron and submicron spacing in integrated circuits, great reservations were expressed over how far optical techniques for processing could go. The capability proved to be much greater than even knowledgeable professionals expected. Thus the path to the future is always murky and full of surprises—both pleasant and unpleasant. The progress of technology is like an endless mystery story, with some clues proving valid and others spurious, and with new puzzles replacing old.

A location high on the maturity curve is not necessarily precarious unless potentially attractive alternatives exist. Tungsten has been the material of choice in incandescent lamps for seventy-five years, and despite many exploratory excursions, no acceptable alternative has been discovered. Ceramics have attractive strength, stiffness, and thermal characteristics for use in jet engines, but lack of ductility remains a formidable barrier to their replacing "mature" refractory metal alloys.

A search may not turn up a single potential invader, but on the other hand, it may turn up more than one candidate, without any clear indication of which might win. In most cases, there is insufficient information to resolve the uncertainty about the potential of new technology by further study and analysis. It is necessary to invest time and money in generating information from actual attempts to apply the technology, but resources are rarely available to pursue all possibilities. Thus even with the finest strategic planning for technology, success is a matter of probabilities.

In evaluating the vulnerability of technology, it is important to consider whether one is looking at a material, a component, or a system. The vulnerability of a system is considerably less. A system comprises a carefully articulated group of technologies. Consequently, its performance is determined by a variety of features, and its vulnerability to attack is distributed among the group of technologies. Furthermore, any technology seeking to penetrate the system must meet twin tests: improved value compared with present technology and compatibility with other elements of the system. Those criteria are apparent in the current thrust to introduce opto-electronics into telecommunications.

In contrast, a material or component represents a concentrated target, and therefore is more vulnerable. Even so, if the figure of merit that integrates the value of the technology encompasses a variety of attributes, the comparison is complex. For example, even in something as mundane and apparently simple as substituting aluminum for copper in the field winding of a motor, the comparison involves more than the differences in conductivity and in price. More aluminum is needed to achieve a specified performance, not only because it is a poorer conductor, but also because aluminum wire is springier and does not pack as densely in the slots. Furthermore, aluminum is difficult to join; it also flows at room temperature and thus may loosen in a mechanical connection and create a high-resistance electrical connection, which can be a fire hazard. Concern over the possibility of such unreliable connections prevented use of aluminum long after it was economically attractive and otherwise functionally acceptable.

Technology substitution. Substitution of one technology for another rarely involves a simple one-for-one replacement. A newcomer typically brings along some disadvantages that either must be overcome or the customer must be induced to accept. It also includes some new strengths, and ability to capitalize on them usually determines whether the attempt at substitution will succeed. DuPont's Corfam had the great advantage, in comparison with leather, of close control of properties for different applications, consistency from piece to piece, and availability in large sheets, but these proved inadequate to compensate for poor conformation to vagaries of foot shape, heat buildup in the shoe, and high price.

Comparing the value of a conventional technology and a new one also involves some important economic and psychological burdens as well. One could argue that investment in present equipment represents a sunk cost that should not enter into evaluation of the attractiveness of a new technology, but it sometimes does. The conventional technology also provides employment for people who are familiar with its use both in manufacture and in application to customers' needs. People responsible for investing in the conventional technology, as well as those employed by it, will perceive the newcomer as causing nothing but trouble. Consequently, their calculation of cost comparisons of the old versus the new are likely to reflect their preferences for the old. Despite the apparent objective quantification of a cost comparison, all such comparisons require a set of assumptions that can strongly influence the result.

Similarly, projections of customer response by salespersons may be unfavorable and unreliable. General Electric's marine turbine business competed successfully for years on the basis of offering the world's most efficient drives, each custom-designed to the specific requirements of the customer. This approach inevitably created problems and delays in maintenance, because it was impossible to stock all needed parts. Eventually, cost pressures forced the business to reconsider its strategy and it switched to a carefully configured family of products that sacrificed something in customized performance, but that greatly improved repair time all over the world. The salesforce, who had feared the deemphasis of customized design, came to prize even more highly the excellent reputation for on-line availability that the new approach permitted.

Spurious product attributes may be invoked. When semiconductor memories were being touted to replace magnetic cores, the "volatility" of semiconductor memories, i.e., the fact that memory was lost when the machine was turned off, was highlighted as a fatal weakness. In fact, as it turned out, the permanence of magnetic-core memories was not a feature that was actually used in practice. The information in memory was downloaded onto tape before machines were turned off, anyway.

The psychological factors favor present technology. Technical people whose careers are associated with a given field always see continued opportunities for improvement. They are slow to accept that a field may be maturing, because that threatens their own feeling of self-worth. I once participated in and wrote the report for a National Academy of Engineering study of the future of civil aircraft manufacture. Distinguished aviation experts on the panel waxed enthusiastic about the bright promise of further progress in powerplants, in sophisticated materials, and in advanced electronics for aircraft control and flight management. One panel member pointed out that none of the improvements in aircraft economics and reliability that these advances would lead to would be visible to the passenger—a comment that was not received with enthusiasm. Technical people whose lives were intertwined with aircraft technology resisted any suggestion that the field was maturing.

In consequence, the task of comparing old to new is complex. The amount of information available is rarely adequate to support a clear-cut determination. Proponents for both old and new technology have strong psychological reasons for overstating their case. People whose careers and status are threatened by a newcomer will resist change. People whose careers are committed to the new will, in their sincere enthusiasm, tend to overstate virtues and underestimate problems. Perhaps most unsettling of all, statistically most attempts to replace a technological capability will fail.

Additional sources of value. The process of evaluating the competitive viability of a technology when it is embedded in a particular industry is even more complicated than the preceding discussion would suggest. The maturation of technology, and its application over time, is not a straightforward, single-dimension process. The S curve of the technology life cycle carries the implication that there is a configuration of physical attributes of the technology-performance features—such as dynamic range, efficiency, speed, stability, strength, corrosion resistance, ductility, life, power, and so on—that together with cost, remain relatively invariant as a set in determining value to the customer. The S curve concept postulates that improvement in these attributes increases value, but improvement is ultimately limited by what nature permits.

As we noted in Chap. 2, the characteristics of technological progress tend to change as a technology matures. Improvements in product attributes tend to be replaced by improvements in cost and quality and to become increasingly capital-intensive. Figure 9.2 suggests this sequence.

Thus even a qualitative understanding of where a technology stands in its life cycle can help guide planning. During the early stages, efforts to reduce costs are probably not an ideal use of resources unless costs are the major barrier to broader application. Conversely, a continued focus on physical properties to the exclusion of process improvements when a technology is maturing is also nonoptimal. When improvements become increasingly difficult and expensive, technologists should become more aggressive in searching for and evaluating quite different approaches.

Even this representation of technological maturation is inadequate. The configuration of technology-related attributes that provide value to the customer changes over time in ways which are very much under management's control and which become central to competitive strategy. In order to acquire the proper perspective, we must shift attention from technology as it affects a single product to technology as it affects entire product lines and the operation of the complete business. Table 9.1 indicates a list of elements, each of which influences customer perceptions of value and all of which are strongly influenced by decisions and actions regarding technology. This shift from a focus on

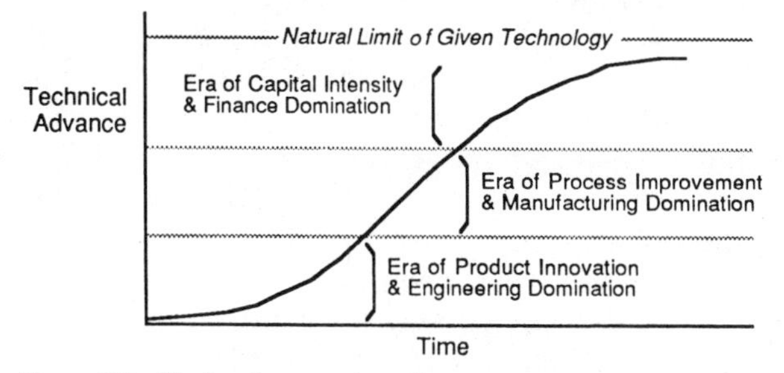

Figure 9.2 Technology maturation.

TABLE 9.1 Sources of Competitive Advantage

- Breadth of line
- Product line structure
- Service
- Application information
- Commonality of parts
- Intergenerational compatibility
- Ease of upgrading
- Ease of maintenance
- Consistent quality
- Faster response
- Flexibility

technology vis-à-vis the development, design, and manufacture of products, to technology vis-à-vis the business as a whole and the way it offers value, is one of the important changes needed in managing technology.

In addition to raw performance and cost, a number of other dimensions can become important. First, the technology can be represented in a growing *family* of products, which cover a wider range of application and give the customer greater choice in selecting the particular product that best meets his need, provided markets have been segmented skillfully. Porter notes this segmentation as one of the principal options for competitive strategy.[1] For example, the broad product line of major appliances for Sears is a powerful element of its strength in the market.

With proper discipline in design, a product can be made easily modifiable to meet special customer requirements. If various members of a product family are designed to use common components and similar configurations, they facilitate repair and maintenance and also reduce training time and errors, as operators switch equipment. These factors have been crucial in the selection of jet engines for aircraft and in Caterpillar's power in its market. But there are other aspects to customer value as well. Are products designed for easy maintenance? Do succeeding generations retain compatibility so that training and maintenance are minimized and systems integration remains easy? DEC has capitalized on this opportunity by imposing rigid discipline on software compatibility among succeeding generations of VAX machines and has achieved customer loyalty in return. Still additional values can be created. Do products allow easy upgrading so that future investment is reduced? Have capabilities been established to supply spares quickly and to assist in installation, training, and maintenance? The maintenance and support that Xerox provided for the early renters of its copiers were crucial to creating the image of reliability necessary for the business to grow. This list could go on, but the central point is that the effective application and management of technology involve far more than single-minded devotion to the raw elements of technological capability for a specific product.

These features can seem mundane and unchallenging to the advanced technologist, but they can be decisive to the customer when he is judging value. In fact, they can become powerful determinants of differentiation for commodity-like products.

Caterpillar long ago adopted metric measurements to facilitate its global operations. Its excellent logistics and effective, worldwide capability for maintenance, both grounded in technology, have been powerful competitive weapons. They not only generate customer loyalty, but also force a competitor either to price items so low that they compensate for these missing services or to invest in competitive products, as well as the logistical and maintenance infrastructure needed for support. This enriched menu of values offered to the customer also broadens the opportunity for further technological innovation.

Unfortunately, these additional dimensions of customer value also serve to constrain technology. Intergenerational compatibility in product lines, ability to upgrade, commonality of parts across product lines, and ease of maintenance all impose requirements that make the job of introducing improvements more difficult. Bell Labs traditionally had responsibility for ensuring the compatibility of any new technology with the installed system. Its rigorous enforcement of this discipline played a critical role in the growth of telecommunications, but it made the task of upgrading the system much more demanding.

Providing the basis for new products

The second role of technology is to provide the basis for new products. This work is popularly labeled *innovation* and is portrayed as the great challenge always facing the technologists. With the amount of attention so continuously focused on this activity, one could easily—and mistakenly—conclude that it is virtually the exclusive task to be done. By looking selectively and over a sufficiently long period of time, one can easily identify a long list of dramatic innovations that have traumatized some companies, glorified others, and in either case affected the structure of competition significantly. Nevertheless, for any given industry or company these events are infrequent, not necessarily well correlated with the effort devoted to achieving them, and much more obvious in hindsight than in prospect. These discontinuities are, by definition, not amenable to prediction. Thus they cannot constitute the chief objective of technological effort. It is, however, important to appreciate their significance when they occur and to anticipate their likely impact on a business. Meanwhile, failure to pursue the much more prosaic work of keeping costs competitive, offering desired product performance and features, and maintaining needed quality can mire a company or an industry in serious difficulty long before dramatic breakthroughs disrupt things. The litany of U.S. industries facing severe international competition—autos, steel, machine tools, textiles and textile machinery, electronics—does not result from major technological breakthroughs developed and exploited elsewhere. Instead it results principally from failure to pursue the prosaic needs and more modest innovations noted above.[2] In some respects, major technological innovation has been elevated into a chimerical Holy Grail that diverts attention from more mundane work. This is not to say that developing new products or dramatic new processes is an activity that can be relegated to a low priority. Rather, its pursuit must not obscure the continuing and virtually endless need for incremental improvement.

Nevertheless, one of technology's critical roles is to be the chief advocate and principal creator of technological innovation. Thus, as we have been arguing throughout, managers of technology must wear two hats: with one, they are members of the management team, providing the technical capability needed to accomplish operating objectives; with the other, they are urging development and introduction of new and better capability. This subject is so important that we will examine it in detail in Chap. 11.

Changing operational conventions

The third role of technology is less obvious and less well recognized than either of the first two. As elaborated in Chap. 3, conventions play a leading role in determining most behavior in an enterprise. Many of these conventions have their roots in technology—technology that was extant during an earlier, formative period. Many of them are concerned with such things as how much time an activity should take, how much things should cost, what level of quality is optimal. Some of these conventions were grounded on physical capability of technology that was available at the time. Others, most in fact, were largely determined by the information processing capability available at an earlier time. As we noted, these conventions frequently are expressed as set points—tolerance limits beyond which corrective action is required.

Technology can change those set points. Failure to recognize advances in technology that make set points inappropriate can make an enterprise traumatically vulnerable. Changing them is one of the most difficult tasks in management, largely because they are transparent within the enterprise even though they may be readily visible externally. Involvement in altering conventions is outside the traditional scope of technology, and yet technology is the principal driving force for changing them. This is the third role of technology in strategic planning: identifying the need for and initiating changes in conventions resulting from advances in technology.

Many of the advances in technology change physical parameters, i.e., they permit more rapid fabrication, closer tolerances, tighter process control, and so on. These advances nearly always have ramifications for other parts of the manufacturing system, but lack of knowledge about the nature of these consequences or difficulty in quantifying their significance often precludes additional adjustments that could be beneficial. Other advances in information processing technology can change the set points that affect an entire business enterprise, both operationally and strategically. Despite the remarkable progress that has been made in applying information technology, we are probably still below the inflection point on its maturation S curve. Great advances have been made in selected segments of business, e.g., in order entry, parts explosion, production planning, inventory control, and payroll. But we have barely begun to capitalize on the potential of information technology in speeding development cycles, permitting product proliferation, enabling faster response to market dynamics, and many similar improvements. Achieving these advances can only come from overcoming the fragmentation within technology—a subject which will be covered in more depth in Chap. 12 on achieving system integration.

Ensuring Effective Performance of Technology's Roles

We have laid out three roles for technology in strategic planning: keeping present products healthy, developing new products, and providing significant advances in operating effectiveness. With respect to all three of these roles, strategic management of technology requires applying two attributes: vision and perspective. We have defined technology as including, among other activities, work that extends capability, typically called R&D. Developing a sense of where technology is going, of what new capability will become available—in other words, demonstrating vision—is central to technology strategy. This applies not only to conventional technology presently in use, but also to new technology not necessarily within the purview of practitioners of the present art.

Perspective requires two things: an understanding of the dynamics of the impending changes, how fast they will occur, and what path they will follow; and a sense of proportion, concerning the significance of those changes for the enterprise, and the ramifications of their introduction.

How can management ensure that technology performs its role; i.e., how can technology management develop and communicate vision and perspective? Just as important, how do senior executives both ensure the existence of and accommodate to the very mixed signals it should be getting from the technical community?

Earlier discussion has postulated that there are rational bases for questioning the inputs from advocates of both conventional and new technologies. And yet sound decisions on the balance between the two must be made. What can management do to improve the quality of decision making? The first step is to recognize that decisions about this balance are management decisions, not technical ones. They require authoritative technical inputs, but those inputs are always colored by the values and perspective of the source. Therefore, senior management, both by its own behavior and by the design of its management system, must encourage bringing to its attention issues involving conventional versus new technology and continuity versus change.

An adversarial process has long been recognized as an effective means of dealing with a complex, murky situation such as this. This approach obviously requires that the various positions be represented by competent, effective expositors. The debate of old versus new must be viewed by management as necessary for survival. Management must organize work, assign responsibilities, and arrange status relationships such that the voices in the discussion are sufficiently comparable in competence and power to command attention. Thus an effective technical operation should be at war with itself—a carefully constrained conflict, to be sure, but a genuine conflict nevertheless. Konrad Lorenz demonstrated that ritualized conflict plays a necessary role in the survival of a species.[3] Perhaps the most important responsibility of corporate management is to nurture and yet control an analogous conflict in an enterprise—conflict is necessary for its survival also.

Inappropriate behavior

Management must deliberately design its system to guard against four kinds of inappropriate (possibly fatal) behavior with respect to technology. First is continuing to focus on technically unique product attributes, to the virtual exclusion of other considerations beyond the point where they provide competitive, attractive value to the customer. Second, at the other end of the industry life cycle, is continuing to build plants based on incremental process improvements when total industry capacity and capacity utilization do not warrant it. At this point, considerations of capital investing should become the dominant parameter.

Third is failing to recognize changes in the way customers derive value from a product or service. There can be important changes in what customers prize in the way of performance. As the use of personal computers has grown, limitations on their ability to exchange information have emerged as a major roadblock to increased usage. The air transport industry is a different example. The civil aircraft industry has made remarkable advances in the comfort, reliability, safety, and cost of air travel. But other more mundane problems are becoming important to travelers and are not receiving high priority. In the mid-eighties the Royal Swedish Academy of Engineering convened an international symposium on the technical details of aircraft design and operation. Toward the end of the session, a representative of a major European airline rose and said, "This discussion is all very interesting, but why does it still take thirty minutes to get your bag?" His question was completely ignored by the panel of world experts. Not a single participant even acknowledged that he had asked the question!

The fourth kind of inappropriate behavior is perhaps most egregious of all: to sit contentedly (or blindly) and assume success is so assured and one's position so powerful that no change in approach is needed. Some have argued that this kind of complacency is what made the U.S. automobile industry vulnerable to attack by the Japanese. A variant of this mistake is to assume that the same "formula" will work no matter what business one enters. Smaller high-technology companies are vulnerable to this syndrome, especially when their original charter was to serve a relatively narrow niche. As they fill that niche and aspire to larger size, they search for new markets and may not recognize the more multidimensional character of value that customers expect in these new and larger markets.

Let us now turn to two kinds of activity that management must foster in pursuit of vision and perspective.

Vision: Technology Forecasting

The formal exercise of attaining vision is usually termed *technology forecasting*. Every practicing scientist and engineer must be, at a minimum, an intuitive forecaster. Whether he is applying the state of the art, seeking to advance it, or developing new technology to supplant present art, he is making implicit judgments about the usefulness of present art, what improvements are possible and what the chances of success are, and the likelihood of success for developing new technology to supplant old. Technology forecasting covers a wide gamut of activities, from science fiction's improbable speculation about possible futures to mechanical extrapolation of trends. Many would argue that science fiction writers have been among our most helpful forecasters. But about fifty years ago technology forecasting moved from the work of individual visionaries, whether scientists or writers, to systematic organized activities. The National Research Council (NRC), an operating arm of the National Academies of Science and of Engineering, sponsored a formal study that led to the publication of *Technological Trends and National Policy* in 1937.[4]

Forecasting can aim at different facets—possible future capability, desired future capability, and expected future capability. In general, forecasting also reflects an uneasy blending of two contradictory approaches: one in which technological advance is assumed to result from the internal dynamics of technology itself (i.e., the intrinsic characteristics of science and engineering determine where they go) and one in which the expressed needs and desires of society determine the direction of technological progress.

The field grew rapidly during the sixties, and considerable effort was devoted to developing more carefully formulated techniques and methods of forecasting both in the United States and in Europe. Technology forecasting became a miniprofession, and a body of literature appeared.[5]

Instead of summarizing this literature, I will first comment on some characteristics of technology forecasting that warrant careful attention and then focus, just as in the case of strategic planning, on process considerations, which largely determine the actual utility of technology forecasting to an enterprise.

Early emphasis on methodology

Technology forecasting has exhibited some of the characteristics that seem to be common to any new intellectual endeavor (operations research showed similar characteristics 35 to 40 years ago). It initially attempted to portray itself as a new and different activity whose contributions were not obtainable from other sources. Its originators and early recruits displayed missionary zeal in proclaiming the value of the contributions and the validity of the intellectual constructs on which it was based. Since there was little in the way of demonstrable value in the early days of this new intellectual activity—forecasting was especially handicapped in this respect—great emphasis was placed on the rigor and sophistication of the methodology being developed. Within the community, high status was granted to those making methodological advances—or what were labeled as advances. These advances tended to make the methodology more complex and in fact often were dependent on computers for their application—an additional subliminal indication of methodological rigor and sophistication. If it required a computer to apply the technique, it must be sophisticated!

Despite the significant effort to advance the methodology and the apparent growth in sophistication, my own experience with much of that effort has raised two kinds of concerns. First, the early questions regarding some aspects of the methodology remained unanswered and worse, even unaddressed. For instance, I had meetings ten years apart with a widely known consulting group, and it was as though I were in a time warp. People discussed, with great enthusiasm, their advances in forecasting methodology over the previous ten years, but during that time they had not addressed the reservations and limitations expressed earlier. My concerns over validity were magnified by the experience.

The second concern is related to the utility of the conventional form of the output. Many forecasts are expressed in terms of probability of occurrence for a given event or given advance within a specified time. Typically, a long list of events is presented, which, in the form offered, has little value because it has no structure. The effort to develop such a structure is considerable, and when there are reservations about the validity of results to begin with, the effort may not seem worthwhile.

Scenarios

Scenarios represent a different species of forecasts. By definition they require some sort of a structure, an identification of the principal factors to be considered and of potential relationships among them. They require a sense of sequence and of connectivity between events. However, their value is closely related to their specificity to the circumstances under consideration. A generalized scenario about space travel or futuristic health technology may be stimulating reading, but of limited value to an individual company. Forecasting should begin with careful consideration of what purpose is to be served. This specification will determine the kinds of people who should participate and the questions they should ask themselves.

The forecasting process

Every strategic plan contains an implicit technology forecast. It assumes either that present technology and extensions thereof provide an adequate basis for competitive success or that some new technology represents a threat or an opportunity that must be addressed. Few, I believe, would argue that it is better to leave the technology forecast implicit and unarticulated. The real questions are what areas to cover in formal forecasts and how to develop the forecast, i.e., what process should be followed. For technologies already being extensively pursued, I believe in general that practicing professionals can be relied on to have a realistic perception of what the future holds—provided they show clearly that they are in good contact with the outside world. If one or two technologies are regarded as critical to the survival of the enterprise, a more formal process of appraising the future should be used. More systematic technology forecasting should be used for technologies that are newer and therefore less central to a company's traditional strengths, or that are changing rapidly. The basic choice is between relying on outside experts and developing it internally. If a firm has no people with knowledge of the subject, it has little choice, at least initially, except to go outside. Wherever possible, my strong preference is to rely principally on one's own people and to make selective use of external consultants as reviewers.

Almost every technical organization I've known has at least some people whose wide-ranging intellectual curiosity leads them to read broadly and to interact extensively with the external technical community. Even smaller companies often have greater internal resources for participating in technology forecasting than they realize. A common mistake is to assume that technology forecasting deals with exotic areas that are far removed from present interests and with projections far into the future. While those might be interesting, if they could be done, much more can be gained by looking into relatively mundane areas that are closer to home. Knowledge of the company and of its technical needs and limitations may be of more value than knowledge of esoteric possibilities remote from the ken of the company.

With an external forecast the problem is how much reliance to place on an inherently questionable document when the level of authority of those providing the inputs is impossible to judge. People offering these forecasts tend to place great emphasis on the sophistication of the methodology used to treat its inputs. Nevertheless, the forecast cannot be any better than those inputs.

Perhaps a more useful approach is to ask what technology forecasting should provide: a series of predictions about the future, whose reliability is inherently uncertain and untestable, or better insights into the forces driving the technology, the level of consensus among those in the field and the strength of their feelings, the critical barriers whose elimination would significantly change the picture, and perhaps most important of all, the likely path and rate of deployment—in other words, a combination of vision and perspective.

It is unrealistic to expect technology forecasting to predict previously unrecognized major breakthroughs credibly. For instance, the 1937 NRC forecast noted earlier did not recognize that penicillin had already been discovered and that Great Britain was already working on radar.[6] A more useful forecast is one that those working at the forefront of a field would probably not regard as a forecast at all. Even though the raw capability to do certain things has already been created, the path from initial discovery to applicability for a particular business is tortuous and strewn with errors and missed opportunities. Usually, new technologies are first applied in some high-leverage applications where a potential advantage warrants the risk. They then diffuse along a path of additional applications as their value is confirmed and enriched and the risks and costs of application are diminished. For most companies it is this path and the rate of deployment that are the crucial questions. A forecast that provides better insights into that path is of great value.

A technology forecast can be regarded as a document where experts predict the future or as an opportunity for systematic learning, not only about the likely future of a technology but also about the capability of an enterprise to deal with that technology. I find the latter more useful.

Content requirements. The type of forecast I have found most useful is short and focused (ten pages maximum, preferably five or six). It is meant to be read by busy people who will not extract kernels of insight from pages of trivia, irrelevancy, or unstructured estimates of the probability of events. Table 9.2 shows the items I believe should be included in a forecast. It begins by bounding the technology, i.e., in succinct terms, what it covers in the forecast and, equally important, what it excludes; that is important for the forecasters as well as the readers. The forecast should identify the key external economic and sometimes social and political forces that both drive and constrain the technology. Technologies do not appear spontaneously totally out of context. By my definition, trying to understand the dynamics of technology development is futile without placing it in a socioeconomic context—technology serves society.

TABLE 9.2 Content of Technology Forecast

- Bound the area to be covered
- Identify key external forces
- Identify driving forces and constraints
- Identify the likely sequence of early application
- Infer predictive events to look for
- Evaluate readiness to cope

Next, the forecast should attempt to identify the internal driving forces and constraints. As a new field of technology begins to emerge, the leading investigators quickly converge on the key problems that impede progress or effort tends to concentrate in areas of rapid progress simply because it is rewarding to work there. It is terribly important to identify the crucial roadblocks to development; frequently, those roadblocks remain untouched while progress on more tractable, but peripheral problems is made and publicized. Those who are not truly familiar with the field may be misled regarding the significance of the progress being made. Thus technology forecasting should include both intrinsic and extrinsic forces shaping the future of the technology.

In many high-technology industries, such as telecommunications, computers, instruments, and pharmaceuticals, market forces and technology progress are clearly closely coupled. Therefore, technology forecasting must take into account both market-driven technology and technology-driven innovation. In cases like these, where technology and market are clearly intertwined, an awareness of customer needs, and of the driving forces that shape and constrain them, should be an integral part of the forecast.

Some examples of work in forecasting that helped to call attention to key barriers follow. A forecasting group looking at local area networks for information systems noted that almost all effort was focused on applications in an office environment. The requirements for communication and control in a factory environment are quite different. In particular, a factory is an electrically noisy place, which inhibits accurate transmission; even more important, it requires a system that can recognize the urgency of different messages. Without the latter, no factory would dare entrust real-time control to a local area network. The work of the forecasting group helped to focus the laboratory's program on the special needs of factory communication systems.

In all of the discussion of robots, there is little reference to cycle time. To replace human hands which can move with remarkable speed and precision, experts in light assembly assert that a robot must have a cycle time of about three seconds. Unless this can be achieved, advances in other areas are to no avail. Thus improvements in cycle time are central in predicting the proliferation of robots.

Much attention is devoted to the promise of biotechnology. Little mention is made of the fact that these processes inherently occur in very dilute solutions; therefore, removing water is a major problem. A world-renowned molecular biologist calculated it would require the entire beer-making capacity of Great Britain to produce the world's supply of insulin—less than 10 tons per year.[7]

Identifying the critical barriers—not just in general, but in the specific applications a business is likely to need—is far more important than achieving convergence on the probability of occurrence of twenty-five different events.

The next element of a technology forecast should identify the likely sequence and timing of the early applications—where they are likely to occur and when. Again, the path of deployment is a means of telling whether or not the technology is proliferating as people expect.

The explosion of interest that followed the discovery of high-temperature superconducting materials stimulated many optimistic assertions about the economic potential of the breakthrough. Although one should not denigrate the enthusiasm of those involved, because it was an important and exciting discovery, *The Economist* provided a useful note of realism in evaluating its potential significance by noting some of the barriers that must be overcome before it can be used.[8]

Sometimes a forecast can help to make inferences about actions that are not visible directly. For example, in the mid-seventies a forecast in materials processing identified reaction injection molding (RIM) as a promising new process. At that early stage the forecast noted that pioneering developers were having to work with materials that clearly were not optimal because no market pull had induced materials suppliers to formulate materials especially for RIM applications. The forecast alerted people that the appearance of such formulations would be confirmation of the growing importance of this new technology.

The final component of a forecast should be an evaluation of the forecast's implications for the business and its readiness to respond. In other words, the forecast should address the if-then question and consider very dispassionately how well prepared the enterprise is to cope. The spectrum can run from adequately aware and responding appropriately, to somewhat aware but lacking resources to respond, to unaware and likely to be blind-sided. Rather than simply point to the level of capability to respond, the forecast should recommend courses of action for various components. Unless the forecast gets down to specific actions for particular businesses, it can easily become an empty exercise.

Process requirements. The process used for preparing a forecast will strongly influence its validity and its impact. The tendency is to rely on the R&D organization since it is most likely to be working at, or familiar with, the forefront of technology. This approach is apt to lead to a forecast that is naive in some of its predictions about application and that encounters problems of credibility in operations. A more productive approach is to identify a mixture of technical specialists, some expert in developing the new technology—or as expert as possible in side the company—and others expert in pioneering applications. My own preference is to have the chairman come from operations. It should be made clear at the outset that the goal is to establish the present state of knowledge that can be extracted from inside the company and not to make a federal case of studying the problem. If it turns out that the level of expertise is thought to be inadequate compared with the potential of the technology, that in itself is important information.

The team should be asked to complete its work in short order and rely heavily on the knowledge people already possess or have readily available. Some will argue such a superficial approach will inevitably lead to a shallow and vulnerable output. I would respond by saying that a more in-depth approach will rapidly generate costs that few managements will tolerate for long and will inevitably narrow the range of subject matter that can be addressed. Perhaps most important, the process will rapidly collapse to the work of a few highly motivated people whose output will be regarded as a questionable, specialists' document.

The purpose is not to produce the world's best forecast. The purpose is to systematically establish the perceptions of the most knowledgeable people in a company in a way that will gain credibility with a large and diverse audience. Since the forecasts are inherently estimates with wide confidence limits, one should be wary of spurious accuracy. Efforts to push a forecasting group to full rigor in specifying events and precision in estimating time intervals are largely wasted energy.

The initial group should contain primarily technical specialists. Where appropriate external experts can be identified, I have found it helpful to have a forecast reviewed on a confidential basis. A review helps to reassure the forecasting team and lend credibility to its output. However, effective reviewers are not easily located. Sheer knowledge of a field is not enough. Many reviewers simply vent their personal prejudices without providing a balanced perspective on the forecast.

I have also found it useful to have an additional internal review by managers representing a different cross section of the company from the original team. Service on a review board gets managers involved in the process, helps make them converts, and broadens the constituency for the forecast. The board adds credibility to the final product. But a review also produces valuable benefits for the content of the forecast. Since managers usually have a different perspective on the business implications of a technology and on the readiness of their company to respond effectively, their inputs can help make the findings more realistic. This internal review should not be construed as approval before release. The original forecast team must feel that the output is its product, but that it has had the benefit of constructive criticism.

Benefits of forecasting. I believe a forecast should be widely distributed internally. Some will argue that the information is highly proprietary, but the competitive threat can be clearly defined and controlled, if necessary. It is difficult to identify the exact audience that should receive the forecast, but I think it is better to err on the side of overdistribution. One survey of usage turned up a spectrum of favorable reactions, which covered the span from top executives to first-level managers in R&D. Uses were much more diverse than anticipated: for example, helping an important customer, assisting a senior executive in an important external presentation, reorganizing an important operating component, or restructuring programs of a component in the corporate laboratory.

The benefits of preparing technology forecasts in-house extend beyond the learning that is gained within an organization and the creation of a forecast that is customized to the local business and culture. The exercise provides a measure of the caliber of internal expertise and of its coupling with the outside. If a forecasting team is unable to identify authoritative external reviewers, that, too, is a warning flag.

The process of preparing a forecast aids in building internal communications networks. This is particularly important for emerging technologies. Typically, the early efforts in a new field are small. People may well be somewhat isolated, and participation in a forecasting effort fosters the sense of community that practitioners of a technology tend to develop over time. It also helps to improve the perspective of the team members on the important trends likely to emerge. Every forecasting team I have known has expressed gratitude for the opportunity to participate and has recommended that the work should be updated periodically.

Technology forecasting in operations can also be helpful in stimulating work by vendors and by the corporate laboratory. One advanced materials group makes a forecast of what improvements in properties should be possible. It then commits itself to achieve those improvements, with the product development people who will use the material. The laboratory selects the avenues of development it feels are most promising, but it also goes to vendors and to the corporate lab and asks for help. For vendors, there is the incentive of potentially attractive orders downstream; for the corporate lab, the forecasting group in operations can speak in precise terms about what is needed, the avenues that are being pursued in operations, and the kind of higher-risk work that might be appropriate in the corporate lab. By stating specifically the properties of materials that must be achieved and the urgency of the work, the operating group is able to exert leverage on the corporate lab.

Forecasting as an ongoing process

Forecasting should not be regarded as a one-time activity. The output should be reviewed periodically with the chairman and members of the panel and reopened when enough has transpired to warrant another look. The ongoing activity should seek a blend of continuity and new blood. I have found it useful to rotate the chairmanship. The task is time-consuming, and rotating spreads the workload. It also ensures a new perspective. Beginning with a completely new panel negates the institutional learning that has accumulated and reduces the additional learning that can occur from asking why an earlier forecast went astray—as it inevitably will in some respects.

In internal forecasting, effort tends to generate its own agenda. Typically, as a team is at work it uncovers a tangential field that is not in its scope or a special subset of its own work that warrants more focused attention. For example, a panel examining the future of applications in integrated circuits identified solid-state sensors, integral with such circuits, as a special field to pursue. Similarly, a panel considering information processing identified local area networks as a field to explore more carefully. The recommendation went further to suggest that for this particular company, factory-oriented networks should be distinguished from office-oriented networks, which were receiving most attention by vendors and new ventures.

If technology forecasting is regarded as a sophisticated way to anticipate the future and to both open unexpected opportunities and prevent nasty surprises, it is doomed to failure. If it is regarded as a systematic way for appropriate members of the technical community to clarify their thinking about the future and to orchestrate their communication, and at the same time, as a means for educating and interacting with management, it can be a useful tool.

As we have seen, effective technology forecasting is not something to be done by a group of experts set aside for that task. Such a group remains objective only until it has completed its forecast, at which time it becomes a defender of its output. A temporary group of people already working on various aspects of a technology is less prone to defending its output, because that output is quite peripheral to their principal work and less important to their own future. The practical realities of budgeting pressures also favor orchestrated but dispersed activity. The visible overhead to provide continuity, to organize a team, and to produce an output is small, and the real work is spread in small increments over many components. In contrast, a special forecasting group is concentrated, visible, and vulnerable when budget pressures appear.

Perspective: Appraising the Competition

Perspective rests partly on a sense of history, which should be an integral part of technology forecasting, and partly on a clear-eyed view of what is going on in the competitive environment. Since technology is an almost universal competitive weapon, it follows that strategic planning, which also requires careful assessment of competitive status and of likely future actions of competitors, must include technology in that assessment. The conventional way to evaluate competitive position is to relate it to market position. In the strategic plans I have reviewed, the usual approach is to say, we have the leading share of the market; therefore, we are ahead technically. That does not necessarily follow. Market position reflects effectiveness in a composite of factors, and competitive technological position, as one of the most effective weapons for changing market position, will often lead or lag behind it. Evaluation of competitive position must proceed along three paths: (1) the status of conventional technology versus present competitors; (2) the potential new competitors resulting from advances in new technology that are emerging; and (3) the potential new competitors resulting from changes in strategy that could create confrontation.

Conventional competitors

Evaluation of competitive status among conventional competitors is relatively straightforward, if one but puts his mind to the task. Their products are in the marketplace available for comparison. The quality of competitive products and of their apparent strategy in using technology are accessible for analysis. Nearly all industrial concerns develop relatively consistent patterns of behavior that are evident if studied over time—the conventions we noted earlier extend to visible external behavior as well. Competitor participation in professional societies and at technical meetings provides additional clues. The patent group often has insights into the activities of competitors from the patent literature. Salespersons can be helpful sources of information and can even probe customers for specific information if it is deemed desirable. Competitors' plants can be visited during "open house" to learn something about their processes. Competitor "house organs," newspapers from communities with competing factories or laboratories, annual reports, and presentations to security analysts can all be helpful. The carefully attuned mind can often learn more from a competitor's publications and comments than expected.

In short, within its own staff a company has a remarkable array of potentially useful information. The challenge, as in any intelligence activity, is to dig the information out of the woodwork and assemble a mosaic that provides some insights. This can never eliminate all surprises, but it does help keep the organization calibrated. Nevertheless, lack of persistent surveillance of competitive position tends to be a continuing weakness in the strategic planning with which I am familiar. Just as with other aspects of strategic planning, conducting this appraisal is probably more effective if it calls forth the knowledge and insights already within the organization, rather than establishes a separate expert group. In many cases management can stimulate more effort, and certainly more attention, simply by making known its interest in competitive information and by continually raising questions about how in-house technology stacks up.

Less obvious threats

The evaluation of offshore competitors is a particularly vexing problem. Unfortunately, most competitive evaluations have a domestic bias. Usually, if potential foreign competitors are not yet established in the domestic market, they are not regarded as competitors. The response I usually get when querying about foreign competitors is, "But they aren't active in our market!" Advances in technology can make it possible for a company not previously thought of as a competitor to become one. The classic examples are in materials substitution where higher-strength alloys enable aluminum to replace steel, plastics to replace both of them, etc. Less obvious cases arise from the substitution of solid-state electronic components and subsystems to replace electromechanical and electromagnetic controls and mechanical watches. The success of video camcorders in replacing Super 8 movies is the first act of a dramatic confrontation between silver imaging and electronic imaging.

This type of technological competition is particularly difficult to recognize, much less evaluate. The advances occur in fields that company experts often do not monitor and are poorly equipped to evaluate. The competitive style of the new threat may be quite different regarding pricing, applications support, pace of change in introducing advanced products, and so on.

No sure-fire system exists to prevent nasty surprises. One can reduce the likelihood considerably by asking insistently if possible threats have been taken into account and if a deliberate search for new threats has been conducted. In single-industry businesses, additional protection can be gained by assigning formal responsibility for monitoring new technologies to key people. The essential criterion in selecting them must be that they are already virtually performing the role. This is not a task that can simply be assigned. (Thomas Allen at MIT has long advocated the role of such "gatekeepers" in keeping in touch with external technology.[9])

In multi-industry companies, internal technical workshops and symposia can provide a mechanism for tracking technologies not being applied in an individual's own component. In large businesses with a separate central R&D component, this group is often charged with monitoring technological advances across a broad front and alerting appropriate businesses as progress in new fields warrants their attention. A central facility for technical information may also be charged with the task.

The mere fact that mechanisms exist to alert the technical community to advances in technology from a potentially new competitive quarter does not necessarily mean that their inputs are automatically included in strategic planning. Only if the technical organization is charged with, or simply assumes, responsibility and performs the necessary work to convert this competitive intelligence into a form useful to strategic planning—how significant the advance is, how close to application it is, what specific threats and opportunities it presents, what possible strategic options are available in response—will the input have an impact on strategic planning.

Another area of competitive appraisal is perhaps most subtle of all. It involves changes in strategy that could bring previously noncompetitive or limited competitors into more extensive confrontation. The changes introduced by Harry Gray in his attempt to diversify United Technologies brought it into much broader confrontation with GE than just jet engines, their conventional field of competition. Conversely, the relatively greater emphasis of Westinghouse on energy-related businesses compared with GE reduced their areas of competitive confrontation. The increased attention of foreign companies to obtaining a foothold in the United States is bringing them into confrontation in domestic markets, where heretofore their impact has only been in foreign markets. For example, the sale of A300s by Airbus Industrie to Eastern changed the competitive equation in the United States. Similarly, in electrical equipment and electronics, the efforts of Philips of Holland, Siemens of Germany, Brown Boveri of Switzerland, and ASEA of Sweden to establish a presence in the United States have affected the nature of competition.

Strategic business planners and line executives are more likely to be first in noticing these kinds of changes. Technical managers should seek their inputs as a dimension of strategic technology planning. A GE vice chairman alerted me to the changing status of United Technologies by saying, "You'd better begin to consider UTC more broadly, because it is becoming a different company and one we'll be meeting in new areas."

Conclusion

It is a mistake to believe that strategic planning in technology requires the creation of an elaborate new planning capability. The foundation should rest on the way line managers of technology go about the task: the questions they ask themselves and others. Understanding the three roles of technology in strategy and the work and the viewpoint needed to fulfill those roles is the indispensable starting point. Similarly, technology forecasting and competitive appraisal are best done by marshalling and focusing in-house skills and interests rather than by installing a group of professional specialists for the work.

Notes and References

1. Michael E. Porter, *Competitive Advantage: Creating and Sustaining Superior Performance,* The Free Press, New York, 1985.
2. "The Competitive Status of U.S. Industry—An Overview," National Academy Press, Washington, D.C., 1985.
3. Konrad Lorenz, *On Aggression,* Harcourt, Brace & World, New York, 1963.
4. W. F. Ogburn et al., "Technological Trends and National Policy," U.S. National Research Council, Natural Resource Committee, 1937.
5. For example, Robert U. Ayres, *Technological Forecasting and Long Range Planning,* McGraw-Hill, New York, 1969; and James R. Bright and Milton E. F. Schoeman, eds., *A Guide to Practical Technological Forecasting,* Prentice-Hall, Englewood Cliffs, N.J., 1973.
6. Ayres, op cit., p. 12.
7. Lecture by Sydney Brenner, Director—Laboratory of Molecular Biology, Cambridge University, at Seminar on Emergence of New Technologies, September 16, 1982, at Chateau De Marcay, Chinon, France, sponsored by Centre de Formation aux Réalités Internationales (CEFRI).
8. "Not-so-Superconductors," *The Economist,* June 13, 1987, pp. 93–99.
9. Thomas J. Allen, *Managing the Flow of Technology,* MIT Press, Cambridge, Mass., 1977, pp. 141–181.

Chapter

10

Integrating Technology with Strategic Planning

Getting Started

Strategic technology planning is an iterative—sometimes parallel—process of thinking about the nature of an enterprise and about the role of technology in that enterprise. Technological inputs should help frame the basic strategy of the business, and then technology strategy should contribute to its implementation. Of course the nature of technology strategy differs depending on whether it applies to a single strategic business unit (SBU) or a multibusiness corporation.

One common mistake in strategic planning is to assume that its success rests on the skillful application of various analytical techniques. It is comforting to assume that strategic planning is a deterministic process in which application of a logical sequence of rigorous screens leads unambiguously to the best strategy for a given enterprise—comforting, but wrong. The actual practice rests on hard work and hard thought in a very indeterminate universe. The objective is to understand, as comprehensively and realistically as possible, the nature of the business under scrutiny as well as the future competitive and economic climate in which it will function. In other words, you try to look at the business from different vantage points to try to ensure that you truly perceive it adequately. The planning techniques, such as industry-attractiveness–company-strength matrices, learning curve, market share, stars–cash-cows–dogs classification, are mere tools to ensure that you have really established the vision and perspective on which sound strategy is based. They can help diagnose a situation and perhaps indicate the direction in which you might like to go.

Developing a response is a creative process of quite a different nature. The task is, on the one hand, to understand both the opportunities and threats that may confront the business from either internal or external forces, and on the other, to recognize the constraints arising from resource limitations, management limitations, external market and competitive considerations, and so forth. The eventual plan that is established is a synthesis of all this information by a process that defies description—one of my former associates used to refer to it as a "transcendental experience."

My own experience suggests that the value of strategic planning goes well beyond the creation of "the plan." I was always struck by the fact that in GE we talked mostly about strategic planning rather than about strategy. The planning process can be a powerful tool for educating management at all levels. It can force consideration of shared purposes, objectives, and priorities. It can make people more aware of traditions and assumptions that are so taken for granted that they have almost become transparent. Planning should make one aware of inadequacies in information and risks inherent in possible courses of action. It can clarify the assumptions which underlie a plan and identify the trigger points which would require its reexamination. Finally, planning can force consideration of alternatives and make people more aware of the spectrum of options available to an enterprise.

The process of preparing a plan requires a management team to develop a common language for talking about its business. Done properly, a planning process and its output can be a powerful instrument for communication within the organization—for saying: this is why this organization exists, this is what it must accomplish, this is how we intend to go about it, and these are the results we expect to achieve. In that same sense, the plan can also be useful in communicating with other components. It can help clarify interfaces, elicit support, and sell clients on what they should expect.

Let us now turn to the *what* of planning, namely, what should the technological component of the strategic plan include? and then the *how,* i.e., how does one go about creating a plan?

Incorporating Technology into the Strategic Business Plan

The tack to take in integrating technology with strategic planning is to create a single instrument of which technology is an integral and indispensable component. Too often this activity is approached as though the objective were to achieve the unification of two separate documents produced by two separate processes. Thus technology managers should think of their task as helping in the creation of the business strategy, with a further elaboration to guide activities within the technical community, not the creation of a technology strategy to be unified somehow with business strategy. Obviously, both the general manager and the other functional managers must share this view, or the technology manager cannot play the role he should.

What are the basic technological concepts of the business?

Every strategic business plan should include, first of all, an articulation of the basic technological concepts of the business, together with a statement of the data and assumptions which underlie the chosen approach to technology; that is, how does the business propose to use technology in providing value to the customer, and why has this particular approach been chosen? Every business has an internal technological system. It begins with inputs at some stage of completion—raw materials, semifinished products, components, or subsystems—and adds further value by fabrication, assembly, modification, and distribution. Therefore, the starting point of the business is itself a strategic technology issue that underlies the rest of strategy. The parts of the technology system are at varying stages of maturity, are of varying degrees of criticality to competitive success, and represent differing degrees of opportunity or threat for the future. This technological basis for the business must be laid out for evaluation. The discussion in Chap. 3 of alternative ways of conceiving of a business provides a means of approaching this task: businesses that are resource-focused, that are technology- or product-focused, or that are systems- or market-focused. Whatever the present technological foundation of a business, will the assumptions and relationships on which that foundation is based continue to be valid for the planning period under consideration?

The forest products industry provides an illuminating example. The lumber companies in the northwest have operated on the assumption that owning timberland and growing trees effectively were central to success, because they also have assumed that the demand for pulp would continue to grow virtually indefinitely, that the northwest's climate provided an inherent competitive advantage in growing trees, and that pulp from those trees had an inherently better quality. If those assumptions are correct, the basic technological character and priorities of the business are more or less determined. Owning or controlling good timberland and effective tree farming remain central to competitive success. If those assumptions are questionable, as may now be true because of the emergence of a forest products industry in the tropics, then the entire concept of the business must be rethought.

Similarly, for decades General Motors operated on the premises that economies of scale would provide an economic advantage, that internal competition among brands would provide incentives for managerial effectiveness, and that a broad product line was needed to sustain the edifice being created. As a result, its management structure provided for an intricate combination of centralized standardization and modularization to achieve economies of scale and decentralized design to achieve market segmentation and product differentiation. It also realized retrospectively that it had focused on one component (the car) in a complex transportation system and left the characteristics of the rest of that system up to others.

Technical management obviously is the key resource in delineating a business from this technical perspective, but it should be done under the leadership of general management. Two particularly egregious management failures often occur at just this juncture in the process. General managers may conclude (often because they are uncomfortable with or feel inept in dealing with technological considerations) that the seminal elements of strategy lie in marketing and finance and that technology strategy should be formulated subsequently so as to support or implement the business strategy. Conversely, technical managers may complain that they cannot develop effective plans until the business managers figure out where they will try to take the business. Both attitudes are failures in leadership. If technology is fundamental to the core features of the company, general managers cannot approach it as an afterthought when the really important decisions about marketing and finance have been made. By the same token, technology managers should view leadership in developing consensus on the basic technological dimensions of the business as one of their major responsibilities.

This assessment of the basic technological foundations of the business should conclude with an evaluation of the future prospects of the present core technologies. How close to their physical limits are they? How well understood are the phenomena they are associated with? What are the key barriers impeding performance that approaches the physical limits? What alternatives can be perceived on the horizon? These questions must be asked even though they are not easily answered. A management must be sensitized to its technological prospects despite the fact that it cannot be certain of them.

What is the projected environment?

The second technological component of strategic planning is an evaluation of the projected environment: the likely change in market requirements for technology, the implications of competitors' probable strategies, and the advances in technological capability that could present threats and opportunities. Our earlier discussion of vision and perspective focused on this component of planning.

Interaction of financial goals with strategy

These two elements—the assessment of the basic technology used by an enterprise and of the environment in which it will be operating–are the two major technology inputs to the formulation of strategic business objectives in terms of growth and profitability. Obviously, technology is by no means the sole determinant of growth and financial performance, but technological considerations should weigh heavily in establishing these objectives. These objectives, expressed within the context of the basic character of the business and the environment in which it will operate, determine the extent to which a trajectory must be changed. For technology this means determining how far applying conventional technology or extending it can go in achieving strategic objectives. Alternatively, will new technology be required to enable the enterprise to achieve its strategic objectives? Obviously, the process of determining strategic objectives for the business and the level of technological capability they will require is an iterative one.

The basic strategic issues we have been considering thus lead to an examination of the full spectrum of strategic options: should the basic concept of the enterprise be changed—its business portfolio or its basic operational approach—or should it continue along its present trajectory? The decisions made should reflect relevant technological considerations, and they will also establish the guidelines for a detailed strategic plan for technology.

Implementing Competitive Strategy

Once these three basic components of a strategic plan that have implications for technology (i.e., technological characterization of the company, technological environment, including specific threats and opportunities, and objectives for growth and financial performance) have been established, issues of implementing specific aspects of competitive strategy become dominant, i.e., what combination of product performance and features, cost, and customer segmentation will be used to create value, and how will technology be used to provide this value?

Now the critical technology issues become the balance between the pursuit of conventional technologies versus development of new ones; the competitive position one will seek in various technologies (leader, equal, or follower—it is often impossible and rarely cost-effective to achieve competitive leadership in all technologies) and the value to offer through technology; and the allocation of resources among various fields. A fourth element is beginning to assume increasing importance: planning and integrating the balance between internal resources and capabilities and external resources around the world that will be assembled to pursue technological objectives and obtain strategic advantage.

So now we have the basic elements of technology strategy—i.e., what strategic technology planning should attempt to produce: (1) a technological characterization of the business and the assumptions which underlie it; (2) an evaluation of the environment—market requirements, competitive factors, and technology; (3) the implications of the basic objectives for growth and financial performance for technology; (4) the technological threats and opportunities which must be addressed in order to ensure that the business has the capability it needs to be viable; (5) the competitive technological position to be sought and the value to be produced for the customer; (6) the balance of effort between conventional and new technologies; and (7) the allocation of resources to the various technical programs.

The Method of Strategic Planning for Technology

The language of planning

What kinds of activities are essential to establish these elements of strategy? First, one must develop a language for planning so that many people can participate usefully. Clearly, form influences substance in planning, and the tools employed influence the output. Classification systems that can group individual technical efforts into larger categories of various types are an important element of the language of planning. The form of classification chosen inevitably highlights a particular perspective; it calls attention to a selective point of view. Consequently, it reflects implicitly a sense of priorities about what is important. The work required to develop classification systems is an integral part of planning. In my experience, as situations change or management learning progresses, the classification system changes.

Display techniques should be developed to portray the total effect of a proposed technology strategy in a variety of ways, so that management can examine and pass judgment on the attractiveness and viability of a given strategy and so that one can communicate plans to others at a level of aggregation appropriate to their role. Although technology must fashion its own language, it should, wherever possible, use language and concepts being adopted for other aspects of strategy formulation, such as the matrix discussed in Chap. 8.

Figure 10.1 displays an array of circumstances that an industry-attractiveness–business-strength analysis could reveal and suggests the appropriate strategic response. The competitive strategic options that Professor Porter developed are an important part of the appraisal involved in selecting the strategic trajectory, and they form the basis for the strategy adopted for each entity.[1]

Technical classification systems

It is important at the outset to recognize that strategic technology planning is not just program planning done in a longer time frame. Earlier we considered at length the nature of program management and the kind of planning that it required. Strategic planning must group programs into larger aggregates. The process is interactive in the sense that strategic planning should influence program selection and priorities, but at the same time, the technical progress promised by programs should affect strategy. When technologists discuss technology strategy, they more or less implicitly use categories that they feel comfortable with. This stance requires a general manager or CEO to "come to them," so to speak—to learn their language and adopt their thought processes. Because the technical community is rife with dissatisfaction about the quality of communication with general management and unhappy about its lack of adequate participation in strategic planning, modes of discourse carefully chosen to improve communication with top management should be sought.

Obviously, general managers need to comprehend the strategic significance of what they are receiving. Just as important, they need to be able to relate it to information they receive from other sources. This latter point is particularly significant. General managers are information processing machines. They receive a flood of information. If the information is presented in such a way that a general manager cannot easily incorporate it into his personal processing system, i.e., if he has to repackage it before he can use it, he often will not bother or he will do so resentfully. The objective of technology managers should be to

STATUS:	Unsalvageable	In Trouble	Limited Growth	Attractive Growth	New Opportunity
ACTION:	Dispose	Fix	Generate Cash	Extend	Diversify

Figure 10.1 Business situation—strategic response.

identify the general manager's concerns and to describe them and the response of the technical community in ways that make the least demand on his time and energy—he is usually impatient and overcommitted. The presentation of the plan also provides an opportunity to educate him regarding impending changes in technology or in the competitive environment.

I would argue that technology-oriented categories are indispensable for planning. But they are not an adequate basis for developing technology strategy, much less for communicating with general managers—especially if they are not technically trained. The classical mode of categorization (basic research, applied research, and development) I indicated at the beginning of this book, is not particularly helpful in discussions with businessmen. In the first place, it is not complete because it does not begin to cover the full spectrum of technical work, which goes all the way to customer service. Second, it is more oriented toward staking out turf than toward identifying contributions to a business. A general manager is interested in what he receives for his investment, what risks he runs, and what options he has, i.e., he is interested in the outputs of technology. If he probes into how and why, then a discussion of inputs can occur to build credibility; however, it is difficult for him to relate technology inputs directly to his business, and he should not have to interpret "what does this mean for my business?" Even more important, choosing to communicate with the general manager about strategy by discussing individual programs in detail not only makes heavy demands on his time but also requires him to provide the strategic framework into which the programs fit. In practice, a strategic plan that is framed in terms of individual programs probably has not really addressed strategic issues at all. A series of classification systems which permit grouping individual programs into larger categories in different ways is an indispensable element of planning; in fact, it is one of the most important techniques for doing strategic technology planning.

The process of categorizing must satisfy two different constituencies: those doing technical work and those paying for it in expectation of valuable output. Those doing the technical work need to be able to see a sensible path from their work to larger strategic entities. Those paying for it should be able to ascertain the expected strategic contribution of the technical effort without necessarily going into the detail of how that impact will be produced. It may be helpful to think of categories for the former, which are more likely to have a distinctively technical flavor, as input categories, and the latter as output categories.

Input categories. The discussion of the technology life cycle in Chap. 2 provides a helpful approach to categorization of technological inputs. First, a list of technologies used in the products, processes, and operations of a business has to be developed. The terms used in most cases correspond fairly closely with conventional labels for areas of technical specialization. They may include, for example, heat transfer, nonsteady flow, integrated circuit design, high-speed circuits, signal processing, low-cycle fatigue, machining, welding, or application software. Second, by processes such as those discussed under technology forecasting, a vision of the future has to be created, together with a sense of how important each technology is to the business, where it stands with respect to competition, and what changes in status must be sought. The resulting pictures of technologies, irrespective of what labels are chosen, are most likely to correspond to the base and key technologies described in the A. D. Little conceptual structure.[2] They represent the present state of the art in terms of the technological capability that is the foundation for current competitive advantage.

To this list of technologies must now be added the emerging or adolescent technologies that represent the most promising candidates to replace existing ones. For example, in aircraft there are full electronic control, reconfigurable structures, and extensions of composite materials into primary structures. A vision of future technical impact must be extended to evaluating the competitive significance of advances in these technologies and any changes in resources needed to provide the required technical capability.

All this work of looking at technologies, and examining the implications of their future prospects accordingly, is in the nature of homework within the technical organization. Although it is indispensable, it may well be invisible in the final product.

Output categories. Eventually, technical categories must relate to businesses or groups of businesses so that management can compare the character and allocation of technical effort across them or across product lines. However, slavish use of current business organizational labels as the basis for categorization can be troublesome because reorganization can require relabeling or shuffling categories for quite extraneous reasons.

The primary audience to have in mind is not the technical community, but the rest of the organization and especially the general manager. He is asked to supply funds and he expects to be told what his funds will buy. The objective must be to display for him the priorities among businesses or product lines in the plan, i.e., how money is to be allocated among businesses and how the emphasis will change in the future. The plan should also indicate what he can expect in the way of technology output—lower risk, improved performance, new features, or new products.

In considering outputs, one immediate problem that arises is how to handle programs which have multiple outputs or whose outputs have multiple potential beneficiaries—a very common circumstance in more fundamental programs. This problem can be dealt with by a combination of allocating benefits on the basis of judgment and double-counting when there are multiple beneficiaries. The important point is that the basis for assigning programs be made known.

There is no simple formula for devising categories. They often use labels that depart from classical names for technical work. Thus one may talk of laundering fundamentals, high-pressure discharge lamps, high-temperature alloys, switching, or power transmission. In telecommunications, examples might be network architecture, switching systems, transmission systems, or optics technology. Aircraft might include aerodynamic design, aircraft control, optics, flight management systems, or materials. The programs grouped into these categories may be base, key, or pacing, but that feature is now less important.

Connecting Output Categories with Business Strategy

Let us review what we have been discussing before going on to the crucial phase: connecting output categories with business strategy. The challenge in planning technology strategy is to accommodate two very diverse intellectual activities and disparate modes of discourse—the thought processes and language of business with those of science and engineering. Technical ideas and technical programs must be couched in technical terms and must relate to technical objectives. But somehow they must be recast into the language and objectives of business. We must get from the world of gallium arsenide and recombinant DNA to the world of lower cost and higher performance or to specific businesses that are the intended beneficiaries of technical output.

Our discussion of technical inputs emphasizes the development of vision and perspective regarding the likely future prospects of particular technologies. This aspect of planning is focused on the needs of the technical community itself, to give it guidance in choosing directions and priorities in technical programs. The key transition point is in preparing each program plan. A plan must indicate the technical objective, of course, but it must also indicate the business objective. This applies even if the business objective is simply to strengthen the technology base underlying technical work, e.g., catalysis or arc phenomena or tools to test software.

Everything in the planning process after the preparation of the program plan should relate output to business strategy and evaluate allocation of resources, seeking optimum use of funds and people to support business strategy. The goal is to evaluate the adequacy of the technology plan in terms of four strategic questions: Does the plan support optimally the competitive strategy of the business (or businesses)? Does the allocation of resources and choice of priorities support the higher-level elements of strategy—for instance, change in portfolio or in the character of the enterprise? Does the plan demonstrate sensitivity to the contradictory need to improve and extend present key technologies and to create radical new technology? Finally, does the plan provide an effective balance between using inhouse capability and obtaining technology from external sources? Obviously, these questions cannot be examined by any single analytic technique. I have found that a variety of matrices are useful in evaluating proposed strategies.

Figure 10.2 enables one to array programs against the spectrum of options for competitive strategy. The labels for the columns should correspond with the terms used in business strategy. I have used a conventional set. The cells could be filled with dollars, person-years of effort, or percent distribution. The choice will depend on the audience. Technology managers preparing a plan will need all three. A general manager or CEO reviewing a plan will want levels of effort, perhaps with an illustrative program or the largest program mentioned by name. This matrix provides a quick visual display of the work planned and the allocation of effort. It enables management to question the appropriateness of both. For example, if the column on new products shows very little effort, is that a danger signal?

Figure 10.3 approaches the problem in a different way. It asks technical managers to consider, at least in a qualitative way, the likely

Program	Cost	Quality/ Reliability	Product Performance	Features	New Products	Total
A						
B						
C						
D						
Total						

Figure 10.2 Distribution of effort by stategy elements (expenditure in dollars)

Cost	→
Quality/Reliability	↘
Product Performance	↘
Features	↗
New Products	↗

Figure 10.3 Changes in level of technical effort for each strategy element.

direction of future allocations of effort, perhaps over the next five years. Generating the display forces management to think about needed shifts in emphasis and is a way of discussing with general management the possible need to shift emphasis in competitive strategy.

Figure 10.4 is a matrix that raises strategic questions of another sort. As I indicated in Fig. 10.1, one important step in strategic business strategy is to evaluate the prospects of a group of product lines or businesses and assign each a strategic trajectory from sell or otherwise dispose of to diversify into new markets or businesses. The matrix in Fig. 10.4 arrays programs against a spectrum of such strategic trajectories. The option of "dispose" is excluded because presumably it should receive little or no effort from technology. The same questions about the appropriateness of the allocations that were suggested for Fig. 10.2 are applicable here. And again, a display such as Fig. 10.5 is useful in raising questions about projected changes in allocations of effort among strategic trajectories.

Figure 10.6 suggests another way in which individual programs can be grouped into larger aggregates that can be related to strategy for an individual business, i.e., strategic program areas rather than individual programs. If I were general manager of lighting, say, I would want to know what relative emphasis was being given to various product lines and how effort matched with basic competitive strategy over the five years normally included in a strategy.

Program	Fix Up	Hold Position	Grow Share	Extend	Diversify	Total
A						
B						
C						
D						
Total						

Figure 10.4 Distribution of effort by stategic trajectory for market segments or business units.

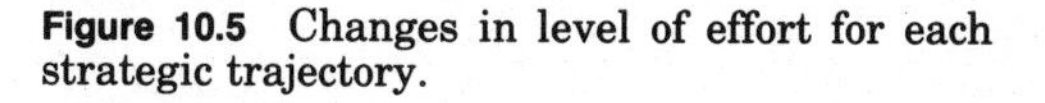

Figure 10.5 Changes in level of effort for each strategic trajectory.

The work the technical organization has to do to answer the question, what can I expect from you in the way of output for my investment in technology? exerts a healthy discipline on the technical organization. Year-by-year expenditures on a single program can seem small and defensible. Aggregate expenditures for five years on a group of programs can become considerable and may well lead a boss to ask, what do I get for my money? If the results seem thin, then maybe the allocation is wrong or maybe the technical organization should rethink what it intends to produce.

Of course, if I were a manager of a diversified company of which lighting were only one part, I would like an additional level of integration so that I could compare proposed effort for, perhaps, consumer products against other business aggregates, as in Fig. 10.7.

The amount of detail one chooses for communicating with management depends on the level of the audience. The CEO of a multibusiness company is most likely to want information at the level of technology output areas, with only selected examples from strategic program areas, whereas an operating manager below him is apt to be interested primarily in strategic program areas, with some exploration of individual programs. Again, it is helpful to display at least qualitative changes in distribution of effort.

With the effort allocated into the output categories, both the technical manager and the general manager can ask two kinds of questions: Is the emphasis correct in light of the expected opportunity in

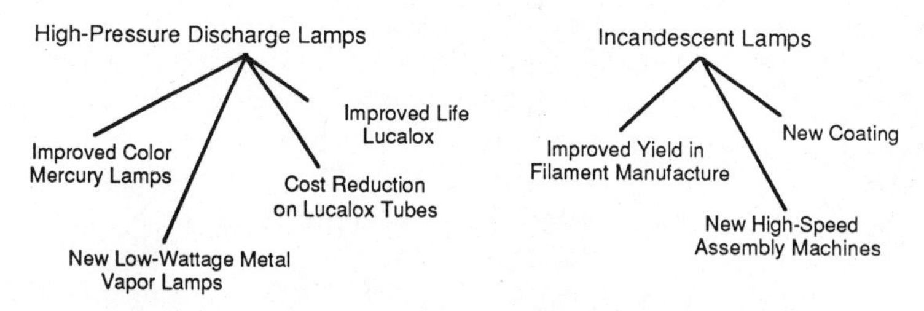

Figure 10.6 Strategic program areas (lighting business)

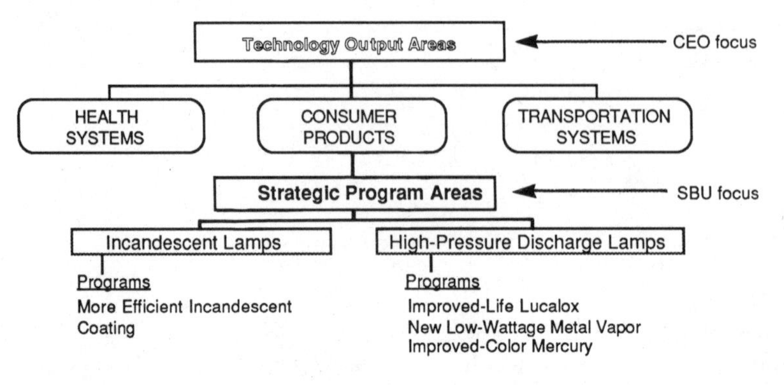

Figure 10.7 Strategic program structure.

various technologies? Is the emphasis consistent with the strategic trajectory assigned to each business or product line? There are as yet no analytical procedures for demonstrating that a proposed allocation of effort is optimal. Certainly, consistency with overall business strategy should get heavy weight.

Use of Technology

The evaluative techniques we have been discussing focus on the support technology provides for competitive strategy and strategic trajectories assigned to various business entities. They address the first and second questions raised earlier. The remaining two questions regarding balance between conventional and new technology and use of internal versus external capabilities are equally important and must be evaluated.

A business has a range of options in the way it chooses to use technology. The allocation of effort among those options is influenced by the stage of maturation of its technologies. But the priority assigned to various options is very much a matter of choice, and those choices define the most basic elements of technology strategy. Irrespective of the mode of presentation, the options run from application of present in-house technologies, to evolutionary improvements in those technologies, to application of existing technologies presently outside the organization, to development of new technologies to supplant existing ones, to (rarely) developing new technologies that provide capability not previously available in any form. That spectrum, as shown in Fig. 10.8, represents the range of options from which technology managers must choose in formulating technology strategy. These options can be applied to all fields of technology and can be used to characterize man-

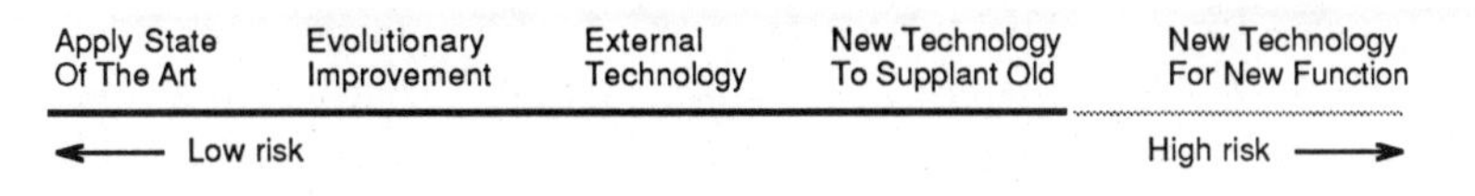

Figure 10.8 Strategic technology options.

agement options independent of the specific features of any particular technology.

Technology Generation Matrix

Those options offer a range of risk from which management may choose. Another spectrum of technology applies to the range of uncertainty associated with any technical work, usually as seen from the perspective of the people doing the technical work. A "technology generation matrix" (Fig. 10.9) can be constructed from these two axes. The idea for this matrix evolved from discussions with dozens of general managers regarding the way they thought about technology and the alternatives they had to weigh in determining strategy. This display of technical work is more descriptive of industrial practice than is the conventional basic research–applied research–development spectrum. It also highlights the balance between conventional and new technology as well as between use of in-house and external resources.

The rows represent the range of options by which technology can be brought to bear on a business. The range is expressed in terms of increasing risk—a concept of great interest to a CEO. Application of the state of the art includes, but is not limited to, mature base technologies that are subject to little change. It also includes those elements of

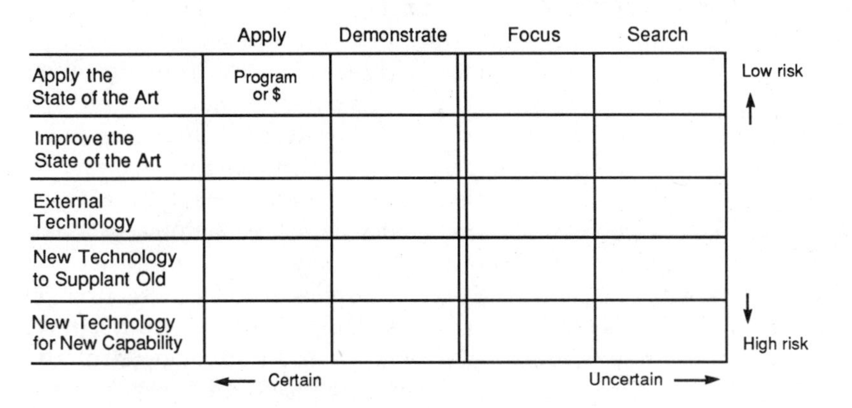

Figure 10.9 Technology generation matrix (SBU level).

key technologies that involve applying proven techniques. The technical risk in this work is virtually nil. A competent technical manager can tell you what he can produce, how long it will take, what human resources it will require, and how much it will cost with little likelihood of error. Risk is still present because the product may not sell or its cost may be too high (even though it meets its target), but technical risk can practically be ruled out.

Some might argue that innovation is precluded in such a circumstance, but that is not so. The food processor and the Black & Decker Dustbuster were new products based on state-of-the-art technology; so was the trash compactor. Even radically new products normally include large elements of conventional technology. For example, the early Xerox machines used power supplies, light sources, mechanical linkages, controls, and so forth, that were no more than state of the art. A good designer consciously seeks to make extensive use of existing materials and components—they are less expensive, readily available, and more reliable.

Evolutionary improvement in the state of the art introduces some technical risk. A competent manager can still estimate with relatively high certainty what he can deliver and at what cost. He needs to explore some alternative approaches to see which is best, and sometimes he encounters unexpected difficulty.

The importance of these two categories of technical work is seriously underplayed: most innovation results from effort on them. Most difficulties that companies encounter arise from failure to do this work well or to invest enough in it. The "microinvention" that underlies such innovation is the bedrock of sustained technical success.

The placement of external technology as having greater risk than evolutionary improvement may seem surprising. Evaluation of external technology, especially if it is in an area where internal competence is low, is uncertain. It may work well, but it can prove very troublesome. There is probably no technology that is so completely specified that no human input is required. The difficulty is that even skilled practitioners of the art may not recognize the extent of their own judgmental inputs. For example, suppose the specifications for a motor or a transformer lamination call for edges free of burrs. Well, how "free" is acceptable? Electropolishing every edge is not economically feasible. Skilled operators "know" when an edge is acceptable.

The next category, new technology to supplant old, represents the pacing technologies noted earlier—those fields, low on the S curve, from which some new winners will emerge. Most innovation in the popular literature is associated with this category—xerography, integrated circuits, and compact disks, for instance.

The final category is essentially sui generis. It comprises those rare advances that enable one to do something heretofore impossible or impracticable, such as television, instant photography, and magnetic recording. Most examples now come from, and are likely to continue to come from, biotechnology, pharmaceuticals, and instrumentation. Space technology has produced important examples in communication, remote sensing, and global weather sensing. No one contests the significance or drama attached to these advances, but they are pretty chancy stuff on which to base the future of a business.

The columns in Fig. 10.9 indicate the stage of technical work and, implicitly, the extent of technical uncertainty for a given program. "Search" means just that—looking for and trying out bright new ideas. Most search programs are small, and they may not even last very long before they prove unproductive. Thus even though the uncertainty is great, the financial risk is small. Since the attrition rate on programs in the search category is large, few survive to receive more extended examination in the focus category.

If an investigation in "search" proves rewarding, further effort is required to learn more about the promising new discovery. What additional attractive attributes does it possess? What are its limits? What problems emerge? The focus column is for work in this stage. Focus programs are larger than those in search, but are still typically two- or three-person efforts. Consequently, expenditures to develop new technical capability inherently must represent a small fraction of total technical effort. This fact is frequently lost sight of in the attention devoted to the high failure rates for such endeavors. These programs are frequently described as high risk, but as we saw in our earlier discussion of managing risk, risk should include the probability of occurrence times the magnitude of consequences. On that basis, these programs should not be characterized as high risk. They contain large uncertainties, but the financial exposure is rarely significant.

The columns for demonstrate and apply shift the focus from the creation of a new capability to learning how to use it. The attrition among programs as we move from column to column is substantial. Fewer programs will survive to the demonstrate phase, but those that do will escalate in size and complexity. Furthermore, involvement with, and contribution from, operations will become more important if programs in demonstrate continue to show promise. Programs in the apply column are in the domain of operations. An R&D group would provide more than backup support for this phase. The information entered in the cells would be similar to that in Fig. 10.2, and management should go through an analogous process of evaluating the appropriateness of the distribution. For example, is it acceptable to have no effort on new technology or very little in using external technology?

Distribution of effort

It is instructive to look at the distribution of effort in this matrix for different kinds of organizations (Fig. 10.10). For a laboratory devoted to R&D, the distribution of effort from one column to another, particularly the proportioning between apply and demonstrate, varies depending on the charter of the laboratory and the cutoff between work in the lab and work done in operations. Even so, rarely is more than 10 percent of effort devoted to search. The striking thing is the approximately equal distribution between creating capability and getting it applied. In other words, for every dollar an R&D organization spends creating new technology, it also spends a dollar on application. Obviously, these numbers are influenced by the dynamics of program effort. The resource requirements of programs escalate substantially as they progress, and it may be necessary to curtail technology creation effort temporarily to complete the development of major advances. Conversely, R&D management may deliberately decide to shift the balance between the right and left halves of the matrix in response to operating exigencies, i.e., it may postpone effort in search and focus to supplement effort in operations during a time of severe cost pressure. The matrix offers a useful framework for providing perspective on the total technical effort that would be difficult to achieve, staying at the level of individual programs.

The distribution of total technical effort for an entire business is affected by the stage of maturation of its industry as well as by the particular technology strategy it has adopted. Figure 10.11 suggests the distribution for a company in a relatively mature industry.

	Apply	Demonstrate	Focus	Search
Apply the State of the Art	↑	↑	↑	↑
Improve the State of the Art				
External Technology	10%	40%	40%	10%
New Technology to Supplant Old				
New Technology for New Capability	↓	↓	↓	↓
	50%		50%	

Figure 10.10 Distribution of technical effort—R&D.

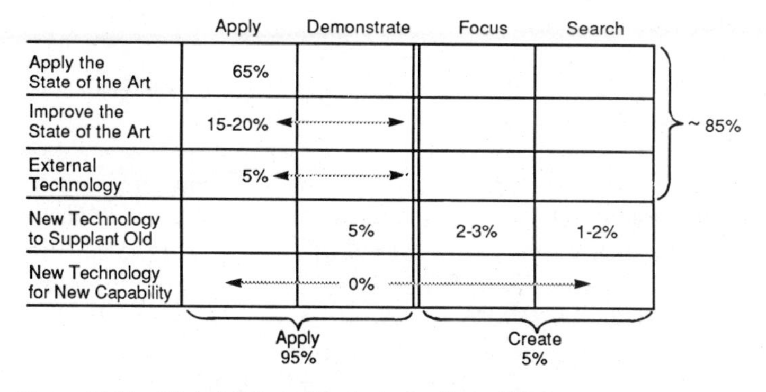

Figure 10.11 Distribution of technical effort—SBU.

Influence of management philosophy. The matrix in Fig. 10.11 reflects a company's basic philosophy about its use of technology. Should it limit itself primarily to application of the state of the art? Many smaller companies have no option other than to focus almost exclusively on the creative use of existing technology to provide value to their customers. I must emphasize this point. Thousands of companies have little choice except to make creative use of existing technology. That does not mean that technology is not important. They can get into trouble for failing to use existing technology effectively just as certainly (and much more frequently) as for failing to adopt new technology. Most foreign companies hoping to penetrate world markets use conventional technology. Hyundai in South Korea, for example, in seeking to penetrate the world auto market, used existing technology. In fact, it relied extensively on external technology in doing so. Note the high percentage of funds, approximately 85 percent, devoted to the use or extension of existing technology. The 85 percent is not based on a formal survey, but on extensive discussions with managers of technology in many different industries. Except in new fields, the existing state of the art (or its extensions) represents the principal technology base by which value is created for the customer. The focus is on supplying value to the customer, not on creating new technical capability. Only approximately 5 percent of effort, even in large companies with a strong technological emphasis, is likely to be focused on creating and applying new technology. In fact, the effort on truly trying to create new capability by fundamental work on breakthrough discoveries is conducted by few companies—some very large and some new ventures. We need to remind ourselves that the sophisticated work to extend microelectronics to ever smaller dimensions is just that—extension of advanced state of the art. People have foreseen its coming for many years and have known most of the problems that had to be solved. In contrast, much work on gallium arsenide is trying to create a new capability; but the totality of that effort is trivial in magnitude compared with the application and extension of silicon.

The nature of the technology and of the basis for competitive success also affects the distribution of effort. Pharmaceutical companies succeed by developing new drugs for new applications; therefore, for those companies the efforts in the last row, instead of 1 to 2 percent, become a significant portion of total effort. Since the new biotechnology companies have little technology that constitutes present state of the art, most of their work, of necessity, is in the bottom two rows.

This technology generation matrix is of value both to the technical community and to general management because it displays a company's basic approach to technology: Is it primarily an applier, or will it now seek to create new capability?

The matrix also makes it possible to display alternative strategies for a company or for a country. For example, until the early eighties the Japanese emphasized extensive use of external technology and extensions thereof. They have been trying to implement a new strategy of more internal development and more effort on new technology to supplant old. Great Britain, on other hand, has been particularly successful in developing new capability, but less so in applying it.[3] The self-image of the United States since World War II has been that it was effective in all parts of the matrix (except using external technology). More recently its competitive success in applying technology has come into question.

Technology Delivery Matrix

The same technology spectrum can also be used in a technology delivery matrix to display technology portfolio strategy vis-à-vis the business portfolio strategy, as shown in Fig. 10.12. In this case, the col-

	Dispose	Catch up/ Fix up	Maintain Position	Grow	Extend	Diversify
Apply the State of the Art						
Improve the State of the Art						
External Technology						
New Technology to Supplant Old						
New Technology for New Capability						

Figure 10.12 Technology delivery matrix.

umns show the range, described earlier, of strategic options for an individual business from sale at one end to diversification into new areas at the other. The dollars associated with each category would be shown in the cells.

Figure 10.13 demonstrates how this matrix can be used to compare the strategies of various businesses (or countries). Intel has a strategy of placing heavy emphasis on developing new technology, whereas Motorola has stressed evolutionary improvement and manufacturing efficiency. Japan achieved eminence by aggressive acquisition and application of external technology and is now trying to move more toward developing new technology. Emerson Electric has focused internal effort on the first two categories and has relied on acquisition to broaden its technical base. On the other hand, GE has traditionally sought primarily to develop new technology internally, as has the Bell System until its breakup. Siecor is in the optical fiber business, which is in the early stages of rapid growth.

An important use of this matrix is to examine changes in trajectory over time and in particular to ascertain where incremental funds are being allocated. Are increases in funds going to sustain business through evolutionary improvement or to develop new technology as the basis for extensions, as suggested in Fig. 10.14?

Displaying information in this matrix does not tell one what to do. It does make evident the distribution of effort in meaningful ways. Perhaps most important, it is a way of treating technology expenditures as an investment portfolio, subject to varying degrees of risk. It also enables one to check the consistency of technical effort with other elements of strategy. As such, it can become a useful mechanism for

	Dispose	Catch up/ Fix up	Maintain Position	Grow	Extend	Diversify
Apply the State of the Art		Steel Industry				
Improve the State of the Art		GM & Ford		Emerson Motorola	AT&T	GE
External Technology					Japan	
New Technology to Supplant Old				Siecor	Dow	
New Technology for New Capability					Intel – DuPont Great Britain	

Figure 10.13 Illustrative strategies.

	Dispose	Catch up/ Fix up	Maintain Position	Grow	Extend	Diversify
Apply the State of the Art	Business A - $200,000					
Improve the State of the Art		Business B $0		Business D + $100,000		
External Technology						
New Technology to Supplant Old					Business C + $250,000	
New Technology for New Capability						

Figure 10.14 Allocation of incremental funds among businesses.

discussing resource allocation with a CEO—who might question, for example, whether zero increase in effort for business B is appropriate.

Limitations of Display Methods

Discussions that focus on particular programs cannot be replaced by these display techniques. Individual programs are concrete and can help create better understanding of the promise of technology and thus generate enthusiasm; however, unless they are placed in some sort of strategic context, as the matrices attempt to do, they can leave CEOs with a frustrated feeling of having examined many trees, but not knowing what the forest looks like. Furthermore, when CEOs are given information only about specific programs, they have no choice except to probe program details if they wish to test credibility or validity. If their own background is a technical one, this practically invites them to intrude into the development process—a step that can be stimulating, but difficult to control and even destructive. Furthermore, it is a perversion of his role as CEO. Technical managers in one prominent electronics company talk bitingly about "board-room invention."

Scope of Technology Strategic Plan

The point of view we have been taking in discussing technology strategy needs to be supplemented by taking an additional approach to technology outputs. In the first chapter we emphasized the importance of viewing technology more broadly by incorporating product technology, process technology, and information processing under the umbrella of technology. To these three, we must now add a fourth—programs to strengthen and extend the infrastructure of technology itself, e.g., computer-aided engineering, more powerful techniques for analyzing heat flow and the behavior of high electrical fields, nondestructive evaluation, low-cycle fatigue... the list can go on and on. These four categories become another important way for displaying the allocation of technical effort:

Product technology

Process technology

Information processing

Technology infrastructure

Again, there is no ideal distribution. The mere act of accumulating the data will raise consciousness and cause questions to be asked. If, as one business found, you are spending seven times as much on product technology as on process technology, you can question whether, in an era of intense concern over productivity and product quality, that distribution is optimal. The choices are to do nothing, transfer resources from product development to process development, or increase total effort and devote the increment to process technology; but unless the present distribution of investment in technology is known, there is no basis for judging whether it is optimal.

The attempt to complete this examination of the distribution of effort among the four components of technical work can, however, raise one critical question: Can the business even assemble the data? Is there some manager (or group of managers) to turn to, who on the basis of his responsibility, can supply the data for one or more of the components? In other words, are the interests of the enterprise in each component of technology assigned as a strategic responsibility to some manager? Is anybody, other than the general manager, responsible for evaluating the appropriateness of the distribution among the four components?

This chapter has placed heavy emphasis on various ways of displaying arrays of information regarding allocation of technical effort. The emphasis was deliberate because gaining a comprehensive perspective of a technological effort (or a business) from different points of view is central to effective strategic planning. I do not know of any methodology that tells what strategy to adopt, but often a need for change is made evident simply by displaying information in new ways. The use of displays that emphasize financial considerations is deliberate. Since changing the allocation of resources, i.e., making financial decisions, is perhaps the crucial step a manager takes in implementing strategy, it is helpful to present him information in the same format that he uses for decision making.

The Planning Process

The information gathering, analysis, and evaluation that underlie an effective planning system obviously require a complex, multitiered process involving the orchestration of both top-down and bottom-up activities in any given operation or laboratory. The bottom-up approach begins by a searching review and critique of present programs: where do they stand in their development, has progress been satisfactory, what technical obstacles lie ahead, how will they be attacked, do the objectives still appear attainable and attractive, what skills and funds will be required for the coming period, are other functions or operating components on board and cooperating...? The review is intended both to help reaffirm a sense of the value of a project and to determine what will be needed to support it for the coming period.

The final step in the bottom-up activity is to consider possible new programs that could be initiated if resources could be provided. This step is absolutely indispensable in calling forth the creative talent in the technical staff. The charter for a technical component can be looked upon as a fence that constrains what it does and that minimizes territorial disputes, or it can be regarded as a hunting license to discover what opportunities lie in a region, largely undeveloped at present. The latter produces an environment more conducive to creative thinking.

While this bottom-up process of program review and evaluation and search for new ideas is going on, a top-down process must also be under way. This process addresses four issues: delineation of business strategy for each business or product line, evaluation of competitive technical position, formulation of technology strategy in support of business strategy, and allocation of technical effort in accordance with those business strategies.

This top-down analysis is intended to provide a broader strategic perspective against which to examine the output of the bottom-up, program-by-program review. The exploration of the four issues noted above can be carried out by line management or by a strategic technology planning unit. The basic purpose is to check the consistency between the emerging technology strategic plan and the business strategy.

If the work is done by a central planning unit, it can have the advantage of a fresh look, unencumbered by commitment to past decisions. However, the review must not be done in such a way as to intrude on the basic responsibility for planning, which must rest with operating managers. The planner's objective should be to raise questions, perhaps make people uncomfortable, but leave responsibility with line managers.

The allocation of technical effort need not track perfectly with business strategy. For example, a heavy commitment of technical effort may go to rescuing a business in trouble because a technical fix is deemed to be central to the salvage effort. Nevertheless, it is important for both the technical managers and the general manager to have examined the correspondence between allocation of technical effort and business strategy and to have agreed on the rationale for "contrarian" thrusts in technology.

Competitive comparison should include an analysis of what can be learned about the technology strategy of major competitors and even their levels of technical effort. With increased line-of-business reporting, more data are available than heretofore in various company reports and surveys. Industrywide data are available from a variety of sources. Again the goal is to check consistency between technical effort and business strategy. If a business embarks on a program of growth based on technological innovation, but its expenditures are only comparable with competitors, the realism of the strategy is questionable.

Presenting the plan

The best plan in the world is worthless if the boss does not buy it or if the functional people who must implement it are not committed to it; therefore, planning must also include effective communication with higher management and other relevant constituencies. This is especially important in multidivisional companies. Although preparing a plan and presenting it are obviously closely related, they are two separate activities. First, one prepares a plan and then asks, how can I best go about communicating this plan to others? A significant part (I would estimate 25 percent) of the total effort should go into preparing the presentation of the plan. The effective planning presentations I have been involved with choose a few simple themes for their structure. These themes are intended to convey one's sense of priorities and emphases during the period ahead and to inform higher management about the most important issues that technology will be addressing. They are chosen with the goal of giving higher management a sense of security that the technological needs of the business are in good hands. This process of developing a planning presentation is also a good quality check on the plan itself. If identifying key themes and priorities proves difficult, if the themes and priorities do not address the full concerns of top management, or if they do not seem sufficiently exciting or promising, perhaps the plan is deficient. The presentation of the plan can also include material to begin alerting higher management to emerging issues it may be asked to address at a later time.

Role of the planner in the process

The role of the planner is important, but it is offstage. First, he should supply tools, such as the techniques for aggregating programs and evaluating them noted earlier. He can also provide data, such as historical analyses within the technical function, data on operations, and data about competitors.

Second, he can critique the planning proposals as they evolve. Here his intent is primarily to educate rather than to change. Given the tight schedules on which planning operates, there is rarely time to go back and redo much of the work. The planner's goal is really the next cycle. He would like to help managers alter their perspective and priorities so that next time around they will do it differently.

In making his critique, the planner should focus on three considerations: vision, realism, and consistency. It may seem odd to suggest that a planner can critique the vision of a strategic plan prepared by competent technical people. However, these people feel committed to accomplishing what is proposed in the plan, creating a certain conservatism. The common wisdom is that technical people tend to be overly optimistic about the near term and to be too conservative in projecting longer-term accomplishments. Progress in technical work requires putting on blinders and focusing all one's resources on a chosen target, but that also may lead to overlooking opportunities outside the area of concentration. It is not uncommon for technical people immersed in one technology to view its replacement on a one-for-one basis (i.e., its properties versus those of its competitor), without asking what else the new technology might make possible. This attitude maximizes the difficulty of pushing aside present technology because it is rare for a new technology to be better than an old, especially in cost, if limited just to the attributes of the old technology. It is the new things that are possible that make the new technology attractive, e.g., small, low-cost transistor radios rather than simply replacing vacuum tubes in conventional radios, or new services to the airline passenger rather than just replacing reservation clerks and telephones with computers and communication networks. A planner may be in a better position to call attention to these features even though he cannot himself provide the necessary modifications.

The critique for realism is focused on resources. Technical organizations tend to spread themselves too thin. They are reluctant to leave some areas totally uncovered. Consequently, they have a tendency to maintain small efforts that may do good work, but that are unlikely to succeed against the much larger efforts of some competitors. The planner can raise questions about the size of competitive efforts and the likelihood of wasting not only money, but more important, the time and creative ability of good people. Such programs are unfair to those working on them. Somebody needs to raise the question: If we really aren't likely to win with an effort of this size, why are we doing it?

The first check of consistency is to compare resource allocation against the rhetoric of the plan. On occasion a plan waxes poetic over an important opportunity or expresses concern over an important threat, but then proposes that few or no resources be devoted to the situation. As we noted earlier, it is especially important to check resource allocation in technology against the strategic thrust of various businesses. Technical organizations tend to build up client relationships with businesses they understand well and are effective in working with. In this circumstance the technical staff has no difficulty discovering additional problems and opportunities to address. Meanwhile, other businesses may be growing rapidly and represent opportunities for technical contributions that are being overlooked. The planner can be alert to such situations and point out the vulnerability that they create. There are indeed businesses for which at certain times technology is a low-order concern compared with other issues. However, it is much better for technical management to have reached that conclusion after careful examination than to ignore some businesses because it has never really examined them. That danger is particularly pernicious for companies with a hardware tradition which are gradually diversifying into services. The hardware-focused technologist is likely to conclude—if he even bothers to look—that technology has no role to play in services businesses.

Conclusion

This chapter has focused on some of the practical problems in preparing a strategic plan for technology. Throughout, the emphasis has been on arraying information in a variety of ways to help ensure a comprehensive perspective on a business and the strategic options being considered. I have stressed the importance of presenting information to the general manager in ways that help him to achieve a strategic perspective on technology and to make his decisions in a framework he is comfortable with, i.e., decisions about the allocation of money.

Notes and References

1. Michael E. Porter, *Competitive Advantage: Creating and Sustaining Superior Performance,* The Free Press, New York, 1985.
2. *The Strategic Management of Technology,* A. D. Little, Cambridge, Mass., 1981.
3. Sir Montague Finniston, *Engineering Our Future,* Report of the Committee of Inquiry into the Engineering Profession, HMSO, London, 1980.

Chapter

11

Nurturing Innovation

As we noted earlier, the fundamental question that must be addressed in strategic management is whether and how the enterprise must be changed in order to survive and achieve its potential. The change may involve any combination of alterations in the business portfolio, in basic technology, and in the internal culture, organization, and management system. In order to convert this question to operational steps, one must ask a series of questions: How much change is needed? How rapidly must it be introduced? How much can we afford, since implementing change requires both human and financial resources? How much can we manage?

Change Is a Universal Ingredient of Survival

Everything we know and are learning about both biological and social systems indicates that change is an indispensable ingredient for survival. The need to adapt can arise from dynamics of a system (or organism or institution) and from changes in the environment. All business enterprises are subject to the same requirement. The challenge is to manage with vigor and foresight so that the dynamics of change can be matched to internal trends and shifts in the environment without having to resort to draconian, catch-up measures, i.e., to be proactive rather than reactive. In this chapter we will focus specifically on innovation as an agent of change.

The ubiquitous impact of change is easy to accept in the abstract; it is almost universally traumatizing when one is himself the victim of change. I have always found it ironic that technical people, those great agents of change, are just as anxious and concerned as others when they themselves are subjected to change. I hope we will be able to show that change is not only necessary for survival, it has many desirable consequences as well.

Although this discussion focuses on technological innovation, it is important to note that innovation (change that creates value) occurs in many other spheres as well. Low-cost, no-frills air travel, air shuttle service, credit cards, cash management accounts, "junk" bond financing, fast-food chains, next-day air freight, just-in-time inventory—these and a myriad of others constitute innovation in which technology plays at most a facilitating rather than a central role.

Management Challenge: Preserving Tension

Probably the most important challenge in managing technology effectively is to recognize the necessity of preserving tension between continuity and stability (i.e., resistance to change) and change, both evolutionary and revolutionary. The central role of this tension pervades this entire book. The present structure of technological capability constitutes a resource of enormous power. It enables us to satisfy a staggering array of human wants, and its power is by no means exhausted. As we noted in Chap. 2, a new technology faces a daunting opposition in attempting to penetrate and become part of the corpus of conventional technological capability.

A new technical capability is at best "unproven" with respect to both its superiority over the application of conventional technology and the timing of its availability. Any aspiring newcomer must be fit into a carefully orchestrated sequence of activities that include already proven technologies, and if it does not fit or is not available on schedule, it wreaks havoc. We too easily forget that the pioneering user of a new technology is often more at risk than its supplier. The pioneering user not only runs the risk of an unsuccessful development, but far more damaging, the derision of "you damn fool, you ought to have known better than to try something nobody else had ever used successfully!"

Thus a manager of technology who contemplates sponsoring an innovation faces a paradox. Given the power of existing technology, the likelihood that any given attempt to introduce a new technology will succeed is quite small—probably 5 percent or less. However, the universal pattern of technological maturation dictates that practically every technology is fated to be replaced—sometime. That "sometime" is the wild card!

This paradox in turn creates a management trap: If a manager says no to all proposals for major innovation, he will be right perhaps 95 percent of the time; the other 5 percent may destroy his business. That is a very uncomfortable position. Unfortunately, the 5 percent of successes is not readily distinguishable in advance from the 95 percent of failures. Luckily, however, the cost of probing and monitoring is relatively small. The real danger is more insidious—relaxing one's vigil in the constant search for technological breakthroughs in the face of long odds that any will be found.

Although the rewards for successful innovation can be spectacular, the fundamental drive for innovation is survival. Management must maintain in one part of the enterprise a constant effort to innovate in the presence of an obviously valid, pervasive resistance to change. All of management, even though its immediate task is to produce certain results, must foster a wariness that present products, processes, techniques, etc., are vulnerable to attack. Thus an enterprise with a high potential for survival is at war with itself. It is continually nurturing and retuning the tensions between those seeking operational efficiency and those attempting to ensure survival by introducing innovation. This inevitably produces discomfort.

Why Is Innovation Cause for Concern?

At the national level, the United States especially, but every other country as well, faces another paradox: The desirability, indeed the necessity, of innovation is universally accepted (certainly very few argue that innovation is undesirable) and yet there is also a universal belief that we do not do enough and that we need to do it more effectively. Both of those statements have been common wisdom at least since World War II. How can it be that an activity which is universally regarded as necessary and desirable is also regarded as a continuous cause for concern because it is not done well or enough?

The economic argument for innovation is couched in terms of capturing its benefits, i.e., appropriability. Economists argue that the results of research cannot be predicted. Consequently, an investor first cannot be certain whether any benefit will accrue, but even more, he cannot predict whether a result will be applicable to his business. At best, it is likely that others will benefit as well; at worst, he might gain no benefit while others do. Under these conditions of shared or zero benefits, the aggregate investment in innovation will be less than optimal for an entire economy.[1]

The management perspective, although it is very different, is not incompatible with that argument. The general management literature did not devote much attention to innovation until the eighties. A number of books have now extolled its virtues. *In Search of Excellence* identifies innovation as a central characteristic of effective management.[2] Peter Drucker, in discussing entrepreneurship with its implicit requirement for innovation, argues that it is a necessary requirement for rational managers.[3] Richard Foster, in *Innovation: The Attacker's Advantage,* claims that innovation is the key to success and that those seeking to innovate have an intrinsic advantage over the defenders of conventional technology.[4] I spent much of my professional life in a globally respected R&D organization chartered to create innovation. I have seen countless promising discoveries fail to survive the gauntlet to commercialization and have seen the careers of competent, dedicated scientists and engineers blighted by innovations that did not make it. From my perspective, these assertions of robust value seem like tempting oversimplifications. If it is so important, so valuable, and so likely to succeed (as some say), why is innovation a source of perpetual concern?

Many attempts fail

Unfortunately, human nature is such that we focus on the successes. It is all too easy to point to Intel, Hewlett-Packard, Texas Instruments, Tektronix, or Loctite, or more recently, to Apple, Microsoft, or Lotus, as dramatic demonstrations of the value and rewards of innovation.

The longer-term, less testimonial perspective is more sobering. Of the 500 firms that went public in 1961 to 1962, most of which presumably were attempting to offer something new, twenty years later only 2 percent were still generating the level of profit that would make them attractive investments, 53 percent had gone bankrupt or completely disappeared, and another 25 percent were operating in the red; less than one-quarter were profitable (Table 11.1).

The remote and usually retrospective views of innovation typically found in the literature remind me of the satellite view of the earth: it shows earth's spectacular beauty, but it is a poor guide for climbing Mt. Everest. One of my cynical associates suggests that a more appropriate analogy to the course of innovation would be to say that the satellite view does not reveal the war, famine, and plague that affect humanity.

A senior vice president of technology for a company that is striving mightily to encourage innovation and entrepreneurship noted with

TABLE 11.1 Status of 500 Firms Going Public in 1961–1962

Status	Percent
Vanished	12%
Bankrupt	41%
Losing Money	25%
Moderate Profits	20%
Attractive Profits	2%

SOURCE: Business Economics Group, W. R. Grace & Co., 1983.

some puzzlement that none of the aspiring entrepreneurs seemed happy. I replied that I had never encountered a group striving to bring an innovation to market that *did* seem happy. Beleaguered is a more apt description! These groups are almost mesmerized by their problems. They worry about intractable deficiencies in the technology. They fret about demanding investors. They strategize over how to find and convince doubting customers to buy. Committed to the promise of the new technology, yes. Determined to succeed, yes. Excited over the brave new world they are going to create, yes. But happy or confident, no.

I am certain that this description is not characteristic of all entrepreneurs or innovators, but a picture of warriors going forth joyfully to change the world is not typical. The sense of exhilaration, of having been more alive and more effective than at any time in their lives, often comes later; they look back nostalgically on the memorable days when they were changing the world—that is, if they succeed, of course. Otherwise, the people scatter, they are hard to locate, and they talk about their experience somewhat defensively. I have tried to do post mortems on aborted innovations with people I knew well, and it was apparent that they were still trying to rationalize their actions and decisions even years after the event.

Innovation is difficult

Innovation is not easy. Uncertainty means you *cannot* know whether something will succeed, and this leads inevitably to anxiety. Furthermore, when it does not work, the personal cost can be high. Consider S., a brilliant engineer nearing the end of his career. Nothing he has ever worked on even made it to production. Another S. devoted fifteen years of the best years of his life to a sophisticated energy-storage device before the program was abandoned. B. devoted seven years to a promising new steel-rolling technique that was abandoned. R. devoted seven years to a process that was only a limited success. Another B. has devoted fifteen years to achieving the application of ceramics as engineering materials of construction, with little to show for it. Both J. and C. have remarkable records as creative inventors with no significant commercial success to show for it. Innovation exacts its toll, and we must not ignore that fact when we attempt to nurture it. Fortunately, the rewards of success—not only financial, but probably most important, of "having made a difference," of having changed the world—call forth the strivers who are among our industrial heroes.

Rewards must be commensurate

It should be apparent that no matter how vital the need, innovation will not occur unless powerful forces are mobilized to make it seem like a gamble worth taking. Innovation will not happen because people sing its praises. It won't happen by pointing to dramatic successes, because they are too easily dismissed as unrepresentative sports. Sustained commitment to innovation requires a clearheaded understanding of what it is, what benefits can be derived from it, why it is difficult to accomplish, what microenvironment is required for success, and what corporate macroenvironment is needed to nurture it.

What Is Innovation?

First, what *is* innovation? Innovation is the creation and introduction of change—any kind of change that results from focused, purposeful action. It calls for a particular state of mind to look for a new perspective, a new solution. From my viewpoint, such change must create value for customers or improve the viability of an enterprise. Change just for the sake of change itself is of course possible, but it is difficult to see why a business should seek it. Change covers a spectrum from small increments to major discontinuities. It is misleading to focus on innovation in products as opposed to process or information processing. The spectrum can be thought of as follows.[5]

	Example
Incremental improvement in product, process or system	• Better color or efficiency in lamp phosphors • New lower-cost catalyst • Improved word processing software
New component, process, or technique in a larger system	• Plastic dishwasher tub • Sintered instead of forged jet engine parts • Automated lamp filament inspection
New product for existing market	• Compact disk • Color TV • Lucalox lamp
Radically new process	• Pilkington float plate glass • Low-pressure polyethylene
New product for new market	• Lexan • Silicones • Personal computer • VCR
New system	• Satellite communication • Doppler radar for wind shear • Fiber optic communication • Computerized passenger reservations
Entirely new capability	• Xerox • Instant photography • Beta blockers for high blood pressure • Interleukin 2 • Satellite earth sensing

It is apparent that the risk, uncertainty, and needed resources associated with innovation at different places in this spectrum differ widely. In general, product innovations involve greater uncertainty because they entail market response as well as technical performance. Process innovations are often invisible to the customer. Incremental improvements tend to be less risky. An innovation such as color TV or compact disks is beyond the reach of a small company. The CD required the development of an entire system—a complex recording technique and a sophisticated process for reproducing the disks, an advanced opto-electronic machine for playback, and of course the capability to generate program material for the new medium. Some new systems, such as satellite communications, air transportation, weather forecasting, and air traffic control, are so large and complex that only close government-industry cooperation makes them possible.

Potential impact is difficult to forecast

Much of the literature and media attention focuses on radical new discoveries—for example, Xerox, instant photography, microprocessors, and beta blockers for heart disease. This has the unfortunate consequence of distorting perceptions about the nature of innovations. It encourages the illusion that effort in innovation should be focused on "going after the big one"—the home run.

At a recent seminar of R&D managers for exploring ways to encourage innovation, the most widespread, common problem was corporate management's imposition of arbitrary minimum thresholds of projected size before candidate innovations would be supported. My experience has been that projections on size, even by avowed enthusiasts, are notoriously wrong—in both directions! If a product is truly new, it is almost impossible to predict the eventual size of its market. Even for ventures that have begun to demonstrate that they are going to make it, the dream of their dedicated participants is to achieve sales of perhaps two to three times their present size. If they were told their ventures would grow by an order of magnitude or more, they would say: pipe dream! Almost certainly, if an enterprise limits its ventures to those that pass some arbitrary minimum-size test, it will be missing many that would more than meet the test within an attractive time frame. Furthermore, if it aims only at markets which are demonstrably large, it is probably limiting its efforts to markets which have also been targeted by others that may well be ahead of it. History is replete with examples of predictions by knowledgeable people who underestimated the potential of new technology:[6]

"Who the hell wants to hear actors talk?"—film mogul Harry Warner, 1921.

"X-rays will prove to be a hoax."—Lord Kelvin, President of the Royal Society, 1903.

"I think there is a world market for about five computers."—Thomas Watson, IBM, 1958.

"They will never try to steal the phonograph—it is not of any commercial value."—Thomas Edison, 1915.

People underestimate the potential of a new technology principally because they do not perceive the collective impact that thousands of people can have in inventing new uses for a new capability after its credibility has been demonstrated. As we have mentioned, the existing system of technology is hostile to aspiring entrants, but remarkably adaptable once a newcomer has begun to be accepted.

Incremental versus discontinuous advances

The kinds of advances associated with major discontinuities should probably be regarded as analogous to genetic sports. They can be dramatic, but they are best viewed as sui generis; they certainly cannot routinely become the goal of every business and every R&D effort.

A focus on major advances not only distorts perception of the entire spectrum of innovation, it leads to underrecognition and underevaluation of the less dramatic advances associated with incremental progress. Home runs may be dramatic, but they are statistically improbable and a dubious basis for ensuring the survival of a business. The path to survival requires a certain wary probing for potential breakthroughs; it requires *relentless* pursuit of incremental improvement. To me, the most outstanding characteristic of businesses that achieve sustained profitability is their implacable search for improvement. In contrast to my earlier observation about technology that "good enough is best," for persistently successful enterprises there is no such thing as "good enough." The exhortation of that great inventor and entrepreneur Thomas Edison encapsulates the challenge: "There has to be a better way—find it."

Our earlier discussion of resource allocation in strategic management emphasized this same point concerning the importance of incremental improvements. Except in industries such as biotechnology or pharmaceuticals, where success is dependent on major advances because of the nature of the technology, resources should be focused first on making better use of the technological capability near at hand. Perhaps the most important lesson the Japanese have taught the world is the tremendous competitive power that can be gained from incremental improvements in technology. It is an unfortunate mistake to depreciate the significance of these advances and to assign lower status to those who achieve them. One could easily make the argument that "high technology" should more appropriately be associated with those fields that have been honed over decades to astonishing levels of performance and reliability, such that we take them for granted—and "low technology" applied to those fields that are able at present to accomplish very little compared with the promise that nature holds. In practice, the term "high technology" is applied to industries in which the technology is changing quickly and rapid advances in performance are being achieved. This is not, I repeat, *not* to argue that one should ignore or exclude efforts to create the big discontinuities. It is just that such home runs are a dubious strategy for competitive survival—home runs win ball games, but it takes more than that to win pennants.

So much for putting innovation in better perspective. Why is it so difficult? To be an effective protagonist for innovation, one must truly understand what it entails—you have to know the product in order to sell it effectively.

Barriers to Innovation

Barriers to innovation can be classed as intrinsic (inherent in the nature of the process itself) and extrinsic (inherent in the interaction between innovation and the situation it will change).

Intrinsic barriers

Intrinsic barriers are all associated in one way or another with uncertainty. When trying to innovate, no one knows, cannot know, whether the attempt will succeed—or even whether the course of action being pursued is worthwhile. It is all too easy to mouth the term "uncertainty"; it is quite another to experience the gut-wrenching anxiety that uncertainty creates. I have known people who worked for a whole year before they completed a single experiment that yielded meaningful results.

Innovation requires such an array of talents—those of inventor, reducer to practice, project leader, marketer, financier—that their appearance in one individual is most improbable. We will explore this subject later, but it is important to realize that uncertainty applies to not only whether or not an invention will succeed but also whether customers can be found and induced to use it.

As we have noted, rarely do those customers have no choice except to use a particular invention. Most of the time, they already have available technology that is much more certain and that can perform acceptably. With new technology, there is not only the uncertainty of whether and how well it will work, but also of its relative cost. Perhaps even more damning is the uncertainty about its availability—will it be ready as promised. I know of cases where customers lost their jobs because they had made a commitment to incorporate a vendor's innovation that was not available as promised. These uncertainties loom so large that new technologies which offer only modest improvements are unlikely to be adopted. Innovation is most likely to succeed because of fear or greed. A customer facing a problem with no apparent solution may risk using an unproven technology out of necessity. Conversely, if a new technology promises a significantly improved performance that has attractive market potential, a customer may "go for the gold."

One particularly distressing aspect of innovation is that it does not easily lend itself to cumulative learning—the experience curve has a low slope. Most innovators live through only one in their careers. I asked the only manager I have ever known who has been closely associated with several important successful innovations, "Are you getting any better at it?" His answer was, "Nope! Each time you are dealing with a new cast of characters, because the people connected with the earlier ones move on. You never know whether you're doing the right thing. You just have to suck in your gut and make the best guess you can." In fact, I think he was being too harsh in his judgments. He was focused on the substance of decision making, where his observation was no doubt true. The process invoked in bringing an invention to market was one he had become more familiar with and was managing more confidently.

Extrinsic barriers

Managerial barriers. The extrinsic barriers to innovation are both managerial and organizational. Managers have good reason to avoid innovation, if they can. As we have emphasized from the beginning, operational management, which comprises the great majority of all management activity, attempts to create certainty. Its entire focus is on maintaining order and discipline in a world that is always threatening to come unglued. An operational manager can only view an innovation as a thoroughly unwelcome intruder. He lives in a very unforgiving world in which, having committed himself to a budget, he knows his performance will be measured by his success in meeting or exceeding that budget. In preparing the budget, he has given high priority to reducing uncertainty to a minimum; innovation just torpedoes his whole approach.

The baleful effect of innovation goes beyond increasing uncertainty: it disrupts routine. New products or processes have to be mastered, new skills learned, and administrative routines altered. The process is very much like a gauntlet with each function having reasons to attack a disruptive newcomer.

Furthermore, these events divert resources. They require people, often the best people, and they require funds. It is not as though a general manager has no place else to make investments. Often he has other opportunities—e.g., in plant renovation, training, promotion, and physical distribution—that he views as more certain, having shorter payback, or having even more attractive return.

In concluding a meeting of technical managers, I noted that while it was gratifying to see more attention being paid to external competitors, I was disturbed to see no reference to the most important competition of all—the internal competition for resources. As I spoke, a representative of corporate finance at the back of the room was nodding his head vigorously. This competition from other functions and other investment opportunities warrants all the creative thought and careful analysis that the technical community can give it. The most successful laboratory in operations I have ever known owed much of its success to its skill and ingenuity in identifying and quantifying the comprehensive benefits that flowed from the innovations it proposed. It was particularly creative in identifying the second-order effects that could make an innovation's impact much greater. Its success in turn, of course, also required understanding internal customers' needs and constraints.

It is a mistake for advocates of innovation to conclude that operating managers lack vision or courage or are too focused on the short run. These managers are simply responding rationally to the environment in which they operate. An effective advocate does not waste time or energy bemoaning the fact that operating managers do not respond as he thinks they should. He tries to understand his potential customer's or sponsor's constraints and acts accordingly.

In addition to these barriers to innovation, a manager also faces the conflicting expectations of the various constituencies with which he must deal. Employees want opportunities for growth in responsibilities and income, but they also want job security and continuity in skills. Stockholders and investment analysts want growth in earnings, but they want predictability of performance. Customers want products that embody the latest technology, but they also want compatibility between generations of products; they want an extensive product line to match their needs, but they also want low cost and fast service. The aggregate of all these conflicting expectations is an exceedingly complex balancing process that gives a manager ample reason to be ambivalent and cautious about innovation.

Institutional barriers. In addition to managerial barriers to innovation, there are also extrinsic institutional barriers. Innovation is disruptive of people's lives and administrative routines. New products and processes threaten careers and status relationships. Skills that support images of self-worth are threatened with obsolescence. Salespersons face the unpleasant task of learning about unproven products and trying to talk customers into taking a chance with them. They may have to make more "cold" calls where they seek to attract new customers. The diversion of human and financial resources is resented because much of it comes from squeezing or canceling other work. A senior operating manager, leaving a meeting that kicked off a new corporate-level ventures program, commented: "I'm all in favor of what we're trying to do, but this business about these programs being ex-budget is a bunch of nonsense. I know perfectly well that these programs will come out of my hide!"

Finally, innovation is just very hard work. The people involved work very long hours under intense pressure. They must cope with relentless anxiety about whether deadlines will be met or crippling barriers overcome, and indeed whether the whole effort will ever amount to anything.

Barriers are useful

Too often in our advocacy of innovation we overlook an important fact—these barriers serve a useful purpose. A business or an economy that is too committed to innovation will be unstable. Changes having only virtues, but no faults, are nonexistent. The disruption and cost that are associated with innovation would be counterproductive if they did not lead to substantial improvement. Tough entry barriers help to ensure that the innovations which succeed do indeed justify the disruption they create. Most new ideas or proposed changes simply are not good enough to warrant adoption, or else they have serious drawbacks that negate their value.

Why Innovate?

Having built such a case against innovation, we could well ask, why bother? The answer is more multidimensional than it might appear. The overwhelmingly most important reason is that innovation is the primary source of growth and of improvements in productivity.

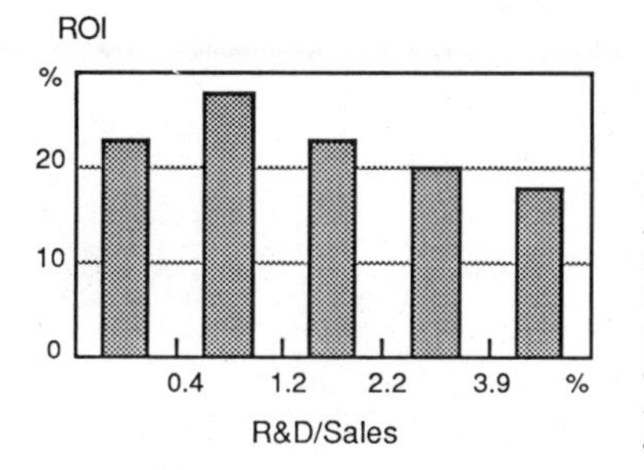

Figure 11.1 Impact of level of R&D on profitability. (*Source: Donald W. Collier, John Mong, and James Conlin, "How Effective Is Technological Innovation?" Research Management, September–October 1984, pp. 11–16. Reproduced with permission.*)

Despite Joseph Schumpeter's seminal work fifty years ago in providing a conceptual structure for economic growth that included innovation at its core,[7] economists did little until the late fifties to develop any empirical data. Robert Solow's pioneering work on the contribution of technology to improvements in productivity and economic growth stimulated additional effort and won him a Nobel prize.[8] Edward Denison, in his extensive examination of the causes of economic growth, attributed a major role to technology.[9] He attributed 20 to 25 percent of the improvement in productivity to contributions from technology—approximately equal to the contributions from labor and capital.

Effect on business performance

Efforts to demonstrate the value of innovation on a more local basis, such as an individual enterprise, are more difficult. It is easy to cite anecdotal evidence from an Intel or Merck, but while such evidence is comforting, it is not convincing. The Profit Impact of Market Strategies (PIMS) data base constitutes a rich lode to examine for the impact of innovation.[10] Unfortunately, the evidence is unmistakable that in a gross sense increased expenditures on R&D reduce profitability, expressed as return on investment (Fig. 11.1).[11]

This aggregate result has two major deficiencies: it does not eliminate the effect of other factors, such as investment intensity, that also affect profitability; second, it treats profitability at a given point in

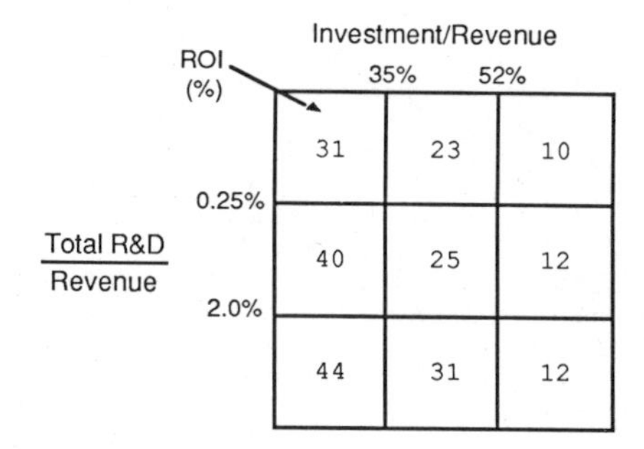

31	23	10
40	25	12
44	31	12

Figure 11.2 Relationship between investment intensity and R&D intensity. (*Source: Donald W. Collier, John Mong, and James Conlin, "How Effective Is Technological Innovation?" Research Management, September–October 1984, pp. 11–16. Reproduced with permission.*)

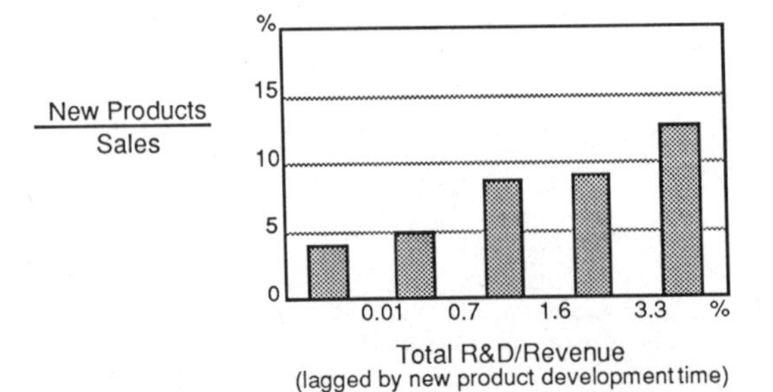

Figure 11.3 Relationship between new products and R&D intensity. (*Source: Donald W. Collier, John Mong, and James Conlin, "How Effective Is Technological Innovation?" Research Management, September–October 1984, pp. 11–16. Reproduced with permission.*)

time as the only measure of business performance. Figure 11.2 shows that if firms are grouped according to investment intensity (defined as investment divided by revenue) and R&D intensity (total R&D divided by revenue), then profitability does indeed increase as R&D intensity increases, i.e., the numbers in the cells in Fig. 11.2 increase, going down the column from lower to higher R&D intensity.

If business performance is considered over time, rather than at a single point, the evidence is also supportive. Figure 11.3 indicates that, after lagging R&D expenditures to allow for the time required to bring products to market, the portion of new products in sales is positively related to R&D intensity.

In turn, Fig. 11.4 shows that increases in market share, a virtually universal goal and one strongly supported by Fig. 11.5, are positively related with the portion of new products in total sales.

Figure 11.6 indicates improvements in margins are also positively related to R&D intensity (lagged as above in Fig. 11.3).

A separate study of the effectiveness of R&D expenditures in producing profits, as compared with investments in marketing, indicated that the time lag between expenditures and benefits was much

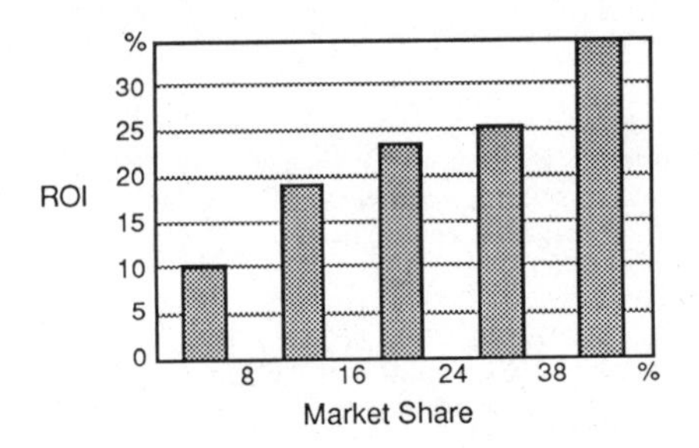

Figure 11.4 Relationship between market share and profitability. (*Source: Donald W. Collier, John Mong, and James Conlin, "How Effective Is Technological Innovation?" Research Management, September–October 1984, pp. 11–16. Reproduced with permission.*)

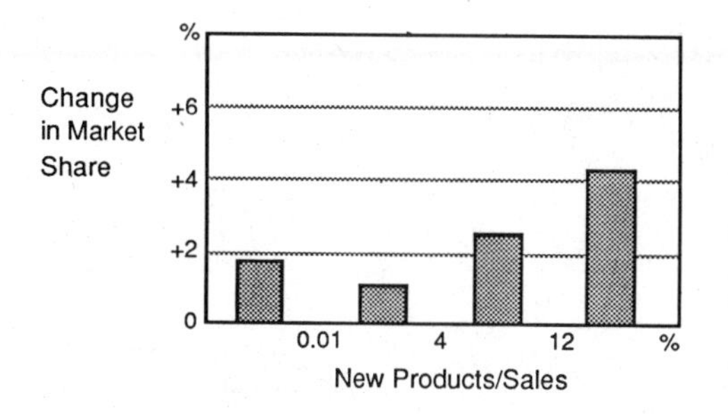

Figure 11.5 Impact of new products on changing market share. (*Source: Donald W. Collier, John Mong, and James Conlin, "How Effective Is Technological Innovation?" Research Management, September–October 1984, pp. 11–16. Reproduced with permission.*)

greater than for marketing, but the benefits were two times greater (Figs. 11.7 and 11.8).[12]

Additional benefits

One of the basic precepts of economics is that competition promotes efficiency. By extension, the competition between the old and the new, between conventional technology and new technology, is a powerful force for keeping conventional technologists on their toes. One could well make the argument that even when innovation fails, it serves a useful purpose in prodding conventional technology to improve.

The benefits of innovation extend beyond financial performance. Innovation is a forced-draft management learning experience. DuPont executives have commented about how much they learned from the extensive efforts to create new businesses in the late sixties and early seventies. 3M, which has a distinguished, decades-long record of innovation, gives evidence of the self-knowledge that comes from continuing efforts to innovate. The small team that created the silicone business for GE produced two senior corporate executives, several division

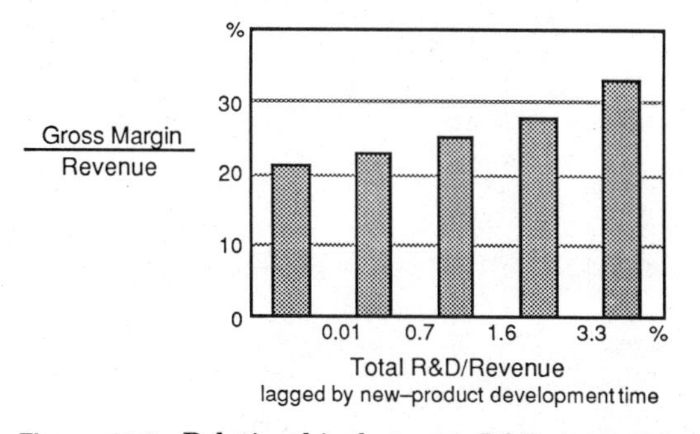

Figure 11.6 Relationship between R&D intensity and gross margins. (*Source: Donald W. Collier, John Mong, and James Conlin, "How Effective Is Technological Innovation?" Research Management, September–October 1984, pp. 11–16. Reproduced with permission.*)

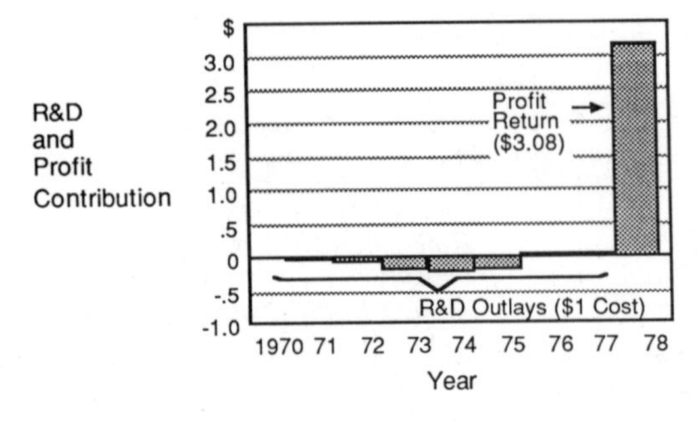

Figure 11.7 Time lag between R&D outlay and contribution to profit. (*Source: David Ravenscroft and F. M. Sherer, "Is R&D Profitable?" PIMS Letter 29, 1982. Reproduced with permission.*)

general managers, and a larger number of department general managers.

Innovation places one in new situations. It imposes stress on, and therefore helps illuminate, the limits of managerial skills and judgment. A management team with continuing exposure to the stress of innovation is much more likely to be adaptive and flexible, to perceive a new situation in a realistic light.

Perhaps another way of saying this is, innovation helps counter ossification. Success not only breeds complacency, it breeds arrogance. Management's styles and practices become accepted doctrine. Innovation acts as a therapeutic shock. The stresses created by GE's undertaking three major innovations simultaneously—nuclear power, commercial jet engines, and computers—stimulated the soul-searching that led to the development of strategic planning.

Nurturing Innovation

As I have insisted, innovation will not occur just by talking about the need for it or pointing to dramatic successes or even by setting up a

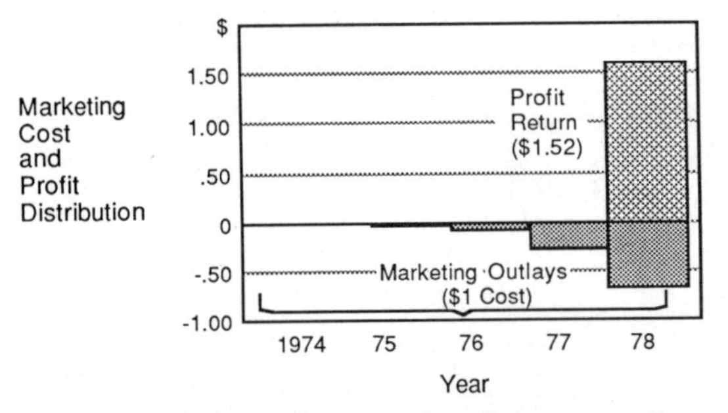

Figure 11.8 Relationship over time between marketing outlay and contribution to profit. (*Source: David Ravenscroft and F. M. Sherer, "Is R&D Profitable?" PIMS Letter 29, 1982.*)

special component or program to do it. Success requires a multidimensional, multilayered approach. Let's start at the corporate level and work down to the ground troops who are sweating it out, and then back off to try to deduce what precepts help nurture success.

Commitment

The key word at every stage of this process is *commitment.* Innovation can perhaps best be thought of as an anxiety-laden but necessary process. Possibly the best physical analogy is a change of state, such as going from a solid to a liquid. Just as a change of state requires a large infusion of energy, an innovation requires large expenditures of energy—human energy. A number of key players at different levels in an organization must make a dedicated, tenacious commitment to innovation, or it will not occur. That begins at the top. Without a senior management dedicated to making innovations happen, they are unlikely to succeed.

Involvement

The second key to innovation is total involvement. Unless the person directly responsible for innovation has no other responsibility to execute, or to hide behind, he will not have the almost fanatical drive to succeed that is needed. Innovation must be set up as a separate activity, with people charged solely with that task. Additional operating responsibilities divert attention, they become excuses for less than all-out effort; often in fact it *is* more important to a person's career to perform well on the operating than the innovative work. Performance on the former is more easily judged, and poor performance is more likely to be deleterious to a career.

Resources

The third requirement is resources. When I hear the comment, "Yes, we have a new ventures manager, but he hasn't been able to come up with anything that is attractive enough, or big enough for us to support," I know I am probably listening to a manager who has not crossed the line from intentions to commitment. The initial stages of innovation rarely require substantial resources. Consequently, "we can't afford it right now" is more often an excuse for not running the risk than a legitimate reason.

Accepting probabilities

The fourth requirement is psychologically the most difficult: accepting the probabilistic nature of innovation, i.e., accepting failure as inherent in the process. I cannot emphasize too strongly the importance of that understanding. The most successful internal entrepreneur I have ever known, by that time a senior corporate officer himself, once said: "The trouble with my associates is that they don't understand probabilities. They think probabilities are something that management can change." From his perspective, these managers, whose careers had been devoted to eliminating uncertainty, if at all possible, before they acted, were reluctant to accept the need to act—even when there were inherent uncertainties that could not be reduced by further study and analysis. All their instincts told them, learn more; require more data and more study before you decide! Those sentiments apply to the intrinsic risks I discussed earlier, but by learning how to innovate better, the overall probability of success *can* be improved.

In the lexicon of operating management, to miss a target or to fail to meet a budget is evidence of incompetence. Consequently, there is a strong aversion to making mistakes, i.e., missing targets. That kind of atmosphere, which equates failure with incompetence, is antithetical to a climate that nurtures innovation. This difference in attitude has a profound effect on the environment for innovation. If the attitude is, if it didn't succeed, then somebody goofed and we will exact retribution, then innovation is patently an irrational act.

In the aggregate, this set of conditions means that in order to succeed, a corporate program to encourage innovation must represent sufficient commitment and sufficiently powerful advocacy to provoke palpable tension with operating management. We have stressed repeatedly the need for tension between the advocates for continuity and effectiveness on the one hand and the need for change on the other. Unless each advocate represents enough power to be a force to contend with, there can be no tension. For example, GE's simultaneous involvement in commercial engines, nuclear power, and computers evoked strong feelings of resentment among many operating general managers, who felt they were being obliged to pass up needed investments in order to support these ventures. Without a strong, sustained corporate commitment to innovation as an important avenue to growth, that resentment might well have aborted all three. As it was, the enormous drain on resources which these three simultaneous ventures created led management to dispose of the computer business.

Implementation—Centralized Approach

Given the necessary commitment to innovation, what choices are available to implement it? The two principal options are to declare innovation a corporate-level function (centralized) or to decree that innovation is a part of operating management's responsibility. The centralized approach has the advantage of simplifying the work of operating managers. It says to them, don't you worry about innovation or growth; you just concentrate on managing your businesses for maximum performance, and we'll worry about innovation at the corporate level.

This approach in fact dichotomizes innovation, between that focusing on growth or diversification and that resulting from incremental improvements. Because businesses cannot survive unless they consistently introduce incremental innovations, an edict for operations to concentrate solely on operating performance and reserve all innovation to the corporate level is misguided and doomed to failure. These smaller incremental changes cannot be made without intimate knowledge of current operations. What must be preserved, whether in operations or at corporate level, is the separation in responsibility and resources. As we mentioned, innovation—even incremental innovation—requires commitment and resources. An operating component that is literally focusing on maximizing short-term benefit from input will not have people with the time or the resources to attempt to innovate. Furthermore, innovation is simply a very different kind of work. It is extremely difficult—I am tempted to say, virtually impossible—for an individual to succeed at both simultaneously.

The centralized approach is more often intended to aim at larger discontinuities that will lead to totally new products or new businesses. It does have the advantage of visibly demonstrating corporate management's commitment to innovation—provided it actually initiates some programs instead of sponsoring endless studies. The centralized approach also has the advantage—especially with an uncertain management—of reassuring it that activities are being properly monitored and that unwise investments or overcommitments are not being made.

I make these statements with tongue in cheek, because in practice what frequently happens is that fledgling innovations are stared to death or are drowned in resources and unrealistic expectations. In operational terms, fledgling activities close to senior management are often very vulnerable. Top corporate managers tend to see themselves as first-class operating managers—after all, that is how they got where they are. Unfortunately, those skills have often grown rusty or, more likely, are inappropriate for the scale and pace of a nascent innovation. The purview of a CEO covers the whole company. Furthermore, he normally deals with businesses that have established a level of activity and a momentum built up over many years.

An unfortunate, but all too common, scenario is played out. A promising invention or discovery is made and a decision is made to implement it. Almost all senior executives are obsessed with the challenge of growth—finding opportunities to place their enterprise on a higher growth trajectory. In their eagerness they begin to amplify the potential of the innovation so that its contribution can have an impact on the entire business during their tenure. Since they are able to commit large resources, they begin to insist that the innovation has greater potential than its advocates have projected and pour on the resources to bring these benefits on-stream faster. In consequence, the fledgling innovation acquires an overhead its trivial income cannot possibly support; the added resources do not significantly speed up the development cycle, and losses skyrocket. Senior managers become disillusioned with the venture's potential and either kill it or cut it back drastically in order to get in the black rapidly.

Even if senior management escapes these traps and sponsors a successful innovation, the problem of inserting it into the operating structure remains. Its reception may not be completely friendly. Even successful ventures nearly always require continuing infusions of resources—both human and financial—to support rapid growth. Hard-nosed operating managers often correctly forecast that those requirements will compete with the needs of their other businesses.

Nevertheless, for all its problems, many times a centralized approach is the only way an uncertain management is willing to proceed, and it should not be precluded from doing so. In fact, some innovations can be so unlike present operations that there is little choice except to initiate them at corporate level.

Implementation—Decentralized Approach

A decentralized approach that assigns responsibility for innovation to operating managers is the other option. This approach obviously diffuses responsibility for innovation throughout an organization. It also greatly expands the number of potential players and in principle should lead to the identification of a larger number of opportunities. On the other hand, this arrangement complicates life for operating managers by asking them to share the burden of tension between operational certainty and change that seeks to create new value through changing the established order.

An immediate problem arises: corporate management will now be trying to gauge remotely the vigor and effectiveness of an activity that is more elusive than operating performance. Mere delegation of responsibility will accomplish little without an environment that provides incentives to innovate and rewards for success—or even for trying.

The requirements for success that were laid out for the corporate level apply in operations as well. Unless a separate component (or at the very least, selected individuals) with no other responsibility is created, innovation will be crowded out by operating demands. Small operations will simply lack the resources to dedicate any to innovation alone. This limitation must be kept in mind in contemplating the delegation of responsibility for innovation. There is no point in laying on assignments that cannot possibly be carried out.

However, some of the problems that exist at the corporate level are mitigated or removed. Operating managers, being immersed in the exigencies of life at that level, are more likely to establish attainable objectives and realistic time tables. They are less likely to overwhelm a fledgling with resources. Furthermore, since the venture was born in operations, there is no problem of insertion. Conversely, operating managers are unlikely to initiate innovations that take them far afield from their base business. Again, incremental innovations are likely to proceed more effectively if they are undertaken close to operations—close both organizationally and geographically.

The obvious question, of course, is how to distinguish between innovations that should be undertaken locally in operations and those that should be attempted more remotely. The key distinctions are skills and professional identification. If an advance in the state of the art is of the sort that could be made by a skilled, creative professional working in a field, operations could reasonably be expected to assume responsibility for that kind of activity. In the light of the historical evidence that big advances come from outside a field, assigning that responsibility to a separate group would be sensible. When Monsanto decided to initiate a major thrust into biotechnology, it concluded—wisely, I believe—that the activities should occur at the corporate level. They required different skills, associations with a different research tradition, and interaction with a different research network.

With respect to motivation, the key issue is the performance targets for operating managers that will be used to appraise managerial success, not only in terms of compensation, but even more, in promotion. If corporate management talks about the importance of innovation, but promotes its hard-nosed operating executives, who improve profit margins and turn around businesses in trouble by dramatic cost reductions and elimination of programs for improvement, the real message to managers about how to succeed is apparent. If the risk-reward ratio is perceived as clear and certain risk, with uncertain or severely constrained rewards, an attempt to be innovative is patently irrational.

It is unrealistic (as well as counterproductive) to expect corporate management to establish specific innovation goals for individual operating components. One approach to motivation that avoids this problem is to make it clear that promotion will require more than simply achieving a level of growth equal to the growth of the present markets the business serves. In other words, the business must grow faster than its present markets, either by increasing share or by going after new markets.

Evaluating performance on innovations is crucial. Since the values of operating management tend to attach a stigma to failure, corporate management must stay close enough to a venture to be able to distinguish between poor performance and an innovation that was fated not to succeed. For example, at one point GE highlighted, in an internal companywide employee publication, the favorable aftermath for three managers associated with ventures that were folded.

Guidelines for Success

In establishing a project to pursue an innovation, what guidelines should be followed?[13]

1. Start with a small effort, i.e., do not smother it with resources. The flexibility and fast response that are required in the very fluid situation of a fledgling innovation mandate a small team.
2. Make the innovation the sole responsibility of the project members. They must have no place to hide, no other responsibility to divert them. I realize that smaller companies will say, we simply don't have the human resources to do that. If such firms attempt to combine the two kinds of work, they should realize the poor chances of success. They might be better off to severely limit their efforts to innovate to those few most-promising things for which they can isolate the needed effort. Certainly, one of the most common causes of failure is not appreciating the intense effort required to innovate and consequently spreading resources too thinly.
3. Provide relentless pressure to achieve goals, and make the escalation of support dependent on the progress achieved. The common wisdom is that innovations require patient money and that one must expect years of red ink before success is achieved. My experience indicates that is rarely the desirable path.
4. Ensure close physical contiguity. A project for innovation should feel a bit like a ghetto, a beleaguered sense of us against the world. After all, the goal of innovation is to change the world and that requires a banding together, an intense sense of missionary purpose, which is impossible among scattered participants. Furthermore, the necessary rapid, exceedingly efficient communication among team members cannot be achieved unless they are close together.

5. Include all the functions on the project development team from the very beginning. The traditional sequential functional handoff from engineering to manufacturing to marketing is apt to be unsuccessful. Success requires a melding of marketing, product engineering, and manufacturing perspectives that must begin at the start of the program. The time lost in the transitions from one to the next is itself enough to make failure probable even if an innovation hits the target, which is unlikely without recycling to "get it right."
6. Anticipate the rapid expansion of skills needed as progress is made. The requirements for information, to create the kind of certainty that must underlie useful technology, escalate rapidly.
7. Obtain market inputs early and continually. As I have said before, the customer determines value. Guidelines from radiologists played a critical role in determining the target specifications for GE's CAT scanner and in evaluating the performance of the instrument as it was being developed. Tait Elder, who was associated with many of 3M's innovations, emphasizes this point.
8. Focus on achieving a pioneering application as early as possible. A use that demonstrates the utility and the value of a new technology does wonders to establish its credibility. One common mistake is to spread effort over too many possible applications and to end up with insufficient resources to realize any of them.
9. Plan for the required financial resources to escalate rapidly. The most serious problem here is a subtle one. As a program succeeds, its escalating demands may outrun the resources or the risk tolerance of its initial sponsor. Thus a successful project may encounter the counterintuitive phenomenon of resistance to larger infusions of funds as it becomes more successful.
10. Staff the innovation project with the various roles required for innovation, not just the functional specialties ordinarily associated with a business. The confluence of people with the skills and interests to fulfill the roles that made Xerox possible is an almost classic illustration.[14]

    The *inventor,* for example, Chester Carlson, who makes the original discovery or who through persistent, creative effort develops a new capability. Innovation is frequently equated with invention, and it is certain that without the inventor's creative act of discovery there can be no innovation. On the other hand, many inventors lack either the ability or interest to perform the roles required to carry an invention through to innovation—the successful introduction of change. Without the contribution of many more people, xerography would never have become visible.

The *developer,* for example, Battelle Memorial Institute in the case of Xerox, who reduces a discovery to a practical form. Unfortunately, this role is sometimes depreciated in comparison with that of the inventor. Often the discovery practically has to be reinvented all over again in order to produce a practical device. A former Kodak executive once told me that he sometimes suspected that Edwin Land showed them the only successful instant color photograph he had ever made when he asked Kodak to develop and produce instant color film for Polaroid. The film Land had developed had terrible shelf life and was very sensitive to temperature, among other deficiencies.

The *entrepreneur,* for example, Joseph Wilson of Xerox, who sees the commercial potential in an invention and marshals the resources and assumes the leading risk in the effort to innovate. The *project manager,* for example, John Dessauer of Xerox, who manages the effort to create an innovation.

The *designer,* for example, Clyde Mayo of Xerox, who actually produces a device with acceptable performance, reliability, life, and cost.

The *marketer,* for example, again Joseph Wilson in Xerox, who had the creative insight to lease rather than sell the famous 914. Although the marketer obviously wants the effort to succeed, he must maintain a certain detachment. His job is to represent customers, to think as they think, to understand what product attributes they might value.

The *sponsor or champion* (a crucial role in internal innovation), who metaphorically flies air cover for the endeavor. He must have sufficient credibility with top management to reassure it to "stay the course" during the inevitable vicissitudes that occur in innovative efforts.

Conclusion

Advances in technology are generally expected to be the wellspring of innovation. Furthermore, the manager of technology is typically the designated spokesperson (the advocate) for innovation. Technologists are expected to have the most informed perspective on what will be possible as well as the capability to convert vision into reality.

On the other hand, innovation cannot be the sole province of technology. As I have noted, many very important innovations arise in other realms. Of even greater importance, successful innovations require the dedicated participation and expertise of all functions. Studies of innovation consistently show that most failures in innovation arise not from deficiencies in the technology, but because of inadequate marketing input.[15] A senior executive, looking back on a distinguished career in innovation, called attention to the great value of a creative, supportive finance man in helping both nurture and protect a fragile innovative effort. The technology manager faces the challenging task of retaining his visionary eye for what technology can offer for the future as well as eliciting the support of other functions that may not share his vision, all the while preserving his image as a results-oriented manager.

The people who have the abilities and motivation needed to become successful innovators are a scarce resource. The probabilities that they will succeed, and even more important, the number of potential candidates who might succeed if adequately buttressed with additional talent, can be significantly strengthened by an environment that nurtures innovation. The creation of this environment is very largely in the hands of management. The bizarre mixture of sustained support and relentless pressure to succeed calls for skill and sensitivity that are themselves rare talents. Sustained success in innovation requires not only supportive management, but a system that motivates and rewards those supportive managers.

Notes and References

1. K. Arrow, "Economic Welfare and the Allocation of Resources for Invention" in *The Rate and Direction of Inventive Activity,* National Bureau of Economic Research, Princeton University Press, Princeton, 1962, p. 622; also, E. Mansfield, *The Production and Application of New Industrial Technology,* W. W. Norton & Co., New York, 1977, pp. 144–166.
2. Thomas J. Peters and Robert H. Waterman, Jr., *In Search of Excellence: Lessons from America's Best-Run Companies,* Harper & Row, New York, 1982.
3. Peter F. Drucker, *Innovation and Entrepreneurship: Practice and Principles,* Harper & Row, New York, 1985.
4. Richard Foster, *Innovation: The Attackers Advantage,* Summit Books, New York, 1986.
5. Brian Rushton, "Strategic Expansion of the Technology Base," *Research Management,* November–December 1986, pp. 22–28.
6. Graham Nown, *The World's Worst Predictions,* Arrow Books Ltd., London, 1985, pp. 11, 67, 71, 75.
7. Joseph Schumpeter, *Capitalism, Socialism, and Democracy,* Harper & Brothers, New York, 1942.
8. Robert Solow, "A Contribution to the Theory of Economic Growth," *Quarterly Journal of Economics,* Vol. 20, 1956, pp. 65–84; "Technical Change and the Aggregate Production Function," *Review of Economics and Statistics,* Vol. 39, 1957, pp. 312–20.
9. Edward F. Denison, *Why Growth Rates Differ,* The Brookings Institution, 1967, pp. 279–295, 298–299; *Accounting for United States Economic Growth, 1929–1969,* The Brookings Institution, 1974, pp. 62, 79–83.
10. Donald W. Collier, John Mong, and James Conlin, "How Effective Is Technological Innovation?" *Research Management,* September–October 1984, pp. 11–16.

11. Collier figures 1 to 6 are reproduced with permission of *Research Management.*
12. David Ravenscroft and F. M. Sherer, "Is R&D Profitable?" PIMS Letter #29, 1982.
13. J. B. Quinn in "Managing Innovation: Controlled Chaos," *Harvard Business Review,* May–June 1985, pp. 73–84, provides an insightful discussion of the environment needed for innovation. *Technology Review,* MIT, Cambridge, Mass., has collected an excellent group of articles from past issues into a monograph, *Innovation or How to Make Things Happen.*
14. Gary Jacobson and John Hillbrick, *Xerox—American Samurai,* Macmillan, 1986, pp. 53–68.
15. *Success and Failure in Industrial Innovation: Report on Project SAPPHO,* Science Policy Research Unit, Sussex University, London Centre for the Study of Industrial Innovation, 1972, was a seminal study in this field. See also S. Myers and D. G. Marquis, *Successful Industrial Innovations,* National Science Foundation, U.S. Government Printing Office, 1969.

Part

4

Crosscutting Issues

The two preceding parts have focused attention principally on the theme of tension: how to ensure it and how to manage it effectively. This section will first address in more detail the other two themes, the cost of fragmentation and the cost of the "invisible hand" of conventions in nurturing effectiveness but inhibiting change. Fragmentation involves an organizational structure that, for quite valid reasons, disperses technology among many functions, but as a by-product, impedes adequate integration of product, process, and information technology. Chapter 12 addresses this issue. Chapters 13 and 14 address evaluating the performance of a technical operation and selling technology to top management, respectively. In summary, this part takes up topics that cut across the whole field of technology management.

Chapter 12 covers the crucial issue of improving system integration. Technology interacts with all aspects of a business. Unless the impact of technology on the other functions is incorporated into technical decisions, the result will almost inevitably be products which do not meet customer needs satisfactorily, costs which are higher than need be, and response times which cannot react adequately to market and competitive forces, and human and capital assets which are not as productive as they could be.

Chapter 13 is devoted to the problem of evaluating the performance of a technical operation. This task is one of almost universal concern, and it is one at which managers who do not have a technical background feel particularly inept. The chapter identifies an extensive series of clues that result from the way a technical operation goes about its work, which give valid indications of its effectiveness without requiring extensive knowledge of the substance of the technology itself. I have discovered that the search for and evaluation of these clues does not require a technical background. It does require commitment of personal time and energy and the identification of technical people whose competence and judgment can be trusted. But then judging people has always been a key ingredient of managerial success.

Finally, Chap. 14 takes note of the need to sell technology to the chief executive or general manager. The task has similarities with all selling activity. It requires knowing the customer, i.e., top management, being certain that the technology you ask it to invest in is truly responsive to its needs, and taking a variety of actions to influence its mind set favorably regarding technology.

Chapter 12

Achieving System Integration

One of the continuing themes of this book is the pervasiveness of technology. With the inclusion of the information dimension, technology—both its present capability and its future potential—affects all functions. So, managing technology effectively requires understanding, in the broadest sense, the nature of the interaction between the various components of technology and all the other activities involved in running a business, how decisions about technology affect these other activities, and just as important, how the requirements and constraints of these other activities influence technical decisions and actions. Inevitably, failure to achieve integration has consequences: product attributes that do not meet market requirements optimally because the necessary market inputs have not been included; costs that could have been avoided; slower responses caused by recycling programs to correct needless mistakes; and time wasted when information that could have been provided was not available as needed. The final result, of course, is profits and growth less than they could have been. In a tough, competitive situation, poor system integration can be life-threatening. Effective attack on improved integration obviously requires attention to both the management system and the organizational structure.

The Value Chain

The value chain offers a way of approaching the problem of improving system integration.[1] The value chain can be regarded as a system in which the coupling between the sequence of activities for creating value and their downstream impact on that value must be taken into account at every step.

In this chapter we analyze the way in which technology has conventionally been used to provide value; then we consider avenues for improving system integration to optimize value on a sustained basis. The word "sustained" is crucial because a value-producing chain is analogous to a thermodynamicist's entropy-enthalpy balance—no matter how well-designed initially, it inevitably tends to degrade. One of the important responsibilities of technology management is to intervene periodically in the value chain to restore optimal integration, as well as to include "system effects" in technical decisions routinely.

The Value-Creating Process

When the creation of value is viewed not as the choice of modes for gaining competitive advantage, i.e., cost, performance, or product differentiation, but as the sequence of activities needed to *create* value, the sequence looks quite different from a conventional organization chart. Figure 12.1 describes the value chain from this point of view. Superficially, this sequence is so obvious as to warrant no elaboration. Nevertheless, no matter what competitive strategy is chosen, this sequence must be used to create value.

The value-creation process begins with the identification of a need—something that customers want or would come to want once they perceived what need it would satisfy. The value-creation process moves logically down a sequence that ends in after-sale service. The necessity for close coupling between each step also hardly requires justification. If the customer need cannot be satisfied or the response does not truly meet the need, for example, the consequences are obvious.

The criteria for success include the following: Do attributes truly meet customer needs? Are costs such that price and volume goals will produce planned margins? Does the entire process operate in a sufficiently timely fashion to meet the market window before competitors do? Finally, of course, does the entire sequence produce a profit?

The operational view

Unfortunately, that sequence is ridiculously easy to describe and excruciatingly difficult to carry out. The objective of operational management is to drive the entire sequence as rapidly as possible by re-

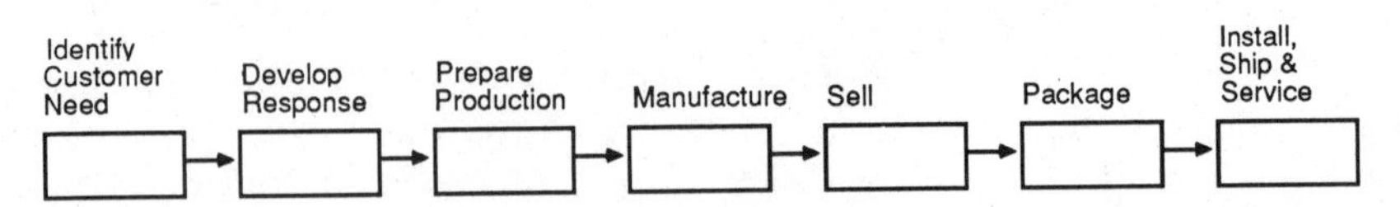

Figure 12.1 Process of creating value. (*Source: William Gutzwiller in a personal communication.*)

ducing mistakes, eliminating wasted motion, avoiding needless delays, and otherwise running a tight ship, as suggested in Fig. 12.2.

Effective operational management, however, requires more than products that meet customer needs and produce a profit. A business that can more rapidly complete this sequence clearly has a competitive advantage. One of the most often-repeated questions I hear is, "How can the Japanese respond so rapidly?" Fast response not only permits quick reactions to market dynamics, it also keeps competitors in a reactive mode.

The profits resulting from expediting movement through the chain can be incremental to the profits generated by individual product introductions because they reduce the basic costs of doing business. The activities prior to manufacture are very people intensive. They require payment of compensation for work that is investment-like in character even though it is regarded as expense. However, the expenditures are not recouped in revenues often until several years later. The products currently producing revenue must not only cover the costs associated with their manufacture, sale, and service; they must also carry the burden of the activities that do not produce revenue until later. If the entire sequence can be speeded up, current revenue-producing products have to carry the burden for a shorter period of time. Alternatively, the choice may be to reduce the resources required to carry out this sequence and keep response time the same. In either case, the business employs its human resources more effectively and therefore increases its profit potential or gains additional freedom in pricing.

Let us examine each step in the sequence from an operational point of view, looking for opportunities for improvement. Then let us examine the functioning of the entire system strategically—not program-by-program—to find opportunities for more fundamental change that could further improve competitive advantage.

Identifying customer need. In the abstract, identifying customer need would appear to be a marketing function. In practice, the process requires a creative interaction between envisioning or discovering a ca-

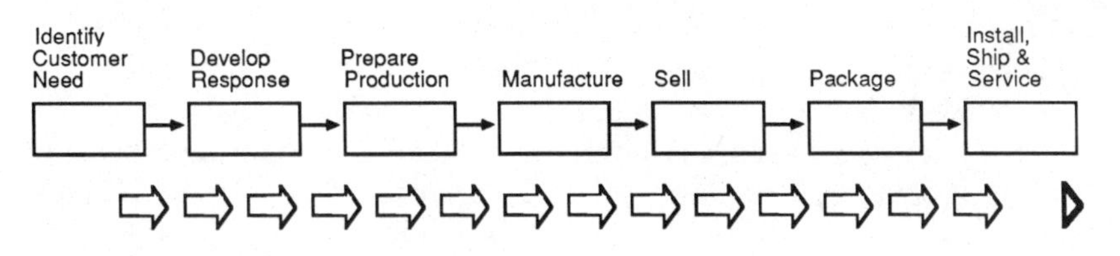

Figure 12.2 Speeding the value creation process.

pability and imagining how it might come to be valued by customers, on the one hand, and observing or querying customers to discover what they might want, on the other. Thus a component of what is feasible or could become feasible (i.e., technology) should always be a factor in identifying a customer need that a business will attempt to meet.

In considering this technology-marketing interface, it is important to identify roles carefully. Three different points of view need to be represented, views that are different and somewhat incompatible. The technologist is an advocate for technology. As we saw in discussing innovation, incorporating new capability is a process that does not occur more or less automatically—somebody has to push. The technologist pushes capability and its value in use. Discovering or inventing something new is a remarkable accomplishment. The posture of the creator is, and should be, this is great stuff—I know the world is waiting for it! Much is made of the desirability of market pull–induced innovation. However, that market pull is much more apparent from an ex post facto viewpoint. In prospect it is usually no more than a barely perceptible tug.

The postures of the other two actors, marketing and sales, are different, and it is important to distinguish between the two. Not infrequently the terms are used interchangeably, or labels are changed from sales to marketing because "marketing" is more voguish, without changing work. Sales represents the existing product line in meeting already demonstrated customer needs. The salesperson has clearly delineated objectives: to establish personal relationships that induce customers to prefer his company's products, to avoid anything that upsets or otherwise threatens the relationships he has so carefully fostered, to maximize the attractiveness of his products in meeting the customers' needs, and to achieve closure on sales. Marketing represents the customers in a more abstract sense: who they are, what they are like, what their needs are, and how they can be influenced. Marketing is focused on the future. Sales is focused on the present. It follows that sales is not a natural ally in innovation, but marketing is or can be. Of course, if a product is not selling because it does not meet customer demands, salespersons will begin clamoring for innovation.

There is an intrinsic tension between these three roles. A company cannot survive unless its relationships with customers are protected—the salesperson's goal. Furthermore, the salesperson's drive for closure, to make a sale, must be present in successful innovation. Innovation does not occur unless somebody extends capability, uses it creatively to satisfy needs, and urges its adoption—the technologist's goal. Innovation also will not occur if it is pushed blindly, irrespective of the customer's preferences. Marketing, in this sense, is the customer's advocate, asking: does our offering truly meet his needs or are we just pushing what we've got? However, marketing must also understand capability, i.e., technology, in order to help imagine uses. This need not, however, necessitate a deep comprehension of the how and why of technology—that is the province of the technologist. Unfortunately, technologists can fall into the trap of insisting that others must understand the details of a capability before they can participate in attempts to uncover possible uses.

In order to improve system integration in identifying customer needs, the initial step in the value chain, a manager must first ascertain that the three necessary roles—technology, marketing, and sales—have been assigned and are adequately staffed; then he should ensure that the representatives of these three activities collaborate in identifying needs. The technologist needs direct access to marketing inputs, but his advocacy for his technology must be monitored lest he damage a relationship with a customer. The salesperson's drive for closure on an actual sale is vital, but he is not necessarily an unbiased interpreter of customer needs. Nevertheless, his legitimate concern with protecting his relationship with his customers must be respected. Marketing considers customers as a class and is concerned with the sustained growth of the business. It mediates to some extent between the technologist and the salesperson—guiding and sometimes restraining the former and prodding the latter. The entire sequence is somewhat like a play in which the roles are specified, but the scenario and script vary from one product development to the next.

Effective system integration, as I stressed in discussing program management, requires continuing attention to "should we" questions as well as "can we" ones. Identifying a customer need is an iterative process of gradual convergence between increasing understanding of what the customer needs and improving comprehension of what the business will be able to offer. Thus identifying customer need is rarely an immutable target set at the beginning.

Developing a response to customer need. This step in the sequence requires a three-way interaction between marketing, product engineering, and manufacturing. Product engineering leads this effort, but it must continuously adjust its work to incorporate inputs from the other two areas. In addition to attempting to meet customer needs, there must also be concern for manufacturability, which includes cost, quality, and yield; for response time (can the requisite facilities and human resources be marshalled in time to meet the market window?); and for investment requirements (can the business provide the financing to fund the product introduction?).

In fostering the participation of manufacturing, it is important to realize that different aspects of manufacturing have different concerns when contemplating the production of a new product. Technologists in product engineering are inclined to regard the fabrication and process elements of manufacturing as most crucial in achieving the desired product attributes. Without questioning the importance of this component of production, manufacturing people place greater emphasis on assembly and materials handling. Complex process and fabrication requirements can typically be achieved by technology embedded in sophisticated equipment. Assembly and materials handling are much more under human control and therefore inherently subject to greater variation. A senior officer of a large company producing mechanical products commented that, in the company's initial efforts to speed product development by having all functions participate on a team, a mistake had been made by having only experts in fabrication involved. He said, "We should have had assembly involved as well, because we built in some problems that could have been avoided."

The bugaboo of manufacturing is the complexity of the entire system, which is determined by how many different kinds of products must be manufactured as well as the variability that is necessary to respond to changing output requirements. These considerations affect the cost of the entire system by influencing inventory requirements, employee productivity and quality, and investment utilization, because change in output disrupts production and reduces capital utilization.

Before going on to examine the problem of system complexity, the system integration requirements with sales and service must be considered. The penalties imposed by failure to pay sufficient attention to this are becoming increasingly common. Traditionally this interface has received the least attention in the product development process. The problem arises in part from organizational separation, but perhaps even more from inadequate information technology. Customers are geographically dispersed and remote, product service is often performed by unknowns, and product performance often needs to be tracked for a number of years to establish a meaningful data base.

Failure to achieve adequate system integration at this interface, however, can lead to product attributes that cause customer dissatisfaction, difficulties in maintenance and repair, and investment in spare parts that is larger than need be. On the other hand, the in-house customer service that Sears has offered through its retail outlets has been a valuable source of information that could be fed back into product design and manufacture. The penalties apply to individual product development, but they loom much larger for the total value-creation system.

Typically the management system provides little help in achieving system integration with installation and service. No mechanism is in place to accumulate field experience systematically. Furthermore, where field sales and service do not traditionally interact with product development, it often is not evident whom product engineering could look to for help. If a business does not have in place the routines and data bases to provide information and a design review process to ensure that this information is incorporated into products, almost certainly it is incurring penalties in customer dissatisfaction and avoidable costs.

Improving Operational Integration

Achieving effective system integration is obviously a complex process involving a chain of people over an extended period of time. Furthermore, the interests of the various participants are diverse and often contradictory. Success calls for a number of ingredients: (1) a management system that mandates integration and then specifies the participants; (2) people charged specifically with representing their various functions in providing the coupling link with the process; and (3) information bases generated from market research on one end to field service experience on the other. The introduction of a new or modified product should require the sign-off of every activity that is significantly affected by its introduction. The producers of large apparatus, such as turbine generators, transformers, or paper machines, were sensitized to the value of system integration decades ago. The curvature of railroad tracks and the dimensions of bridges and tunnels have long been a constraint on the size of apparatus that could be assembled and shipped from a factory.

Perhaps the most important requirement is psychological—a change in viewpoint. A shift in perspective from focusing on individual parts of the value-creating chain to focusing on the effectiveness of the complete system requires a different sense of responsibilities and criteria for effectiveness.

The opportunities for improved system integration are one of the principal motivations for my insistence on improving integration among the product, process, and information components of technology. It is apparent the actions I advocated above are critically dependent on the availability of an integrated information system. This system must not only serve the local needs of each function, but also identify and make available the information needed across functional lines.

The most obvious benefits are financial—lower inventory all across the board, reduction of complexity in manufacture, and products better targeted to serve customers—but these financial benefits are by no means the only ones. An integrated information system permits a faster response time from order to shipment, but also in product development. It enables customers to have direct access to product availability and status of work in process. It can allow integration among diverse sites—and in a world increasingly dependent on multinational operations both in the flow of material to production and in the choice of facility to supply customers, such integration is indispensable.

Strategic Implications of System Integration

In discussing actions to speed the movement of an individual program through the system, we have several times mentioned the general problem of the value-creation chain (system) through which all individual developments are carried out. As we have said repeatedly, the principal factors in determining competitiveness are product attributes, costs, and response time. So far, we have focused primarily on getting the product attributes right and moving a program expeditiously through the system. Let us now consider cost.

Understanding the sources of cost

Direct and indirect costs. The traditional focus of product development has been on the specific costs associated with the product under development. An implicit assumption of product development has been that concentrating on the costs one can readily identify is an acceptable mode of operation. Costs that are not visible or not easily determined are, as might be expected, largely ignored. In practice, development and design engineers have focused on direct costs, especially materials and direct labor, and sometimes primarily only on materials, because those costs could be determined most unambiguously. Engineers have known that there are indirect costs associated with manufacture; but these costs are not easily determined for a specific new product, and for the most part, conventional allocation ratios have been used to "burden" the direct costs with indirect expenses. The primary objective has been to ensure that, on the average, products are allocated indirect expenses sufficient to recover all such expenses in the cost of the products sold and to value goods in process at a level which incorporates all costs already incurred. Occasionally, when a given new product departs significantly from the factory average for indirect inputs, a special "burden" is allocated to more nearly reflect its true costs.

The roots of this practice go back in history to a much more primitive era of information technology, when even keeping track of direct costs was a challenge. We will return to the role of information technology subsequently, but first let us examine the consequences of this method of allocating costs and consider avenues for improvement.

The focus on direct costs has led eventually to a distorted perception of the sources of cost. Given the long, sustained effort to control direct costs, one would expect some success, and that would appear to be the case. Direct labor costs are often only 8 to 10 percent of shop cost in many manufacturing operations. One consequence, of course, is that additional improvements are harder to achieve and the impact on total cost shrinks—even a dramatic reduction in an item that represents less than 10 percent of total cost has limited impact. Materials are more likely to represent 30 to 50 percent of cost and continue to represent an attractive target. That is one reason why reducing inventory continues to receive so much attention. Just as important, costs which receive little attention tend to grow, and that also has happened.[2] Figure 12.3 shows a typical distribution of direct labor and material and all other indirect costs. Indirect costs include unapplied direct labor, indirect labor, scrap and rework, complaint expense, engineering, depreciation, utilities, and carrying charges for inventory, for example. Furthermore, as we have indicated, these indirect costs tend to grow. Figure 12.4 shows a common sequence.

Unfortunately, some indirect costs in fact are increased by intensified efforts to reduce direct cost. The puzzling scenario where several years of successful cost-reduction programs (focused primarily on direct costs) do not actually produce anything like the same aggregate effect on total cost is well known. The adverse effect on indirect costs of many such claimed cost reductions is a major culprit.

The gradual deterioration in cost competitiveness noted in Fig. 12.4 forces practically every business to take periodic corrective action. A common scenario is as follows: The general manager observes a deterioration in margins and urges closer attention to costs; little progress results. The general manager says, we have to reduce our head count and improve on inventory turnover. Operating managers say, "He doesn't mean me. I'm facing these special circumstances that require the people and investment I have in place." The general manager orders everybody to reduce head count by 5 percent

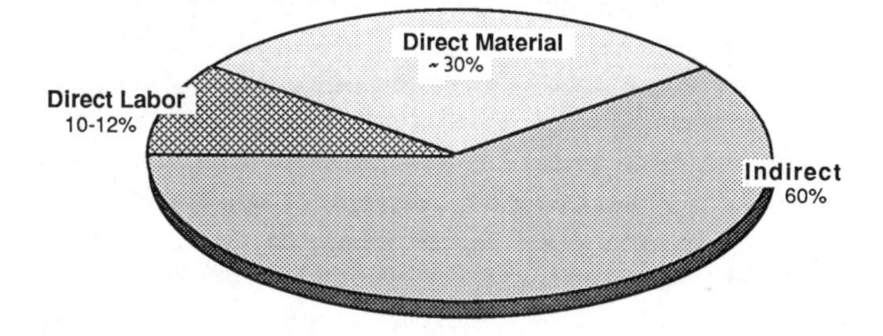

Figure 12.3 Typical cost breakdown.

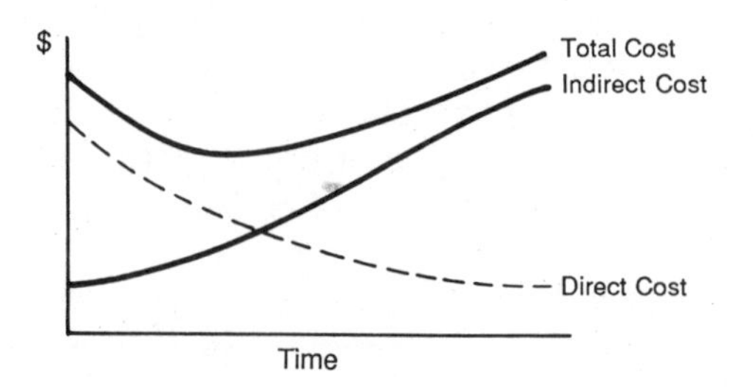

Figure 12.4 Cost history of a product line.

and to reduce inventory by 8 percent by the end of the next quarter. Operations lays off people in a panic and scraps inventory to meet guidelines. The actions that caused the problem in the first place are never identified and rectified because there is not time. Operations is upset at the broadax approach of the general manager. The entire sequence begins again.

Interaction with technology

Actions in the technical community have often contributed to escalating costs without its being aware of the effect. I remember my astonishment at learning in the mid-seventies that the imputed interest cost for carrying inventory in GE's large steam turbine generator business was larger than direct labor costs. This was a long-cycle business in which product performance was paramount. Therefore engineering unhesitatingly sent change orders to manufacturing whenever it devised an improvement, with little awareness of the cost they generated. The goal in manufacturing was to produce what engineering sent it, because engineering was the highest-status function in the business.[3] One U.S. automobile company calculated that the various permutations of the options it offered presented customers with a million market entities from which to choose. This was true despite the fact that a very large percentage of the various combinations was rarely chosen.

Effective management of technology requires a much better understanding of the sources of costs and of the impact of technical decisions on those costs. Furthermore, we need to broaden our concept of cost to include not only monetary costs associated with labor, material, capital, and time, but also response time and flexibility. A business that cannot respond in timely fashion to market dynamics or competitors' moves is just as seriously handicapped as one whose costs, as conventionally defined, are too high. An intense and nominally effective concentration on the conventional narrow interpretation of costs can in fact reduce the overall competitiveness of an enterprise.

Costs arise from three sources: the direct materials, labor, and facilities and equipment for producing a product; the indirect labor, materials, and facilities for supporting the production of a line of products (if a business produces only one model of one product, the distinction between direct and indirect loses much significance); and in addition to these product costs, there are what I would term the system costs—the costs of doing all the other things shown in Fig. 13.1 that are required to produce value, remain competitive, and survive. We have already noted the sustained concentration on controlling direct costs; let us now look at indirect costs and system costs.

Indirect and system costs

The problem of indirect costs is created mostly by the necessity to have one or several lines of products rather than a single product. Thus the sequence of activities for a given product in Fig. 12.2 is in practice an extended array of parallel sequences at various stages of completion, as shown in Fig. 12.5.

The problem of controlling indirect costs is primarily one of understanding and managing the cross impact of each such sequence. The present indirect cost structure encompasses the aggregate of all previous sequences for creating value. When operating management focuses on improving the effectiveness of a single sequence (an entirely worthy objective), it must take into account the interaction between that value-creation sequence and the system that will be used to implement it. Traditionally, the product development process has assumed that the introduction of a new product or the revision of an existing product has no effect on the business that has to produce, sell, and service it. The effects have been largely invisible because they typically occur in small increments, they are dispersed over many disparate activities in different functions, and they are spread out over time. Furthermore, even if one were aware of and interested in controlling this interaction between a new product and the business system, there rarely is any way of measuring the effect.

Nevertheless, a new product *does* interact with the system—usually by raising costs because it increases complexity. A new product affects facilities, complicates production planning and control, increases in-

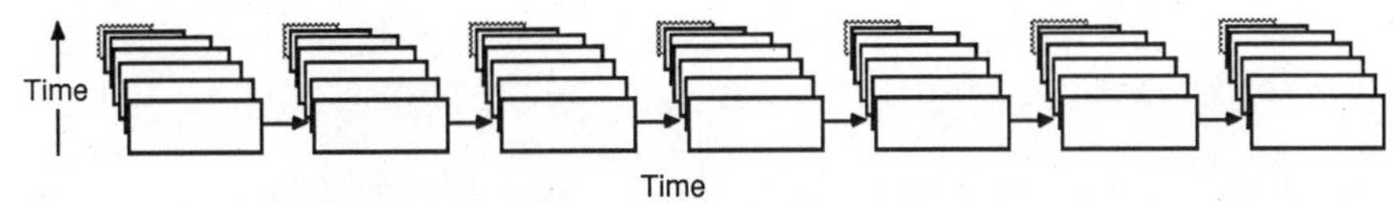

Figure 12.5 Multiple value-creating sequences.

ventory, imposes demands to acquire new skills and adjust to working with a greater variety of products, makes salespersons become familiar with an additional product and spread their effort over more market entities, creates more variety in spare-parts logistics, and requires field service to be able to maintain a greater variety of products. Thus costs that customarily are assumed to be fixed for a given product introduction will tend to increase gradually but inexorably. The individual effects are small, but they are cumulative. After several cycles, the cost structure being used to project costs and margins for a new product no longer depicts the reality of the system it is supposed to represent. This problem should be attacked at three different levels: the product line, the design process, and the entire business system.

Product line structuring. Product lines tend to proliferate. One of the most obvious ways to increase sales is to examine customer needs minutely and to develop products that more nearly match the preferences of a particular segment of customers. Such market segmentation is in the arsenal of every competent competitive strategist. Another obvious way to increase sales is to add products that will add customers. The advantages are easily demonstrated and the results in terms of increased sales easily measured. The pressure to do so is relentless. Unfortunately, many of the costs, I reiterate, are not easily measured and insidiously obscure.

Unless management deliberately creates a back pressure to inject discipline into the product planning process, it will almost certainly end up with too many products that are no longer creating competitive advantage, but are instead imposing a heavy penalty on system costs.

Some businesses even brag about the breadth and variety of their product lines. GE's lamp business used to emphasize that it produced 10,000 different lamps—to the despair of its manufacturing function. When a manager with a background in manufacturing became general manager, he created a product-delisting component in marketing. Every quarter the component was to propose products to be removed from the line. Naturally, the sales force objected strenuously. The procedure called for tentative delisting to measure response, with permanent removal to follow unless a strong negative reaction appeared. In consequence, literally thousands of product variations disappeared—to the great benefit of system costs.

One of the first steps in preparing a strategic plan is to put the product line under a microscope to determine how much each market entity contributes to profits. Technology should play a leading role in this activity. If technology achieves the integrated perspective I advocate, it should become more aware of the system penalty imposed by proliferation of the product line. Once again, I say that effective management requires nurturing and balancing contradictory forces—the value of market segmentation and an extensive product line against the system penalty imposed by the complexity of such a product line—but both points of view must be presented forcefully enough to command attention.

The challenge to managers of technology is even more complicated than that. Technology is in a tenuous position if it turns its role into being an advocate for drab sameness in products. A legitimate fear of a business is that its products will come to be regarded as indistinguishable commodities, and a stubborn refusal to permit product line proliferation can lead to that. The loss of brand differentiation among General Motors cars in the early eighties aptly illustrates the danger. Chrysler, in contrast, managed to achieve remarkable product differentiation from a restricted array of basic platforms.

Technology can be the principal force in reducing the system penalty from product line proliferation. The primary objective of flexible automation is just that—to increase the capability of a system to handle a larger variety of market entities without penalty. Thus effective management of technology requires balancing an additional tension: contributing to the rigorous analysis that will discipline product line proliferation while at the same time trying to identify and change the characteristics of the system that limit its ability to handle diversity in output.

The product development and design process. A process parallel to the tendency toward product line proliferation also exists in design. In this case, it involves proliferation of materials, components, manufacturing processes, and so forth. Traditionally, product development and design has been considered as one of the cardinal acts of technical creativity. It has been accepted as a highly individual process in which a creative engineer invents and extracts from the existing corpus of capability a unique response to a perceived need. If done well, with no external constraints, it presumably represents the optimum solution to that need, drawn from the almost infinite variety of bits of technology available to the designer. If a large number of designers are each given freedom to exercise unfettered creativity in the process, the result will inevitably be a mind-numbing variety of materials, components, configurations, dimensions, etc.

In practice, a large percentage of those individual choices do not signify a uniquely preferred selection, but rather simply the necessity to make a choice. In other words, any one of several materials could work, or several different components could perform the function. The particular choice combines the criteria the designer believes he must satisfy, his past experience with various materials or components, the menu of options he perceives to be available to choose from, and even his diligence in optimizing the process.

Probably the key item in the list is the criteria designers believe they must satisfy. If they believe lowest direct costs for materials and labor are the goal, they will create increasing variety in both materials and human responses because they will tailor materials and labor requirements to the narrow constraints of the specific product they are working on.

Product structuring is a natural follow-on to a strategic review of the product line, which determines what portfolio of products will compose the market offering. Product structuring starts by identifying all the components (even the most elementary ones such as nuts and bolts), materials, and processes currently used to manufacture the newly characterized and more constrained product line. All of these elements are then grouped together in subsets by function. For each subset, each item is subjected to the question "Why?" Why do we have to use this item? Are there other similar items already more extensively used that could serve this function adequately? Every effort is made to modularize design and to standardize a restricted list of elements.

In addition, the product line is examined to see whether a rearrangement of the sequence of fabrication and assembly could delay differentiation among products until later in their manufacture. "Late differentiation" not only contributes to longer production runs with attendant economies but also improves response time to changing market demand. The earlier a production process commits to product differentiation, the sooner finished goods inventory is locked in.

It follows that once this product structuring process has been carried out, management routines have to be established to maintain discipline. A customary routine is for any departure from the approved list of materials, components, and processes to be approved at a higher level.

Clearly, information technology can be a powerful factor in both instituting and maintaining discipline in product structure. The work involved in introducing a computer-aided design (CAD) system accomplishes many of the same objectives—forcing an examination of all aspects of the design process and an answer to the question, do we need this? Furthermore, CAD provides a convenient mechanism for discipline, even though the actual imposition of discipline requires both persistent and consistent management. With all the approved materials, components, and so on, lodged in computer memory, the system can easily raise a red flag on any departure from routine.

Initially, a program like this necessitates a sustained effort of considerable magnitude. It requires a multifunctional team comprising relatively senior representatives from engineering and manufacturing, plus contributors from finance to quantify the costs and benefits of options being considered. Senior people are needed because only they know both the total system and the impact on product attributes well enough to perceive the implications for the system of any proposed changes. In other words, they can judge whether a proposed standard would restrict the design in such a way as to impair product performance. It is not unusual for a product structuring program to take two to three years to complete. Fortunately, the benefits begin to accrue much earlier because implementation is generally piecemeal.

Examining the total system

The problem of the competitiveness of U.S. industry did not materialize just in the eighties. Concerns over productivity appeared at least fifteen years ago. The initial response was to improve each function independently. Gradually managers have come to realize that the problem was a system problem. The actions I have been proposing up to this point still deal only with pieces of the problem: rationalizing the product line and imposing discipline on design. Now let us turn our attention to the total system and examine its functioning.

In Fig. 12.1, we laid out the sequence of the system for creating value. Rather than focus on the operational question of how any given sequence of product development, manufacture, and sale can be made more effective, let us now ask how the aggregate of all such sequences behaves. As I noted before, the present system cost structure is a consequence of the cumulative effect of all previous product development, production, and sale programs. These system costs are more important to the total competitiveness of a business than the specific shop cost resulting from an individual development program.

In discussing this subject of cost, we must be clear about the terms we use. I include in cost not only the conventional inputs (both direct and indirect) for producing a product, but also the system costs for the business to function, i.e., the cost structure determined by all its past decisions. But I go beyond that to include response time and flexibility. A mode of operation that causes a business to react too slowly to market requirements is a cost barrier just as surely as labor or material. A mode of operation that makes a business too cumbersome to accommodate a request for a special product also is a barrier to growth and profitability.

The entire value-producing system should be analyzed from this perspective. How are the human and financial resources of the business deployed? How is the time required to produce value distributed among the various kinds of work? Assembling data of this sort is not easy. It requires cooperation from all functions and some ingenuity by finance in allocating expenditures. The results can be startling. For example, it is not unusual for managers to be unaware of the mismatch between where money is actually spent and where management has focused its attention in controlling costs. A manager of manufacturing once expostulated to the labor relations manager who was bragging about having saved 10¢ an hour in labor rates in contract negotiations: "Big deal! You saved 1 percent on an item that represents 10 percent of our costs." It is not uncommon for operational appraisals to reveal inventory levels six to eight times what is essential. Safety cushions to ensure uninterrupted operation are pyramided up the chain. The costs of quality are scattered throughout the system: inspection, rework, and scrap in manufacture; warranty costs in sales; and repair costs in service. Unless these are aggregated, the true costs of quality remain invisible and their sources unknown. An inadvertent choice of a dimension for a refrigerator which makes it impossible to load a boxcar to capacity is an avoidable cost created by technologists. A design that is not easily revised generates costs downstream. The possibilities, unfortunately, are almost endless. Figure 12.6 indicates the questions that should be asked about every phase in the value-creating chain.

Then comes the crucial question "Why?" Why does it take this long? Why are so many people required? What can be done to shorten time or reduce expenses? What effect would that have on the business? Going through this exercise highlights the rate-limiting activities in the sequence—why fight for minutes on the assembly line and ignore weeks or months at earlier stages? Why focus all effort on a 10 percent reduction in an item that represents only 10 percent of cost and ignore other elements of larger magnitudes?

Obviously, an exercise of this magnitude cannot be done continuously, but a periodic reexamination of this sort can lead to dramatic benefits. One business, facing a severe shortage of capacity, asked the manager of engineering to take several weeks off to reexamine how

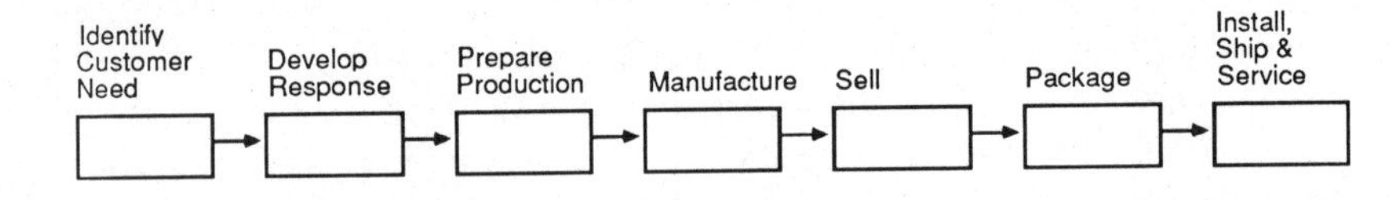

- For each step in the sequence, what is the range of times for completing that step, and what is the average?
- How many people, in what categories of skill, are required to carry out this work for the business?
- How much direct and indirect expense is created at each step?

Figure 12.6 Evaluating the value-creating chain.

the business should be split between two sites and where the second plant should be located. At that point, delivery schedules were slipping, and the factory was filled almost to the danger point with people and material. The manager concluded that a new plant probably was not necessary if they rethought the entire business. By carrying out an exercise such as described here, the business tripled output over the succeeding five years without expanding the plant or enlarging employment. It cut delivery times in half, cleaned up the factory, and dramatically improved profitability in what was already a profitable business. In the conventional sense, this business had not been poorly run, to begin with. It had technological and market leadership and was growing rapidly. However, it also obviously had the potential for greatly improved performance. Later the manager who initiated this effort said, "I'll bet I could go back in there and do it all over again."

We *can* rewrite the book, and the results can be even better than we expected. The 50 percent reduction in cost that my manufacturing consultant friend pleaded for in Chap. 3 can be attained and other benefits derived as well!

The Potential of Information Technology

Despite the impressive results already achieved, there is far to go in exploiting the great promise of information technology. Nowhere is it more important than in the field of improving system integration. From the beginning, I have emphasized the powerful role of conventions in determining our perspective. Many of these conventions are based on the level of knowledge it was practical to achieve on a sustained basis about the current status of a business. The conventions often served to provide a safety margin to cover insufficient information. Two problems exist in using these conventions. First, to the extent that changes in information technology greatly enlarge the information that can be obtained about the status of a business, the conventions must be changed or a business, if operating well below its optimal performance, is competitively vulnerable. Second, to the extent that the conventions have not even addressed some of the important parameters of operating performance, the conventions, even if updated and followed carefully, can lead to performance well below optimum. If a manager is not even attempting to control all of the necessary variables, he is very unlikely to be managing his business optimally.

It is important to recognize the motivational sequence. Managers have little incentive to study situations such as their business system unless they are convinced it would not be significantly improved as a result. The promise of information technology provides such an incentive. In the great majority of cases, managers assert that the benefits that the study led to would have made it worthwhile, even if no more advanced information technology was introduced. *But* information technology provided the motivation to undertake it. The twin goals should be to use the study to improve the management system *and* to introduce more sophisticated information technology.

The roots of information technology lie in accounting, the traditional function charged with keeping a manager informed. My accounting friends emphasize that the initial, and still most important, single role of accounting was to determine profitability. As such, it has focused on credible and consistent determination of value of the inputs incorporated in a product. Its view, to some extent, has to be retrospective: What performance has been achieved by past actions? Inventory, especially the value to be attributed to raw and in-process, has been a matter of particular concern. Much of the pressure to allocate indirect costs has resulted from the need to attribute to goods still on hand, in varying stages of completion, their appropriate share of expenses already incurred.

The sheer physical magnitude of the task of keeping track of a torrent of details is daunting. In those earlier times, allocation of indirect expense demanded the use of averages—nothing else was possible. Accounting was, and is, focused on a business, not a particular product development. With the passage of time and the availability of advanced (precomputer) technology, accounting was extended to provide information useful for operational control—again retrospectively. Cost accounting, with its use of standard costs, in effect said, if we operate as effectively as our criteria assume, these are the costs that should be incurred for producing a given volume of output. When monthly reports (the size of the task precluded anything more frequent) indicated a variance between standard costs and actual costs, managers were alerted to look more closely to determine whether corrective action was needed.

Accounting has more difficulty in providing guidance for decision making regarding the future, quite apart from the uncertainty in forecasting future direct costs of labor and material. Indirect costs, those that are clearly attributable to current production, but especially those attributable to changes in the entire value-producing system, are the major source of difficulty. Working with averages of past burden allocation often leads to erroneous conclusions for an individual product and even more dubious projections for a product under development. Since costs largely determine the fate of a product being developed, the manager of technology is vitally concerned with not only the accuracy of the numbers, but also their comprehensiveness in reflecting results. Some managers have asserted that the validity of cost projections constitutes one of the most serious impediments to innovation. The determination of what to include in costs and in benefits, and even to consider the costs of doing nothing, will obviously control the results of an analysis. Managers of technology would be well advised to become knowledgeable concerning the applicability of accounting principles and to work with finance in making them more suitable.

My continuing theme of the problems generated by fragmentation emphasizes the need for improved linkage between technology and other functions, but until now I have not included finance.[4] The traditional relationship between technology and finance is often one of distrust or even hostility. "Just a bunch of bookkeepers" is an unkind label sometimes applied to technical people who are not demonstrating sufficient vision and creativity. In turn, financial people traditionally are inclined to picture technical people as perennial visionaries, who are either chronically overoptimistic or, worse, may even dissemble concerning schedules and costs until they can no longer hide the facts.

This mutually destructive relationship must be changed, and the key lies in information technology. The growing complexity of operations is a major factor in escalating costs. Most of those costs are associated with processing and communicating information.[5] The growing impediment of uncontrolled indirect costs and the barrier created by invalid cost projections can become life-threatening. Technology managers have no choice except to come to understand the limitations of the present system. As I indicated earlier, the interface between physical reality—people, equipment, materials, components, and processes—and information is peculiarly the province of technology. Indirect costs are uncontrolled because they are very difficult to identify and quantify. The sort of detailed evaluation of the articulation of the entire value-creating system that I advocated earlier in this chapter is a large and intrusive undertaking. It must be occasional rather than continuous. However, if approached properly, it should provide the basis for developing a model of how a system functions, so that one can simulate the dynamic effects of product development and innovation on the system itself. Strategic analysis of the state of the business depends on understanding the true sources of costs as well as the impediments to faster response and greater flexibility. That understanding, in turn, can be generated only by an information system which is integrated across all operations and which can illuminate the impact of any proposed course of action on the business viewed as a system. This capability to simulate a business from the perspective of its general manager represents one of the great unexploited opportunities in information technology.

My own experience is that approached properly, financial people can be willing and helpful partners. They have no interest in producing invalid numbers, but they do resent being thought of as the enemy. I know one lab manager who so distrusted the financial community that he insisted on having his own financial accountant (not a member of the fraternity) and his own private accounting system. This step only exacerbated his problem because no one in operations trusted *his* numbers, either.

Conclusion

System integration presents one of the great opportunities for improving competitiveness. The recycling of programs to get the product "right" not only increases the cost of a product development program dramatically, it also shrinks the market window available. Even worse, it creates frictions and arouses suspicions about the trustworthiness of promises and estimates made by various functions—engineering blames poor market data, manufacturing blames engineering for being late or releasing a design with serious problems in manufacturability, and so on.

The opportunity can be addressed at the operational level by assembling development teams that include the appropriate participants for ensuring that uncertainties are identified and addressed. This issue was covered earlier in managing risk and program management. The opportunity can also be addressed strategically by laying out the present value-producing chain and asking why: Why do we need so many components in manufacturing? Why do we have so many market entities? Why does it cost so much and take so long at each step in the process?

Information technology is giving us the power to answer these questions. Progress depends on a new level of cooperation between technology and finance. It is also contingent upon formulating more sophisticated models of a business and challenging the information system's designer with providing the capability to exercise the models. This process is iterative between the growing comprehension of the kind of "mind" needed to operate the business and the kind of "brain" needed to permit the mind to function. In order to do that, we must move toward an arrangement in which people work from a common data base and use a uniform system of notation, but are free to develop applications that are tailored to the specific needs.

Notes and References

1. Michael E. Porter, *Competitive Advantage: Creating and Sustaining Superior Performance,* The Free Press, New York, 1985.

2. J. G. Miller and T. E. Vollman, "The Hidden Factory," *Harvard Business Review,* September–October 1985, pp. 142–150.
3. Robert H. Hayes and Kim B. Clark, "Why Some Factories Are More Productive Than Others," *Harvard Business Review,* September–October 1986, pp. 66–73.
4. "Accounting Bores You? Wake Up," *Fortune,* October 12, 1987, pp. 43–53.
5. Miller and Vollman, op. cit.

Chapter 13

Evaluating the Technical Operation

Probably no aspect of the management of technology causes more discomfort than evaluating the effectiveness of the technical components. Quantitative measures tend to focus on what can be quantified, not necessarily what is important—"obscuring ignorance with arithmetic," as one disgusted observer noted. The quality of the output is crucial, but so intangible as to discourage measurement. Perhaps most difficult, technical output (for instance, a product design) inevitably incorporates a complex series of trade-offs in a context of incomplete knowledge. Only a virtual reenactment of the design process would enable a completely valid assessment of quality—even by knowledgeable professionals—and that is patently impractical.

In this chapter I will examine some of the complexities of the problem and suggest an approach based on an evaluation of process that can be utilized by both technically and nontechnically trained managers.

Focusing on Inputs

Most attempts at measuring technical activity end up concentrating on inputs expressed in terms of money and human resources. Measuring inputs does at least suggest the importance attached to the activity and, when tracked over time, indicates trends in level of effort, but their correlation with outputs remains an imponderable. Patents and publications in professional journals are most extensively used as surrogates for output in R&D organizations, but their deficiencies are widely accepted—patents and publications vary greatly in their significance, and especially in industry, quantity of both is influenced by local management attitudes about their worth. If managers decry the value of patents or express deep concern over disclosures of proprietary information, a technical staff is unlikely to demonstrate high productivity in patents or papers. Furthermore, their significance in areas of technology downstream from R&D is much more tenuous. They are useful, but limited, surrogates for output.

Problems in Measuring Output

General managers quite properly put continuing pressure on managers of technology to develop acceptable measures of output. But since these are not readily available, they are left—and are likely to long remain—with the task of evaluating performance where much time elapses before an economic impact can be verified and where many others contribute downstream to both success and failure. Sorting out the unique contribution of technology virtually defies solution.

For example, the manager of marketing for the fledgling man-made diamond business always insisted that the invention of a pricing strategy was the crucial step in making that business successful—a point of view that was not popular in the laboratory that invented the technique. The existing market for natural diamonds consisted of "showings," held by the De Beers syndicate, at which prospective customers were offered a predetermined mixture of industrial diamonds on a take-it-or-leave-it basis. This strategy enabled De Beers to clear its inventory of all diamond grades irrespective of the needs of its customers. The manager of marketing for man-made diamonds decided to offer no price advantage, but rather to offer customers the choice of diamond grades they wanted. This tactic provided customers with a degree of control they prized highly without initiating a destructive price war with the syndicate. Although the manager of marketing presumed a lot in ignoring the original invention, he was right in noting that success results from a sequence of creative acts beyond that first step.

Judging technical work is particularly vexing for managers with a nontechnical background. In general, they have been led to believe that technical concepts and techniques are beyond the ken of the layperson. Many are uncomfortable with their lack of knowledge and tend to avoid involvement. Yet they are responsible for results and wary of condoning poor performance. The usual unease would be expressed as follows: I don't know whether my technical people are doing a good job or not. It seems to me that everything costs too much and takes too long. We seem to miss a lot of deadlines, but my people tell me you have to expect that in technical work. After all, you can't schedule invention. I know there is a lot of uncertainty in technical work, but I worry that I'm having the wool pulled over my eyes or that we'll be blind-sided by a competitor.

What those managers may not realize is that many first-class technical people are—at a different level of sophistication—also uncomfortable in judging work outside their own field. I remember one experienced middle-level R&D manager, who also was a superb physicist, declaring that the job of vice president of R&D was, by definition, impossible. Nobody could know enough to make the decisions he had to make. What this manager was really saying was that he personally was unwilling to exist with that level of uncertainty—he required a more complete data base before he would or could make a decision.

A Solution: Focusing on Process

The difficulties of measuring technical work directly are not readily resolved, so managers must resort to other approaches. When they are unable to evaluate an output directly, they can frequently gain valuable insights through examining the process by which the product is created. I have often been impressed with how rapidly experienced technical managers begin to form judgments about a technical operation they are visiting for the first time, even though they are not directly versed in the technology itself. More or less subliminally, they recognize the attributes of an effective technical operation almost independently of its specific output. I am convinced, on the basis of my own experience and from discussions with many managers, that a number of the relevant clues are accessible to the layperson if he but looks for them and develops the judgment to evaluate them.

External indicators of quality of output

Two objective measurements of the quality of technical output are available: market share and complaint expense. Market share is an important indication of effectiveness in technology, but it must be used discriminately. High or growing market share is almost certainly an indication of effective technical effort—it is difficult to construct a scenario where inadequate customer value is rewarded with clear customer preference. However, low share is not necessarily indicative of poor performance by the technical community. Inadequacies in other spheres of operation, such as supervision on the shop floor, sloppy purchasing, poor maintenance, poor pricing, and indifferent distribution, can negate the advantage of excellent product and process technology. And, of course, even when these negative considerations do not apply, the market does not necessarily reward excellent technical performance and sound insights into market opportunities—witness the experience of Sony with its pioneering and technically superior Betamax VCR.

Complaint expense—the amount of money and person-hours of effort required to investigate and satisfy customer complaints—is a useful diagnostic tool. Its level and trends in comparison with key competitors again are revealing. The nature of the complaints can also be illuminating. Difficulties that persist in specific components or parts of a system can suggest a deep-seated problem in technical work. Random problems, provided they are stable in magnitude and compare favorably with competition, are more reassuring.

Complaint expense is somewhat analogous to bad-debt losses: a complete absence is also cause for concern that a technical organization is not being aggressive enough in applying technology. Ultraconservatism can mean a vulnerability to more aggressive moves by competitors or a costly delay in introducing improvements. One astute general manager said, "We do some of our best customer relations work when we get in trouble. If we hadn't pushed the technology, we probably wouldn't have received the order. By going all out to solve a problem, we reassure the customer of our resolve to be a dependable supplier, but retain our image as a technical leader." Although one should not push this scenario too far, the point is well taken.

Measuring output

An effective technical operation puts high priority on defining its objectives in measurable terms. These objectives include not only products redesigned and new products released to manufacturing, but also cost reductions achieved in products and in its own operation, complaint expense reduced, and documentation kept up to date. In other words, its objectives include improvement in its own operations and impact on the businesses as well as in products developed or designed.

The manager of an effective technical operation knows how his operation is doing. He knows not only about the operation's technical success, but also about its own internal operations, and he has records readily at hand to support what he says. His knowledge and concern become apparent with only a little probing, which itself offers a clue to his performance as a manager.

An effective technical operation knows and can demonstrate whether it has achieved its objectives. Its management team has its score card ready at hand and uses the objectives in describing and evaluating its effectiveness. Managers do not have to be prodded to develop information and search for data. It is apparent that these objectives are kept in front of them and that they are a continuing part of management discourse.

Meeting schedules

An effective technical organization knows, and will tell you, whether it is on schedule. It takes its time commitments seriously. Nearly all competent technical organizations are performance-driven in the sense that they take pride in pushing technology to the limit. But the really effective ones also give great weight to meeting time commitments. They are rigorous in appraising the risks and uncertainties inherent in what they are committing themselves to. They include time contingencies to cover uncertainties, or if there is severe time pressure, they back off on their technical targets. A veteran technical consultant, who had reviewed many operations, once told me, "If it takes me some time and some digging to discover that a schedule has slipped, I know I'm dealing with an organization in trouble. The really good ones may get behind schedule, but they'll be upset and irritated that it has happened and they'll bust their butts to get back on schedule."

Schedule slippage is insidious. Everybody is aware of the intrinsic uncertainty in technical work, so it is easy to slip into the pattern of regarding missing target dates as natural, even inherent in technical work. Soon people do not even feel guilty when they miss a target, and before long they are unable to tell whether a schedule is realistic or not, because nobody takes it seriously, anyway.

In a similar vein, an effective technical operation knows and can tell how much and how often project performance or cost goals have been revised. The set of pernicious influences described above for meeting time commitments applies to project objectives. There is a fine balance between the importance of a target and the level of commitment to achieve it. Being committed to unattainable goals is devastating to morale. Sometimes an inadequate or insecure technical organization lets itself be pressured into impossible targets, and the results are catastrophic. Some general managers take the point of view that project plans always have large safety factors in them and their job is to squeeze—"do it in half the time." However, if a general manager tries to make a federal case of every project, he just invites cynicism. An effective technical operation is as stubborn in resisting unrealistic targets as it is relentless in pursuing committed goals.

An effective technical operation measures its output. The tradition of science and engineering is that without measurement there is no knowledge. Although a technical operation may be wary of measurements imposed from outside, because their deficiencies may not be adequately recognized, it strives unceasingly to establish its own measures. They may be lines of software or square feet of drawings completed, or number of specimens tested, or other partial and imperfect techniques—but an effective technical organization is true to its traditions and values and attempts to quantify its own effectiveness.

An effective technical operation knows how much engineering rework has to be done. It has developed tracking systems that establish personal accountability for quality of output. It imposes a discipline that says, you caused the problem—you fix it. The operation also analyzes the causes of rework and attempts to reduce them, because they often result from a breakdown in communication or an inadequate specification of requirements rather than from personal error. Thus part of its annual operating objectives consists of targets to reduce rework.

Adequacy of technical management systems

Test results. An effective technical operation insists on extensive test data and field trials to validate the soundness of its design. Those data are readily at hand, the results are widely known within the operation, and the limitations of the data are explicitly recognized in the design decisions. This observation may seem obvious, but the ceaseless requirement for supporting data can all too easily atrophy. A good technical operation is taut, is ever conscious of the possibility of error, and is unforgiving in its face.

Equal attention is paid to the analytical techniques that have been used to test data against known physical principles; to ensure that attempts to achieve optima have been made; and perhaps most important, to establish the limits of the operating conditions within which the equipment will perform as specified. In other words, an effective technical operation knows the limits of its knowledge and the gap between empirically derived data and fundamental understanding.

Documentation. An effective technical operation is religious in maintaining documentation that is current and promptly distributed to all who should know—especially documentation for marketing and for customers. It is not caught in the position of saying: oh, we changed that a while back, but we haven't had time to change the records. Similarly, when trouble occurs in the field, an effective technical operation is able to produce adequate records of the data underlying design decisions and specifications for manufacture. Sometimes I discover an organization that does not keep its internal documentation fully written up, but the pertinent data can be retrieved readily. Where this is true, the "failing" need not be disastrous. Particularly in the sale of technical products, documentation of product attributes and field performance can be a powerful sales tool. For industrial customers, the process of selecting material is strongly influenced by the quality of the data describing the attributes of materials under consideration. One important user told me that they had chosen a particular vendor because its data were the best, most credible, and most relevant to their decision process.

The practice of the principles advocated in my discussion of project management is also evident in an effective technical operation. The management system keeps the status of projects current, highlights variances in schedules and costs, and monitors the action being taken to get back on track. I know one manager who went so far as to start each day with a brief stand-up staff meeting to share knowledge of the status of all projects. Although I do not advocate going this far, there is no doubt that his managers knew the importance he attached to maintaining schedules.

Uncertainty and risk. An effective technical operation deals openly and systematically with uncertainty. As the discussion on managing risk pointed out, uncertainty is inherent in technical development, and dealing with it directly and systematically not only produces more effective technical development; it also both reduces the possibility of nasty surprises for others and helps control the anxiety that uncertainty tends to generate. Consequently, an effective technical operation routinely makes rolling estimates for achieving cost, performance, and time targets and informs others of the status of projects that affect them. In doing so, the operation need not necessarily be defensive about changes in estimates. It accepts them as inherent in programs that have enough technical "stretch" to preserve competitive leadership. Therefore, an effective, self-confident technical operation projects a posture that says, this is our current best estimate of where we stand. This is what we are doing in the way of corrective action. If the situation changes, we'll let you know. Please let us know how this present estimate affects you.

A senior financial manager once told me the thing he most disliked about relationships with engineering was its tendency to delay informing of slippages in schedules and goals. He said, "We know changes are inevitable; that's why we make rolling estimates of business performance. Why can't engineering do the same thing? I *know* they know things aren't on track; why hide it?"

Interactions with other functions

Obtaining market inputs. An effective technical operation understands what is required to meet market needs and places high priority on obtaining market inputs. In one business I was reviewing, marketing and engineering presented different profiles of what product attributes were important. Even worse, neither gave any evidence of recognizing that the views of the two diverged. When I encounter technical people waxing enthusiastic about the power, sophistication, or elegance of a technology without reference to demonstrable value to the user, a warning flag goes up. Similarly, when I find technical people declaring themselves to be their own market experts and saying in effect, I value this advance in technology, therefore there is a market for it, again warning flags go up. I am well aware of the great innovations that are created by fanatical persistence in spite of no or even negative market inputs, but the exception should not become the rule. We hear of the rare spectacular successes; we hear very little of the hundreds of failures. A number of studies have demonstrated that inadequate coupling with marketing is the most common single cause of failure for innovations.[1] 3M, which has a remarkable record of successful innovation over several decades, insists on early and continuing contact with potential customers. The insistence on contact with customers is crucial. Innovations are not bought by markets; they are bought by customers who perceive sufficient potential value to risk committing to a new advance. Unless *they* perceive the new capability as having value, the views of the innovator amount to little.

Sometimes the effort to identify and validate valuable applications can be far more extensive and time consuming than that devoted to the original discovery. The path to usage can be murky indeed. In the early seventies GE's Corporate R&D invented a low-cost method for fabricating silicon carbide, a ceramic with very attractive properties, but with some severe handicaps to widespread use, especially lack of ductility. Many people are convinced that ceramics will be important materials in applications demanding extreme resistance to wear, high-temperature capability, stiffness, and light weight. However, one of my former associates has spent a good portion of his entire professional life attempting to identify customers and guide development along lines that lead to commercial usage, with very modest success.

This kind of market development cannot just be left up to the sales force. People involved in the technical work must participate. They need the feedback to guide further development, and only they understand the fledgling technology well enough to discuss modifications and extensions with potential customers. If I find a wall of resistance and mistrust between marketing and the technical developers, I see portents of trouble.

An effective technical operation understands the uncertainty inherent in specifying market requirements and in forecasting price-volume relationships. The operation does not regard as immutable the initial marketing inputs that help determine product attributes and cost targets. It arranges, even insists on, a continuing dialogue with marketing about the validity of marketing assumptions on which a development plan is based. It is a demanding partner of marketing, refusing to rely solely on estimates based on past experience any longer than absolutely necessary. It recognizes a lack of customer-based inputs as a danger signal. However, it also appreciates that it is unrealistic to expect completely reliable inputs from marketing and does not accuse it of incompetence for being unable to produce better data to guide development.

Manufacturing. An effective technical operation recognizes that producibility and quality are inherent requirements of the design process, along with performance and cost. If I find manufacturing being taken for granted or, worse yet, being treated as subservient to engineering, responsible for producing what engineering hands it and obviously incompetent if it is unable to do so, I see a technical operation that is headed for problems. If I find that engineers are seldom on the shop floor and that manufacturing people rarely sit in on engineering meetings, that there is little interchange of assignments as a part of career development, or that there are no joint training programs for engineering and manufacturing people set up, I sense trouble.

An effective technical operation understands the impact of its decisions and output on logistics and product service. Is product engineering informed about product service costs and the causes of those costs? Are cost reductions balanced against possible increases in product service costs? Is discipline imposed on the selection of materials and components to ease parts inventory and simplify field repair? Are design features that impede field service sought and eliminated? At one time repairing torn linoleum floors in kitchens was the largest expense for servicing GE appliances. Design engineers for refrigerators had paid insufficient attention to the design of castors, which froze from lack of use. They were simply unaware of after-sale service problems because the information system had no way to accumulate the pertinent data and make it available to them.

An effective technical operation monitors and seeks to minimize the engineering changes sent through to manufacturing. Technically trained people place a high value on extracting maximum performance from a particular field of technology. Therefore, if they discover a way to improve performance, increase life, or reduce cost, for example, their professional instincts are to introduce the change when it is still physically possible to do so. In sending through engineering changes, often insufficient weight is given to the disruption the change will create in manufacturing. The change will affect both cost and delivery and should be subject to some sort of review and trade-off before it is imposed on manufacturing. This issue is particularly acute in businesses that emphasize the quality and performance of their product. An effective technical operation knows how many changes have been sent through, has in place a procedure for imposing discipline on the process, and has ongoing programs to reduce such changes in the future.[2]

Finance. An effective technical operation understands the sources of business costs and the impact of its own behavior and its own output on those costs—the system integration problem discussed in Chap. 12. Failure to recognize the consequences on the entire business system of decisions made in a particular product development is a failure in technology management. One good indication of effectiveness in this area is the relationship of technology managers with finance. Wariness and inhibited communication are a danger signal. Active partnership in identifying both full costs and benefits is reassuring.

Understanding these system effects can provide additional evidence of the value of a technological innovation. One R&D operation identified and took credit for a modest, but clearly demonstrated, cost reduction resulting from a sophisticated materials substitution it developed. But the R&D people failed to talk to anybody in marketing and thus completely missed the fact that the substitution also virtually eliminated the largest source of field failures—a savings many times greater than the one noted in manufacturing, not only in expense, but also customer satisfaction.

Competitive awareness

Existing competitors. An effective technical operation maintains keen awareness of its competitors. It knows who they are, how they behave, and appraises its own position realistically. The most persistent and widespread weakness in strategic plans for technology is the scant attention paid to competitors. Often they are never mentioned; even worse, the plans may assert competitive superiority with no evidence to support it. An effective operation brings up the competition, points out in specific terms its strengths and weaknesses and its apparent trajectory in technical work. Perhaps most important, such an operation postulates its own strategy in competitive terms. Few enterprises engage in competitive gaming in which they ask, if I were a competitor, how would I respond to what I am proposing to do? Effective technical operations buy competitors' products and take them apart if they can. They talk to customers and observe competitors' products in the field. They talk to vendors about competitors' actions and behavior. In short, they actively and systematically pursue all legitimate sources of information about competitors. They also cultivate an environment of awareness and vigilance, which ensures that the threat of competition always receives high priority. When I encounter an operation that says, "Our technology is so specialized and sophisticated that we really can't learn anything from others—you have to have grown up in this business in order to understand it," I forecast an impending trauma. It may take some time to arrive, but the corporate ethnocentrism implicit in such a statement courts serious problems.

Emerging competitors. An effective technical operation recognizes the potential of emerging competition and attempts to monitor it. Weakness in this area is almost universal. Even in technical operations that are aware of competition, the tendency is to focus on traditional competitors, especially domestic competitors. The point of view is typically domestic until international competitors appear locally—even for multinational companies. The assumption seems to be, if we aren't meeting them in the marketplace, they aren't competitors.

I discussed the importance of competitive appraisal in strategic planning. It is a key indicator of the effectiveness and alertness of a technical operation. At least four sources of potential competition should be monitored. First, important customers may decide to integrate backward and make what you have been selling them. An effective technical operation is aware of the technical competence of customers and is alert to opportunities to make backward integration unattractive. Second, an important vendor may decide to integrate forward and add the value you have been supplying. This threat is endemic in electronics, where suppliers of components or subsystems may see a much larger market potential if they can operate in *your* markets as well as theirs. When GE's appliance-control business was contemplating a switch to electronic-based controls, it came to understand the major impact that change would have on the nature of its competition. The business was not large enough to justify establishing its own electronic processing capability. If it purchased the key element in its product from vendors, its value-added would be greatly reduced. But even worse, it concluded there would be little to stand in the way of having the vendor of this key component integrate forward and simply take over the whole business.

This threat is magnified by the third source of potential competition: changes in technology. The continuing advances in integrated circuitry make it possible to place more and more capability on a chip. In so doing, they absorb value formerly supplied by manufacturing processes performed downstream. Similarly, the introduction of reaction injection molding, with its lower capital requirements and fit with intermediate volumes of manufacture, reopens make-buy issues in many applications for plastics, especially for medium-volume usage. Of course, technological substitution by invasion from the outside is now recognized as a classic pattern of innovation. The growing use of electronic controls in autos and appliances is but a reaffirmation of this pattern.

The disruptive effect on competition caused by the introduction of new technology is dramatically evident with the CAT scanner. Who would have thought that a musical instrument and record company could revolutionize the X-ray imaging business? Unfortunately, the fact that the competition seemed so improbable made it easy to deprecate, even when knowledge of the development first became available—well before the product itself was introduced.

Trends in technology can generate a competitive confrontation among companies not previously in competition. The almost total merging of information processing and communications into a single combined field has been creating just such a situation between IBM and AT&T.

Finally, the fourth source of competition arises when a change in strategy brings companies into competition or leads them on diverging paths. The decision of Philips to purchase the lighting business of Westinghouse, to give a base for penetrating the U.S. market, increased its competitive confrontation with GE. Conversely, GE's increased emphasis on service and high-technology businesses, such as medical systems and jet engines, has reduced its range of competitive confrontation with Westinghouse, whose divestiture of appliances further reduced the competitive interface.

Comparison with external peers

An effective technical operation fosters participation in external technical communities to validate its competence, to check its sense of trends in technology development, and to provide early warning of new advances. The ever present pressure to protect proprietary information and to avoid disclosing competitive plans must not be allowed to isolate a technical group from its external community. One can attend meetings and present papers and still protect proprietary information. I well recall a manager's comment about some work whose value he questioned: "Let him try to publish it; he'll soon find out how good it is." Without the continuing prod of professional comparison and peer judgment, a group can easily lull itself into a dangerous state of complacency. One clue that I look for is whether a group even knows who the outstanding authorities in a field are. Evidence of interaction with some of them is a still better signal.

As a supplement to external peer contact, a formal review process by outside experts can be a useful tool. A competent, self-confident technical operation sees the need of such an input to "keep it honest" and welcomes it. The whole structure of integrity and validity in science and engineering is based on a peer review of work, and an organization that cuts itself off from such a review is vulnerable. The review need not be continuous or all encompassing, but the question, how are we doing? should be addressed explicitly. A response by a technical manager of "that's what *I* am being paid for" (i.e., to make these technical judgments) suggests a kind of defensiveness that is a warning signal.

Managing resources

An effective technical operation categorizes its technologies in terms of competitive leverage and selectively allocates resources to the key ones. It is realistic in recognizing that resources are limited and that it cannot, and does not need to try to, maintain competitive superiority in every detail. It understands the danger of spreading itself too thin. It identifies those technologies that are critical to competitive success and ensures that they remain sources of strength.

It seeks to use external technology aggressively where its competitive posture will not be compromised. This practice, more than any other, goes against the grain of technical tradition. People become engineers and scientists to do engineering and science. Their value system and criteria for professional status place highest worth on advancing the state of the art. In the light of that tradition, the idea of buying and selling technology almost like an asset may seem unnatural—even distasteful. Furthermore, many regard a decision to acquire technology as a personal affront. The famous "not invented here" (NIH) syndrome popularizes this deeply held value.

An effective technical operation recognizes that an enterprise is not in the business of creating technology—it *uses* technology and creates it when competitive advantage can be obtained. A CEO who never hears his technical operation propose an acquisition of technology has reason to fear that some of his funds are not being spent wisely. External technology cannot only be obtained less expensively in some cases, it can save much time, which may be even more important. Certainly the Japanese have demonstrated that aggressive, creative use of technology developed elsewhere can be a powerful weapon for obtaining competitive advantage. Again the illuminating principle is tension. I acknowledge the powerful competitive advantage of new proprietary technology. Yet one cannot realistically seek it to the exclusion of all else. Unfortunately, pursuit of that "all else," i.e., creative use of existing technology and shrewd acquisition of external technology, is discomfiting to those imbued with the traditional values of technical work. However, an effective technical manager does both, along with selective effort to develop new technology.

Fostering innovation

An effective technical operation recognizes that the development of totally new technology is a separate task incompatible with the extension of the present state of the art. It assigns separate responsibility for such work or makes some other arrangement to ensure that innovation receives concentrated attention. The manner in which a technical operation goes about fostering innovation is, of course, strongly influenced by its size and the stage of maturity of its technology. Small operations are not able to provide a sufficient specialization of effort to assign separate responsibility for innovation. Of necessity, unless they are involved in innovative technologies, they emphasize applying and extending the state of the art.

There is, as we have noted repeatedly, an intrinsic competition between conventional and new technology. Unless people have been put in the position of having no place to hide, of having performance judged by their success in developing new technology, it will not receive the dedicated attention it must have. An operation that asks people to create their own revolution is only paying lip service to innovation.

Performing the Evaluation

Table 13.1 provides a checklist of the items I have been discussing. The most noteworthy characteristic of the items is that none of them require knowledge of the specific technologies being developed or even knowledge of technology in general.

How can you judge these items? one may ask. They seem awfully soft. The process is not very quantitative, but it does rely on hard evidence that is obtained mostly by asking questions and listening with a trained ear. One must seek a variety of inputs, and one should not hesitate to seek external inputs as well. The final synthesis must, of course, be that of internal management. But this procedure is used in all evaluations of performance.

TABLE 13.1 Technology Evaluation Checklist

1. Quality of output
   - Market share
     - Level and trend
   - Complaint expense
     - Level and trend
     - Localized or random
2. Measuring output
   - Are objectives measurable?
   - Are internal measures of performance used rigorously?
   - Are time commitments taken seriously?
   - Are project slippages and recycling kept visible?
   - Is output measured?
   - Is rework monitored and controlled?
3. Adequacy of technical management system
   - Are test data required and used in design?
   - Are rigorous analytical techniques used?
   - Is documentation complete and current?
   - Is adherence to project goals emphasized?
   - Are risk and uncertainty dealt with openly and honestly?
4. Interactions with other functions
   - Are market inputs sought and used?
   - How effective is the interaction with marketing?
   - Are manufacturing requirements well integrated?
   - Is the interface with finance wary and hostile or constructive?
   - Are engineering changes carefully controlled?
   - Does the management system address total system cost, not just product cost?
5. Competitive awareness
   - Are competitors known, monitored carefully, and their technology and products evaluated objectively?
   - Are emerging competitors given attention?
   - Are potential changes in competition considered?
     - Backward integration
     - Forward integration
     - Technology bringing new competitors
     - Change in strategy
6. Comparison with peers
   - Are external peers known and compared with internal competence?
   - Is reaction to external evaluation defensive?
7. Managing resources
   - Are technical priorities made in light of resource priorities?
   - Does the operation spread itself too thin?
   - Is use of external technology given objective consideration?
8. Fostering innovation
   - Is technology a persistent and effective advocate for innovation?
   - Has it separated innovative work from operating activities?

The unfortunate aspect of judging technology is that too many nontechnical managers believe the task is beyond them, so they renege on their own responsibility by leaving technology to the technologists and concentrating their own efforts on the rest of a business.

Ask to see the operation's objectives. Ask how it intends to measure its results. Do its objectives include improvements in its own performance? Does the manager of technology have at hand records on the status of his operation? Does he know the trends in its performance? Ask questions about competitors. Is the manager knowledgeable in some detail? Can he cite both strengths and weaknesses and describe how his strategy responds? Or does he simply assert a generalized superiority over competition? Ask how he judges the competence of his people. Can he compare them with external authorities? Talk to people in marketing and manufacturing. Do they present a perception of objectives and priorities which is consistent with that in the technical operation, or at least an awareness of inconsistency? Does the operation suppress its problems or talk about them? A good operation does not hide its problems from itself. It is irritated that it has blundered and works furiously to get back on track.

A general manager seeking to evaluate a technical operation can either sit down with its manager and go through these items, or he can keep a mental checklist of questions about which he will seek evidence and simply listen for inputs in his normal interactions with people.

Conclusion

Managers with a nontechnical background often assume that managers with technical training rely on their knowledge of science or engineering in judging a technical operation. I do not mean to suggest that is not the case, but technically trained managers are often required to judge operations where their knowledge of the substance of the technical work is not much different from that of an informed layperson. What they rely on heavily is knowledge of the process that is required to produce valid technical results and of the steps that a competent technical operation follows. That knowledge is accessible to laypeople if they learn to ask sensible questions and trust their capacity to recognize inconsistencies and unnecessarily pseudo-esoteric language.

Notes and References

1. *Success and Failure in Industrial Innovation: Report on Project SAPPHO,* Science Policy Research Unit, University of Sussex, London Centre for the Study of Industrial Innovation, 1972; Sumner Myers and Donald Marquis, *A Study of Factors Underlying Innovation in Selected Firms,* National Science Foundation, Washington, D.C., 1967.
2. Robert H. Hayes and Kim B. Clark, "Why Some Factories Are More Productive Than Others," *Harvard Business Review,* September-October 1986, pp. 66–73.

Chapter

14

Selling Technology to Top Management[1]

American business is not investing enough in R&D, especially the longer-range work that is vital for the future. That complaint is virtually constant and universal in the technical community. Note that this is an assertion of the technical community. Businesspersons might respond, if adequately provoked, that they are not investing enough in a lot of things—market development, improved distribution, internal education and training, more efficient communications systems, quality control—you name it!

As advocates for an important activity, technical managers *should* feel that it warrants increased support. They have been put there partly so that they can be effective spokespersons. However, a chief executive who finds himself in the novel position of hearing his manager of technology say he is satisfied with his level of expenditures on technology, or even any segment of it, will probably conclude—with great validity—that he is spending too much. The chief executive's job is to allocate limited resources, and that process should leave even good committed advocates dissatisfied!

The challenge for all advocates is to be perceived as responsible ones. If they do not push hard enough, technology will not be well represented and will not receive the resources it really ought to have for the good of the business. If they push too hard, they will come to be regarded as "the high priests of technology." Everything they say will be discounted, to the detriment of technology's role.

The nature of the problem the CEO (or the general manager of a division of a multibusiness company) faces is different in large and small companies. In the former, the decision is primarily whether resources should be reallocated so as to support a particular development in lieu of higher earnings or perhaps at the expense of some other investment. For the smaller firm, the problem is more likely to be how to obtain the necessary resources—both money and skills—a problem beyond the scope of the manager of technology.

The need to sell investment in technology comes into focus with specific programs. When an advance in R&D reaches the stage of being scaled up to be incorporated into operations, its benefits, risks, and resource requirements have to be laid out, buttressed by evidence, and packaged for communication. However, it is a serious mistake to focus the selling effort on individual programs, more or less in isolation. The CEO is a customer, if you will, and any particular "transaction" occurs in a continuing context of customer relations.

The goal of technical management in its dealings with the chief executive has to be the same as that of all other representatives of a particular business activity: to influence his objectives and priorities in such a way that technology makes an optimal contribution. Perhaps the first step in selling technology is to understand the difference in viewpoint between technologists (mostly scientists and engineers) and chief executives or general managers when they look at technology.

Perspective of Technologists

Despite the important differences between scientists and engineers, they share many important characteristics. For both, professional rewards come from outstanding original work. Both accept risk and uncertainty as intrinsic in technical work. Consequently, both come to accept the likelihood of failure, either because nature may not cooperate or because some critical element is missing. The professional values of both scientists and engineers foster extending the limits of capability. In their eyes, the potential reward for an important advance is large compared with the penalty for failure. This is contingent on openness and rigor—late recognition of failure, after others have expended large effort thinking the advance was genuine, is not well received. Technical people do not treat kindly those who lead them astray.

Finally, although lower-level management positions in technical work are ones that oversee the creation or application of technology, they rarely deal with the interface between technology and the other elements of a business until they get to higher levels of technical management.

Perspective of General Managers

The perspective of general managers is different. The rewards go to those who not only consistently achieve stated goals, but who also achieve persistent growth in earnings. The penalties for failure are high, because they are regarded as either unrealistic in establishing objectives or ineffective in accomplishing them.

A general manager lives in a very unforgiving world in which one of his most important tasks is to create certainty. Even consistent pleasant "surprises" come to be suspect: they are thought to result from deliberate choice of unchallenging targets. In this environment the toleration of surprises is low; heavy emphasis is placed on stability and predictability. Consequently, great effort goes into identifying and minimizing risk. An important reason, of course, is that the economic stakes are much higher when one begins to bring technology to bear on a business. Much larger resources are involved in production, distribution, and sale of products and services.

A general manager has no personal stake in creating new technology. Even if he is a former technologist, in his new role he must view technology as the capability that enables him to offer goods and services. A business rarely has a proprietary interest in creating new technology for its own sake. If more timely, lower-cost, less risky solutions can be found in conventional technology, the general manager will, and should, prefer them. In other words, he must seek the most available trouble-free technology that is adequate for his needs. A general manager is charged with obtaining at the lowest possible cost all of the inputs needed by a business—including technology. He must be a practicing microeconomist, adjusting the inputs until all marginal returns are equal.

Influencing a General Manager or CEO

Given these two different viewpoints, plus the fact that in many businesses CEOs and general managers do not have a technical background, it is very important to approach the task of selling technology by understanding the environment in which they operate, the constraints they believe they must satisfy, and the criteria they use in making decisions. There is little to be gained in labeling them as short-sighted, or lacking in vision or courage, if in their view they are acting rationally—and usually they are. They do not have the resources to do all the things their staffs ask. Somehow they must balance the allocation of those resources to get the most out of the options available. They need to feel that those seeking support are applying the same criteria.

Establishing credibility

Chief executives are professionals at generating numbers—that's partly how they got where they are—but also know their fallibility.

Therefore, the development of convincing facts and figures for specific programs is unlikely in itself to persuade them. The real tasks are to establish credibility, to get the CEO's attention, and perhaps to change his mind set. Thus the problem is as much psychological as it is cognitive.

Social scientists have learned a great deal about how people gain influence, and among other factors, it is based on establishing two kinds of credibility.[2] In dealing with CEOs and general managers this means:

1. *"Safety" credibility:* A belief that this person's values, beliefs, and priorities are enough like my own that I can trust him, I can trust him to act in such a way that I can accept the consequences.
2. *Technical credibility:* A belief that this person knows what he is talking about.

A focus on facts and figures emphasizes the technical credibility. In matters involving technology, it is often difficult for the chief executive—or most other executives, for that matter—to judge the validity of the numbers. And as we have seen in our earlier discussion of risk and project management, the numbers are by no means infallible. Consequently, "safety" credibility assumes greater importance. Unfortunately, the coupling between the two types of credibility is loose, so that increasing validity and sophistication in the numbers does little to assuage a CEO's concern over whether "this person has looked at all the things I would want him to look at, is reaching the conclusions I would want him to reach, is balancing the considerations I believe are important."

The CEO's questions

When he is hearing a presentation, the kinds of questions that run through a chief executive's mind—even if subliminally—go beyond the specific subject under discussion. Obviously, different executives have different pet peeves, but the list of intruding questions includes the following:

1. Are the issues being raised the ones I think are important? If not, is this person aware of our obviously different perspectives and thus trying to "educate" me? One general manager, fed up with the details his lab director foisted on him, instructed his secretary to schedule his future appointments with the director no earlier than 4:15 p.m. so that he could always gracefully call a halt at 5 o'clock.

2. Are they the sorts of issues I as chief executive should be involved in? in the manner he asks? CEOs at an earlier stage in their own careers have learned that one of the best ways to elicit cooperation or acquiescence from others is to invoke the endorsement of "the chairman." And it does not take them long to learn at first hand that they are just as vulnerable to being "invoked" as their predecessors. When Ralph Cordiner was CEO of GE, he became increasingly selective in signing any internal documents. He had discovered the escalation in importance that his signature created and the ingenious ways that people then tried to subvert the subject matter for their own ends. Reginald Jones, another chairman, probed to be certain that proposed strategic guidelines really warranted his imprimatur. A veteran company watcher once commented about GE, "You'd better be careful what you ask this outfit to do, because you're going to get it, in spades!"
3. Does this person give evidence of recognizing the constraints I am operating under? A common reaction of technical managers who are promoted to a position that brings them to a general manager's staff meeting is one of amazement at the variety of issues and problems he must address. A very successful manager of engineering with many years of experience in a business was promoted to general manager. He commented with surprise, "Boy, I'm learning a lot about finance that I never knew before."
4. Are this person's objectives and priorities compatible with my own? If not, does he or she give evidence of recognizing the difference and of responding accordingly? The inexorable pressure of financial performance, of having to reconcile the unsolicited projections of performance published by investment analysts with the lower forecasts of internal management, is rarely given full due by the technical community.
5. Why am I being told now? Why wasn't I told earlier when there was time for a considered response? or, Why wasn't this delayed until there was enough information for me to be useful? Judging *when* to bring a subject to the boss's attention is one of the most critical features of establishing safety credibility.
6. I wonder what others think about this situation? Did finance sign off on his numbers? Does marketing agree with his projections? Has manufacturing had its say?

If the answers to these questions are not satisfactory, no amount of numbers and backup data are likely to carry the day. Selling technology, just like any other kind of selling, requires knowing the customer and knowing the product.

Know the Customer

Let us focus on four aspects of knowing the customer: the world he lives in, his concerns about his performance, his expectations regarding technology, and the infrastructure that influences him. Later we'll turn to the product, that is, technology.

The CEO's world

The first step in knowing the CEO is to appreciate the world in which he operates. Not doing so will lead him to regard any of his immediate subordinates as part of his problem rather than part of its solution.

The chief executive must live with invidious comparisons—with competitors, with past performance, with his own projections, and with the expectations of the financial community. In particular, the norm of the financial community is quarter-by-quarter improvement in earnings, or failing that, no disappointments and especially no surprises. If financial analysts are surprised, they feel that makes them look bad and they tend to question a CEO's competence, not theirs. The chairman of a major company, expressing consternation at the budget proposal of his R&D director during a downturn, said in all seriousness, "Another quarter like the last one and we'll both be looking for a job." While he was no doubt being melodramatic, the outburst did reflect his deep concern. The expectations of the financial community are taken very seriously, and a switch from "buy" or "hold" to "sell" by an important investment house is very unwelcome.

Chief executives find themselves expected to produce attractive but certain results. As I pointed out in discussing operational management, one could almost say their job is to create certainty. In such an environment it is not surprising that they view an opportunity to invest in an uncertain, destabilizing innovation as a mixed blessing. Ambivalence about innovation is not irrational, even for a CEO who must also be concerned about growth.

Another important aspect of a chief executive's job, which we have already alluded to, is resource allocation. Virtually everybody who enters his office has an ax to grind, is an advocate for some position, is a special pleader for a particular cause. His task is to determine which to accede to and which to refuse. His job is perforce a lonely one, and he tends to seek the counsel of people whose purview of the business is similar to his own. I eventually concluded that that is an important reason for the strong influence of many financial officers—they are perceived by a CEO as being concerned with the performance of an entire business, not some particular component or some particular cause. To the extent that a manager is seen as being aware of the total business equation and as seeking an optimal resolution of all pressures for committing resources, rather than as blindly in love with his own cause, he builds safety credibility.

CEOs may admire entrepreneurs, and even be glad they have them, but still feel a certain wariness and underlying concern that such managers must be watched carefully lest their tendency to become enthusiastic about something new lead them into difficulty. I know one successful entrepreneurial general manager about whom three successive bosses complained (even while recognizing his entrepreneurial success) that they had to spend too much time keeping him on the track.

The CEO's concerns

The second major dimension in knowing the CEO as customer is to identify his major concerns and to respond to them. This goes far beyond pandering to his particular predilections. It requires identifying the concerns that are inherent, or should be inherent, in the CEO's job at this time.

It goes without saying that CEOs are always committed to having a competitive product line and having new products in the pipeline that give promise of protecting the competitive posture of their business. This commitment is classically central to the responsibilities of the technology function. Too often it has appeared to be almost the sole concern of technical management.

In the increasingly global competitive environment we now face, the CEO is equally worried about total business costs and the productivity of all assets and resources employed by his enterprise. The U.S. market is the target of practically all the world. U.S. labor costs are at or near the top of the pyramid. As steel, agriculture, and forest products have demonstrated, presumed advantages in natural resources are no longer adequate to establish competitive advantage. Capital equipment is more or less equally available to all. South Korea is seeking to become a global competitor in autos by buying equipment and know-how from Japan. Visitors to Japanese auto factories note that equipment is similar to that in U.S. factories—a significant amount even purchased here. The education and work ethic of some other labor forces are equal or superior to ours. Technology is perceived as the key ingredient in competitive survival, but it is a technology that goes well beyond the attributes of products or even the direct costs of their manufacture. Even if he cannot verbalize the exact basis for his discomfort, a CEO is likely to feel uneasy with the stewardship of a chief technical officer who presents too narrow a concept of the role of technology. We will return to this subject when we turn our attention to "the product" of technology.

Another of a CEO's concerns is the flexibility and responsiveness of his enterprise in an increasingly volatile environment. A Maginot-line mentality that seeks competitive security in natural resources, capital investment barriers, "keep out" technology barriers, or market franchises with customers has to be shifted to more of a chess mentality of position and movement and momentum. A technical community that appears unaware of this shift in emphasis will not increase the confidence of the CEO in its perspicacity.

The CEO's expectations of technology

The third dimension of understanding a CEO is to decipher his expectations regarding the performance of the senior officer for technology, even when he does not verbalize it sufficiently. It goes without saying, in order to respond to the concerns discussed above, he expects outputs from the technology he pays for: i.e., competitive products, contributions to lower costs of doing business (not, I emphasize again, just lower costs of products), and capability for flexible, timely response. These are the concerns the CEO is most likely to express, but it would be a mistake to assume that they are all that need to be addressed in order to gain his confidence.

Because attaining "safety" credibility demands more than the soundness of technical proposals, which a CEO often cannot judge and which are indeed often difficult to evaluate even by skilled professionals, he will be looking for a number of additional characteristics to satisfy his comfort index. First, as a provider of resources to an activity whose effectiveness and even whose output he often cannot gauge very satisfactorily, he would like continuing evidence that its manager is attempting to be a wise steward of those resources. What constitutes such evidence?

Stewardship. Rudimentary, but all too frequently violated, is adherence to budgets and schedules. Again, the uncertainties in technical work unfortunately make it easy to fall into the trap of accepting slippage in schedules and budget overruns as inherent in the nature of the work. In turn, this can degenerate into an attitude of not even appearing to take cost and time seriously. The laments I hear most frequently from CEOs are, "Why does it take so long to do things?" and, "Why do we miss so many targets on product introduction?" Safety credibility requires visible concern over meeting schedules and budgets and also evidence of vigorous action to meet them.

Beyond these indispensable actions, three additional steps are helpful. There must be a clear indication of awareness that selecting technical programs and evaluating results are intrinsically difficult tasks to which careful, continuing attention is being paid. A continuing process should be put in place to improve the effectiveness of choosing and managing programs and to develop measures of the output of technology. This work must be performed and discussed with top management periodically even when the results are modest. CEOs know the difficulty of the task, but they want evidence that their manager of technology cares. The director of research for a major consumer goods company commented, "When I got this job, I realized that my boss would expect a continuing stream of successful new products. But after a few months I began to realize that, even though he didn't express it directly, he also wanted to see that I was trying to improve my operation."

In addition, the technical effort must be managed as an investment in which priorities change as circumstances dictate, and these changes lead to reallocations of effort. Awareness of this kind of vigorous adjustment of priorities cannot be achieved simply by talking about changes in individual programs. In discussing strategic management, I pointed out that a CEO often has no mental framework for aggregating individual changes in programs into a perspective on the activity unless he is given one. Finally, a CEO should be told of programs stopped and new programs started. I have frequently heard top managers complain, "All you do is ask for more money for new programs. You never stop anything."

Many of these activities may seem peripheral to the basic task of simply producing significant advances in technology. They are, however, central to the need for establishing safety credibility in a field of activity where establishing technical credibility is difficult to evaluate by an outsider.

Realism. The second thing a CEO expects from his manager of technology is realism. Realism can best be demonstrated by thoughtful assessment and articulation of the risk and uncertainty inherent in a particular program. A program proposal and a project management plan that focus solely on objectives and the means of achieving them without articulating the probabilistic element in every component of the plan is fatally flawed. Since converting uncertainty into risk is a major aspect of innovation, a dispassionate explication of probabilities and uncertainties helps the CEO understand the full implications of what he is being asked to decide. Further, it reassures him that the technical community has indeed both done its homework and attempted to protect him. This careful discussion of probabilities and uncertainties should also include a dispassionate evaluation of various options that have been considered to reduce or control risk. The various options I presented in Chap. 5, "Managing Risk," give the CEO an opportunity to become involved, because many of them require a corporate evaluation, not just a technical one. This careful exploration of risks and alternatives also increases the likelihood that the work necessary to ensure success is being undertaken. Too often all attention is focused on achieving the profile of product properties and costs that were selected at the beginning, usually on the basis of inadequate information, only to discover much later that they must be modified.

Realism requires continuing surveillance of program targets and rolling estimates of results and completion dates. A boss expects the same "no surprises" performance from others that he must subscribe to. Otherwise, suspicion and distrust will color all judgments regarding technology.

Vision. The third thing a CEO has a right to expect is technical vision. Technical progress is achieved by putting on blinders and pursuing objectives relentlessly. A technical manager cannot afford the luxury of the narrow focus and the personal commitment that go with it. He must have a historical sense of the maturation of technology and an awareness of the stage in the life cycle of current key technologies. He must perceive their likely future advance, foresee early their possible replacement, and increase surveillance for candidates.

An important part of this vision is the changing competitive environment for technology. Technology is perhaps the most important competitive weapon U.S. companies can employ. They have had more successful experience with the complete spectrum of technology, from basic research and discovery to commercial application and diffusion throughout industry than those in any other country. However, the superior technological position the United States occupied for decades after World War II has clearly disappeared. Furthermore, the rate of change is accelerating, the number of fields of technology that are relevant to a particular enterprise is increasing, and the pace of application is speeding up. A technical manager has a powerful instrument for attracting the attention of his CEO if he keeps abreast of these changes.

Advocacy of innovation. A manager of technology must never forget that part of his job is to be an advocate of innovation. I do not believe it is necessary to oversimplify and oversell in order to be an effective advocate—that elicits painful recollections of the 1950s. A common belief in those days was that if one but did good scientific research in fields that were relevant to a company's interests, the benefits would soon accrue—apparently more or less automatically. The difficulty of moving things into production and on into the market was badly underestimated, as was the likelihood of success. Unfortunately, many technical people then blamed business managers for the poor performance on innovation. "They" were antitechnology, too shortsighted, lacked vision and courage!

The loss of credibility and the misperception of the nature and rewards of innovation engendered by that behavior of the technical community were so severe that U.S. managers of technology have spent much of the last twenty years working their way out of it. Fortunately, they now have both a more realistic understanding of the nature of the innovation process and more credible evidence of its benefits. I do not know of any manager who thinks innovation is a bad idea. I also do not know any who would not like to do it more effectively.

In stark terms, innovation does not happen unless people make a major commitment to it. Part of the manager of technology's responsibility is to make and to encourage such a commitment—it goes with the territory. The long-term survival of the firm is threatened by his failure to be an effective advocate. Somehow or another, those themes must come to life in his words and his actions. One chief executive, exasperated by his vice president of R&D's repeated emphasis of the heavy expenditures of competitors, felt obliged to respond by saying, "We make more money than any of those guys." *But* he went on to tell his general managers, "I notice Tom always does this to me at budget time and it always works. And if you people don't have somebody doing that to you, you don't have the right man in your engineering slot!"

Part of successful selling is courage—the courage to be forceful and persistent even while being realistic and sympathetic to the plight of the general manager. Nobody else has the unequivocal franchise to be the spokesperson for innovation. Its benefits and the risks of not doing it make a powerful argument. At the same time, there are rational creative things that can be done to limit, share, or reduce risk, and the CEO is justified in seeking evidence that they are being pursued.

Breadth of view. Finally, the CEO expects a breadth of view that keeps technology in perspective. The famous "good hands" slogan of Allstate Insurance is the image to be sought. The manager of technology must strive to project an aura of breadth of view that rises above persistent requests for "more" for technology. He must recognize that in some business situations technology is more or less irrelevant, in others it is central. An R&D manager, back in the lab after a stint in operations, told me he was having great difficulty convincing the vice president of R&D that for some businesses, at that point in time, technology simply was not a relevant issue, given the problems they faced in other areas.

The manager of technology should demonstrate awareness that in some circumstances technology is best acquired externally, in others it should be developed in-house. He must be part teacher, part conscience, part advocate, part team player, and always careful steward of the resources entrusted to him.

The CEO's Influence Infrastructure

The task of selling technology requires more than knowing a customer and establishing credibility with him. Marketing professionals refer to "dynamic positioning" for a new product—building a desired set of expectations in the customer's mind. They increasingly emphasize the influence infrastructure that strongly affects the customer's expectations. This concept provides a largely untapped opportunity in dealing with a CEO. He is enmeshed in a network of interactions and messages that largely determine the set of issues he regards as important and that help alter his perspective and priorities.

Successful technological innovation cannot be an isolated activity: it requires inputs from, and cooperation with, a large and diverse cast. This disparate group must have time to come to terms with the innovation, to identify their welfare with it in some measure, and to help shape it to their lights. Failing to sign up this cast of participants and supporters can doom even a promising development.

Know the Product: Packaging the Technology Message

So far, we have been focusing on the need to establish credibility with the chief executive and to work through the influence infrastructure to create a realistic set of expectations toward technology. Obviously, all this is to no avail if the message to be conveyed is unattractive or one which the CEO thinks does not warrant his attention. Attention to four aspects of managing technology, which we have discussed earlier, can play an important role in helping to sell technology: aggregate technology programs and technology expenditures to a sufficiently high level to warrant the attention of the CEO; include all of the fields of work that properly belong under the umbrella of technology, lest the CEO conclude that some of his key concerns are not being addressed; present a balanced picture of technology programs, which includes both incremental improvements and discontinuous changes; and understand the psychology of change in order to elicit support and to provide a mixture of change, but also continuity.

Aggregate programs to larger entities

First, the need to aggregate. Technical people naturally tend to discuss their work in terms of specific programs. Progress in science and engineering demands a careful delineation of the area of investigation and a rigorous specification of objectives and approaches. Unfortunately, that mode of discourse leads to severe problems when attempting to communicate with a CEO. Given his responsibilities and the demands on his time, he understandably resents being presented issues that he believes do not warrant his attention. Rarely is an individual technical program of sufficient import to warrant the attention of the CEO—particularly in the earlier stages when the process of beginning to build support is started. A manager at Exxon Research and Engineering once commented, "Would you want to try to get the attention of the Jersey board, even on the most exciting discovery of this century, on the morning the board was told of the discovery of oil on the Alaskan North Slope? They wouldn't pay any attention to you—and they shouldn't!" This is especially true if the item is presented devoid of any strategic context to help him perceive its significance, which is difficult to do credibly in the early stages. Furthermore, a CEO certainly will not permit the intrusion on his time to discuss the dozens of programs that constitute a typical technology activity. Nor will he have any framework for aggregating them into larger strategic categories.

An exclusive focus on a program-by-program exposition, with no aggregation into larger categories, invites micromanaging by the CEO. He normally tests information by probing for more detail. If he is presented information of a detailed technical nature, he has little choice except to seek still finer detail—even to second-guessing the technical experts. General managers or CEOs who have a technical background are often especially prone to such behavior. An important aspect of every effective manager's work is "managing" the relationship with his boss.

Aggregating into larger entities also serves as a healthy discipline for the technologists. When they see the substantial expenditure proposed for a category and compare it with the outputs expected, they may find the results not as impressive as they would like.

Just as important, expenditures in these larger aggregates can be treated as investments to be managed as a portfolio, and the CEO can see how resources are being reallocated both within them and among them and can discuss longer-term portfolio objectives. Participating in resource allocation is clearly appropriate for him to do.

Ensure scope matches his concerns

The second source of difficulty in selling technology to a CEO is the tendency to define its scope too narrowly. Doing so shortchanges the true contribution that technology can make and leads the CEO to wonder whether some of the needs of the business are receiving the attention they deserve. Much of the discussion of technological innovation focuses on the development of new products. In fact, many equate the role of technology primarily with the creation of new products. Again, however, that focus responds to one important and continuing concern of the CEO but neglects others.

If technology managers do not attempt to provide leadership in achieving integration across the three dimensions of technology—product, process, and information—that we introduced at the very beginning, a CEO may well have reason to question whether his technology is in good hands. Only by achieving such integration can technology make the contribution to the major concerns of the CEO that he has reason to expect.

Present balanced picture

The third element of the message to a CEO is to present a balanced picture of the total technical effort. Businesspersons, appropriately, want the least risky path to success. Innovation comprises two components—incremental and discontinuous changes. The discussion of strategic management highlighted the division between these two. From two-thirds to three-quarters of work labeled as R&D in most industries is in fact application of the state of the art, where the technical risk is virtually nil if it is being used in a similar type of application. Evolutionary improvement in the state of the art and external technology used by others represent approximately another 15 to 20 percent of R&D effort. Thus, about 90 percent of all R&D is probably associated with incremental innovation through the ingenious use of existing technology or evolutionary improvement. Only about 10 percent of R&D, and an even smaller percentage of total technical work, represents effort to create major breakthroughs. And yet most of the discussion of technology focuses on these major breakthroughs and discontinuities. Emphasizing investment in them leads a CEO to conclude that investment in technology inevitably involves high risk. Even worse, it distorts his perception of the role technology should play in his business, encouraging him to believe that its major thrust is to create major breakthroughs when in fact its task is a much more complicated balancing of incremental and discontinuous advances. Equally important, its role in assuring the viability of the enterprise is far more extensive than the occasional home run that generates excitement.

I believe this distortion has two sources. First, the education and value system of technical people emphasize the role of discovery. To be the first to discover or to invent merits the highest accolades. The highest kudos go to creating technology, not just using it. But the CEO, again, just wants the least risky solution that meets his needs.

The challenge of technical management is to provide him that solution at the lowest cost, even if it sometimes means buying it, not developing it in-house. That change in point of view is profound.

Understand the psychology of change

The second source of distortion is the approach usually taken to generate enthusiasm and commitment: to emphasize how new and different and exciting a discovery is. Naturally, an inventor perceives his invention in this light, and he or she believes the path to commercialization is to encourage others to share his vision. What he does not realize is that one person's vision can be another's nightmare. What one perceives as improvement and progress, an exciting new capability, another can perceive as an unsettling change, disruption, uncertainty, and unsuccessful expenditure.

Recent studies have indicated that much "selling" in innovation is counterproductive. More emphasis needs to be placed on the stabilizing elements that will still be present, on areas of continuity, and on fit with existing systems and structures, i.e., on reducing anxiety rather than on stimulating enthusiasm. Better understanding of the psychology of change can lead to a more balanced approach between visions and nightmares.

Conclusion

Selling technology follows the rules of all marketing: know the customer, understand his environment, and seek to influence him toward a set of favorable expectations regarding your product. One must also know that product and be certain that it is indeed what the customer needs.

The CEO lives in a very unforgiving world in which resistance to change is quite rational. Because he often cannot judge technical proposals in depth, he must resort to judging the proposer. Do actions and values correspond enough with those of the CEO for him to be willing to trust him? Does he give evidence of appreciating the CEO's principal concerns and of seeking to address them? Does his stewardship of technology produce not only the kinds of outputs the business needs, but also reflect a concern for effective management and for reallocating resources in the face of changing circumstances? Does his advocacy of programs include a realistic appraisal of risks and uncertainties, and does the program plan address all elements of uncertainty?

Selling technology is more effective if the CEO receives reinforcing messages from multiple sources. Are strategic alliances with other potential sources of influence deliberately nurtured? The "message" on technology must address issues of sufficient import for the CEO to believe they warrant his attention. Individual programs presented with no strategic context rarely do.

Technology frequently emphasizes product engineering with insufficient attention to the opportunities arising from better integration of product engineering, manufacturing processes, and information processing.

Technology is also presented as though the most important goal is to achieve the big breakthrough. In fact, most of the effort is in achieving incremental improvement, and most difficulties arise from failing to perform that work well.

Finally, understanding the psychology of change will lead to more effort on reducing anxiety in conjunction with a vision of a brave new world.

CEOs clearly want to succeed. They are almost universally convinced of the vital role of technology in the success and even the survival of a business. The challenge to technical people is to help CEOs to see the way in which technology can help them achieve their own goals and to make them comfortable with the stewardship of technical management.

Notes and References

1. The material in this chapter first appeared in the *TAPPI* (*Technical Association of the Pulp and Paper Industry*) *Journal,* March 1986, pp. 56–60.
2. Dorothy Leonard-Barton and William A. Kraus, "Implementing New Technology," *Harvard Business Review,* November–December 1985, pp. 102–110.

Epilogue: The Technologically Effective Enterprise

We have now traversed the terrain of technology throughout a business. What will a business that is truly effective in using technology look like? How will it behave? It will not be a comfortable place to work. Complacency will be a cardinal sin. Tension will be regarded as a necessity that must be deliberately nurtured rather than avoided or suppressed. The appropriateness of the present balance between contradictory needs will be subject to continual probing—continuity versus change, long term versus short term, centralization versus decentralization, in-house development versus external acquisition, going it alone versus partners, hard-nosed discipline versus "soft" encouragement of risk taking... the list is long.

Our effective enterprise will take a systems view of technology. It will recognize the need for and encourage the development of specialized technical expertise; however, it will not tolerate technical disciplines or a technology management viewpoint that does not make decisions in the context of the total business. Conversely, it will give due weight to the pervasive impact of technology and expect managers in all functions to incorporate technological implications into their decision making.

A technologically effective enterprise will exhibit great self-awareness, a well-articulated sense of what businesses it is in, how it got where it is, how it behaves, and the keys to its success. At the same time, it will not regard these features as given or immutable, but rather as potential sources of dysfunction, subject to constant assessment.

A technologically effective organization will know who its competitors are, both traditional and nascent; will have a realistic, nonchauvinistic appraisal of comparative status; and can articulate the competitive rationale for its technology strategy.

Such an enterprise will be explicit about the way it uses technology to achieve competitive advantage. At the same time, it will recognize that external considerations can necessitate change—advances in technology, changes in competition, alteration of the market environment. It will have resources explicitly devoted to monitoring these externalities and procedures for considering the need for modifying internal responses.

Our enterprise will accept as indispensable the triad of technology emphasized throughout this book—product, process, and information. This understanding will be buttressed not only by organizational structure but also by internal coordinating mechanisms, technical training programs, and career paths.

A technologically effective enterprise will be relentless in its pursuit of excellence. It will be intolerant of error or ineffective performance with regard to product attributes, cost, time schedules, or response time. Contrarily, it will recognize that seeking competitive advantage often requires pushing the limits of technology. It will do so, but only after vigorous examination of alternatives and realistic appraisal of sources of risk. If it embarks on a risky course, it will ensure that work is initiated to reduce risk wherever possible.

An effective enterprise will be comfortable with uncertainty and ambiguity. In particular, its ability to evaluate managerial performance will include the capacity to distinguish between ineptitude or error on the one hand, and an unfortunate outcome of the probabilities on the other.

A technologically effective enterprise will be aware of the dynamics of maturation of technologies and of industries. It will alter its emphasis in a technology as it matures and will staff managerial positions with people whose styles match current needs.

A technologically effective organization will regard parochialism as an ever present danger. It will insist that its managers in the aggregate have active external interaction—with customers, with professional societies and trade associations, with vendors, and with governmental and academic institutions.

Perhaps most important of all, a technologically effective enterprise will be aware of the changing nature of competition. It will know that in addition to the traditional factors of product features, cost, and customer focus that have provided the basis of competitive advantage, two broadly based secular changes are under way. First is a relentless shift from selling "tools," which customers can use to solve problems, to selling "solutions" that are based on specific knowledge of customers' needs and customized responses to those needs. This shift reflects partly the power of technology to provide product diversity at acceptable cost, even for smaller-scale markets, and partly the worldwide competition from developing countries in commodity materials and mass-produced components.

The second shift is that the concept of a wholly domestic market has become an anachronism. Global markets, global competition, and global operations must become the starting point for all management planning. Since very few businesses have the scale to operate throughout the world, this emerging environment requires a major change in the conception of what a business is and in how resources will be marshaled. The change is analogous to a move from a closed system in which the boundaries of the system are sharply defined and the internal components have been carefully designed to work well together (like the original Macintosh computer) to an open system that is designed to easily accommodate coupling with other components or subsystems. Enterprises will have to become more "open." Instead of defining resources in terms of present employees, in-house skills, and facilities, managers will have to regard locating and gaining access to global resources—whether they be technology, plant capacity, skilled people, access to markets, or capital—as one of their most important tasks. The need for responding faster, for pooling risk, for controlling costs and investment, and for obtaining local presence in markets all over the world dictates the use of resources outside the traditional boundaries of an enterprise. This change also requires the managerial skill to operate effectively in collaborative arrangements with partners, even partners who are also competitors in other realms. It requires learning how to couple much more closely with vendors. Thus the concept of what *is* a business, what are it boundaries, is becoming fuzzy and subtle.

No, the technologically effective enterprise won't be comfortable. But it will prize its alertness, its adaptability, and its realism in evaluating itself and its environment. It will be confident of its ability to succeed and to survive. Its technologists will feel secure in the knowledge that technology plays its needed role in all aspects of the business, even as they compete with others for resources and for the emphasis they believe technology warrants. Meanwhile, they will never lose sight of the necessity to achieve integration between technology and the other elements of a business and will accept both the strengths and constraints that integration produces.

Index

ABOUT THE AUTHOR

Lowell W. Steele's 29 years with General Electric saw him rise to Staff Executive for Corporate Technology Planning in a career that centered on the improvement of business and technology synergy. In the past 6 years as an independent consultant, Dr. Steele has worked across the industrial and governmental spectrum from oil to aerospace. Dr. Steele received his M.B.A. from Harvard and his Ph.D. from M.I.T. in industrial economics. He lives in Fairfield, Connecticut.